THE FOSSIL RECORD AND EVOLUTION

Readings from
**SCIENTIFIC
AMERICAN**

THE FOSSIL RECORD
AND EVOLUTION

With Introductions by

Léo F. Laporte
University of California, Santa Cruz

W. H. Freeman and Company
San Francisco

For Gosia

Most of the SCIENTIFIC AMERICAN articles in *The Fossil Record and Evolution* are available as separate Offprints. For a complete list of articles now available as Offprints, write to W. H. Freeman and Company, 660 Market Street, San Francisco, California 94104.

Library of Congress Cataloging in Publication Data

Main entry under title:

The Fossil record and evolution.

    Bibliography: p.
    Includes index.
    1. Paleontology.   2. Evolution.   I. Laporte, Léo F.
II. Scientific american.
QE711.2.F67   560   81-17401
ISBN 0–7167–1402–7   AACR2
ISBN 0–7167–1403–5 (pbk.)

Printed in the United States of America

1 2 3 4 5 6 7 8 9 0   KP   0 8 9 8 7 6 5 4 3 2

PREFACE

Earth has not always been as it is now, nor will it remain the same in the future. Rather, our planet had a beginning and will have an end: It formed from cosmic gas and dust almost five billion years ago, and it will be consumed by the expanding sun billions of years hence. As far as we know, life is unique to this planet, and its fate is therefore inextricably linked to it. Life thus had a beginning; it flourishes now; and it will eventually end. Each of us, in our own individual existence, bears for a fleeting instant the fire of life that has come directly down to us through countless generations over billions of years. Think of it. You and I, quite literally, trace our origins directly back through those who came before us: from primate, therapsid reptile, Paleozoic amphibian, lobe-fin fish, primitive echinoderm, late Precambrian wormlike creature, to the simplest of single-celled organisms that floated in the primeval seas.

This collection of readings provides an overview of this evolution and history of life—our life—as recorded by the sequence of fossils preserved in the earth's crust. As with all history, the record becomes dimmer as we reach far backward in time, yet we can see some of the major steps in the origin and early diversification of life. As we come closer to the present, we are able to trace and interpret particular lineages quite clearly. The expansion and contraction of life can now be better integrated with the geological evolution of the planet, thanks to the recent development of the theory of plate tectonics.

The readings have been chosen to give a broad representation of the questions and issues that intrigue paleontologists—people who study ancient life. Some articles are concerned with the big picture, others with specific details; some are interested in fossil organisms for what they tell us about their biology and ecology, others for what they tell us about the inanimate world in which they lived. I have also included articles that explain the phenomenon of evolution itself, which provides the theoretical basis for interpreting the historical documents that the fossils represent.

A few of the articles go back almost a generation, yet they are as relevant today as the day they were written. Of course, where appropriate, I have provided any necessary updating in the introduction to each section. Virtually all of the articles are popular classics written by the very people who made many of the original contributions to evolutionary theory and the history of life.

We begin with a discussion of the phenomenon of evolution (Section I) and proceed to earliest evidence for life on earth (Section II). We next consider how fossils of various sorts—invertebrates, vertebrates, tracks and trails, and microscopic single cells—can be interpreted (Section III). We then turn to some of the major patterns seen in the history of life (Section IV), and we conclude with the fascinating evolutionary history of humans as recorded by fossils (Section V).

I believe that this reader will be useful not only in elementary paleontology courses but also in evolutionary biology courses, in that students can see some of the *historical* evidence for evolution. Paleontology is a historical science that has complementary aspects of both theory *and* history. Just as paleontologists need to understand evolutionary theory to interpret fossil history, biologists are well served in their consideration of evolution by examining what in fact happened "way back when."

Finally, I hope that some of the sense of wonder and awe that the evolution and history of life have given me will also grow in you as you use this reader. For, after all, it seems part of the human condition to ask "Who am I? From where have I come? And where am I going?"

Léo F. Laporte

CONTENTS

V   HUMAN EVOLUTION

Note on cross-references to SCIENTIFIC AMERICAN *articles*: Articles included in this book are referred to by title and page number; articles not included in this book but available as Offprints are referred to by title and offprint number; articles not included in this book and not available as Offprints are referred to by title and date of publication.

THE FOSSIL RECORD AND EVOLUTION

I

THE PHENOMENON
OF EVOLUTION

I THE PHENOMENON OF EVOLUTION

INTRODUCTION

Before discussing the fossil record itself, we must first consider the theory that explains the phenomenon of organic evolution. We need such a theoretical framework, because the mere existence of shells, bones, tracks and trails, and silicified wood entombed in crustal rocks can be interpreted in various ways. True, most people nowadays realize, even if only vaguely, that fossils are the remains of ancient animals and plants that once lived on the earth. Some know, too, that these fossils record in some imperfect way the unfolding of life over eons. But these notions, as commonplace as they may be now, are fairly recent in human history. In fact, for most of civilization, fossils were interpreted very differently—as the Creator's mistakes, sports of nature, results of spontaneous generation, tricks of the Devil, or victims of catastrophes that periodically shook the world, including the biblical flood—if they were thought of at all. Like any other historical documents of a bygone age, fossils are objective facts whose interest and significance can vary greatly, depending on the particular theory or conceptual framework in which they are viewed.

Philosophers of science tell us that theory informs us about what is possible and what is not. History, however, goes beyond a consideration of what is possible, to a consideration of "what actually happened." The theory of organic evolution, as developed by Charles Darwin and his successors, gives us as paleontologists a conceptual framework of possibilities within which we try to tell what actually happened in the course of life's history on earth. George Gaylord Simpson, an American paleontologist, has put this another way, by referring to "immanent principles and processes" that are abstract and time-independent (e.g., $E = mc^2$) and "past configurations" that are specific, time-linked examples of those principles and processes (e.g., Hiroshima). Thus, theory explains *how* certain principles and processes operate; history informs about *what*, *when*, and *where*. Together, theory and history combine in a historical science such as paleontology to tell us *how come*.

In its simplest outline, evolutionary theory states that variations in organisms arise from mutations, which are spontaneous alterations in the genes found in all living cells, and from sexual reproduction, whereby gene contributions from parents are randomly segregated when sex cells form and randomly recombined when sex cells cross-fertilize. These individual variations in genetic structure lead to variations in the way organisms look, grow, behave, reproduce, and a thousand other things, ranging from the trivial to the significant, that permit animals and plants to live out their lives and reproduce themselves. Given the limited resources in nature—especially food and living space—some individuals will be better able to exploit the environment, however slight the advantage, owing to particular genetic variations and, consequently, to rear more offspring than others with less favorable genetic attributes. Over successive generations, the more favorable genes increase proportionally, while the less favorable decrease. It is this shift in gene frequencies that defines evolution. And it is the accumulation of these small changes in gene frequency over the millions, even billions, of years of earth history that has led to the great diversity of animal and plant life. By implication, then, all forms of life today—ants and butterflies, daisies and dandelions, sparrows and horses, jellyfish and whales—share a common ancestry, however far back in time.

Before Charles Darwin, there were those who had come to realize that there was, indeed, a unity to life that was particularly apparent from the way organisms were constructed. In fact, by the early nineteenth century, such concepts as the Scale of Nature and the Chain of Being recognized the graduated links from the simplest to most complex organisms as well as the basic architectural, or morphological, similarities typical of one group of organisms as against another. Such concepts, however, were believed to reflect the overall design in Nature or the handiwork of God, rather than the result of common ancestry of species over the ages. Some scientists of this time, such as Erasmus Darwin, Charles's grandfather, and Jean Lamarck, a French biologist, did suggest that these similarities among organisms came about from progressive change, in one way or another, over "millions of ages." But the actual mechanism of change eluded them and others who considered the problem.

Charles Darwin's quintessential contribution to evolutionary theory, therefore, is not the idea of evolution, but rather his statement of the mechanism by which animal and plant species change into other distinct species. In fact, the full title to his masterwork succinctly states this: *On the Origin of Species by Means of Natural Selection, or the Preservation of Favoured Races in the Struggle for Life*. How was it, then, that Darwin succeeded where others failed in recognizing the key mechanism in a theory that, on the one hand, explains life's unity and, on the other, life's diversity?

First, Darwin was an excellent naturalist. All during his formal studies, initially in medicine at Edinburgh and later the ministry at Cambridge, Darwin actively pursued his enthusiasm for hunting, beetle collecting, and bird watching, as well as for cultivating friendships with professional scientists such as John Stevens Henslow, a botanist. He also "gloried in the progress of geology," which indeed had been developing rapidly through the efforts of such Scottish geologists as Hall, Hutton, and Playfair, efforts that culminated with his British contemporary, Charles Lyell, whose three volumes, *The Principles of Geology*, Darwin read during the long voyage of the *Beagle*. In fact, it was Darwin's accomplishments as a naturalist that led Henslow to recommend him to Captain Robert Fitzroy who was seeking a naturalist for the *Beagle* voyage.

Second, the voyage of H.M.S. *Beagle* certainly came at a most propitious time for Darwin. Having just completed his undergraduate studies at Cambridge and a month's field work with Adam Sedgwick studying Cambrian strata in North Wales, Darwin undertook the voyage at age twenty-two. The voyage included stops on major oceanic islands and the continents of South America, Australia, and southern Africa. All these places had sufficiently exotic floras, faunas, and geology (including fossils) to raise many important questions for Darwin. Among these were: How is it that the giant, extinct armadillos of South America have the same ground plan (general design) as living armadillos? How is it that the ostrichlike species of rhea in Argentina and Uruguay are replaced by a distinctly different species in Patagonia? Why do the birds on the Galápagos Islands more closely resemble the birds of South America than they do of other oceanic islands, such as the Cape Verde Islands, whose birds are African in character? Why are the finches of the various Galápagos Islands so different, when the physical conditions of the islands are so similar? As Darwin wrote toward the end of the voyage, the facts that pose these questions "would be well worth examining: for such facts would undermine the stability of species." That is, it was beginning to dawn on Darwin that species might not, in fact, be fixed entities, but that they might alter, change, or evolve one into the other.

A third contributing factor to Darwin's eventual success was that he was of independent means; the family fortunes enabled him to spend all his time and energy concentrating on his biological and geological researches. Settling down in Kent, he spent the next several decades reading and experimenting, thinking and writing. In 1838, three years after the end of the *Beagle* voyage, Darwin read Malthus's *Essay on Population*, written 40 years before, and "being well prepared to appreciate the struggle for existence which everywhere goes on. . .[in] animals and plants, it at once struck me that under these circumstances favourable variations would tend to be preserved, and unfavorable ones to be destroyed. The result of this would be the formation of new species" (Darwin, *Autobiography*, 1876).

The articles in this section review the current status of our understanding of Darwinian evolution based on the discoveries—experimental and fossil—made in the century and one-quarter since the publication of *On the Origin of Species*. It is a measure of Darwin's genius that the fundamental tenets of his theory still hold up well today, despite (or really because of) the evolutionary research that has occurred since his day.

The first article, "Evolution," by Ernst Mayr—himself one of the major contributors to our present understanding of biological evolution—enumerates and explains the chief postulates of Darwin's original theory, namely that species change continually; that the change is gradual and continuous; that all life is interrelated by common descent; and that the mechanism of these temporal changes over endless generations of ancestors and descendants is natural selection. Mayr shows how each of these propositions has been received by biologists and some of the doubts raised about them; he also indicates the degree to which they have been verified. Mayr's article demonstrates, too, the rich complexity of evolutionary theory with its historical antecedents and intellectual roots as well as its on-going ramifications and implications for today (e.g., sociobiology or eugenics). Mayr thus succinctly outlines the justification for his opening claim that evolution resulted in the most consequential change in our view of the world, of living nature, and indeed of our own selves. (If you're doubtful of this, consider the current uproar over the teaching of evolution in schools.)

Richard Lewontin in "Adaptation" pursues in some detail just what is meant by an organism fitting into its environment and how that is effected by natural selection, or survival of the fittest, to use Darwin's phrasing. He argues that we mustn't oversimplfy the notion of adaptation by imagining it as the mere adjustment of an organism to a pre-existing way of life, or more specifically to an ecologic niche. Neither organism nor environment, in fact, is static: Organisms by their very existence modify the niche, and the niche itself varies continuously in its physical, chemical, and biological character. As Lewontin notes, adaptation is better viewed relatively rather than absolutely, as a slowly changing ecologic niche accompanied by a slowly changing (evolving) species, always slightly behind—and if too far behind, it becomes extinct. Lewontin then explores how the immense variety of life has resulted from the occupation of a huge diversity of environments. He cautions us to be wary of inventing detailed adaptive value to everything observed in an organism, whether morphology, behavior, or growth pattern. For some phenomena exhibited by organisms may be present because of differential growth, genetic linking with another, more adaptive character, a different solution to the same problem, or even chance.

Section I concludes with "The Evolution of Multicellular Plants and Animals" by James Valentine, in which we see the broad results of the evolution of life over the last three-quarters of a billion years. It is not enough to know something of the evolutionary process to understand life. We must also trace its history by looking at past organisms and their comings and goings as recorded by animal and plant fossils. Valentine provides functional reasons why different sorts of organisms have appeared and flourished. Although necessarily speculative to some degree because we can't actually observe how these ancient creatures lived, Valentine shows the kind of reasoning needed to go beyond mere description of what lived where, and when. He highlights, too, a number of milestones and breakthroughs in life history that permitted further diversification and specialization. This article thus gives the larger context within which subsequent articles in the reader may be set.

In summary, the readings in Section I provide a historical and theoretical background for organic evolution, against which we can then see particular episodes in the history of life played out, as revealed by the fossil record.

SUGGESTED FURTHER READING

The following annotated references will allow you to pursue further some of the ideas and information included in the individual readings. The references themselves also have good bibliographies.

Darwin, C. 1859. *On the Origin of Species*. Facsimile of first edition, Cambridge, Mass.: Harvard University Press, 1964. *The* book, of course, that all students of evolution must eventually read to understand Darwin's fundamental contribution to human knowledge.

Darwin, C. 1876. *Autobiography*. Reprint, New York: Crowell-Collier, 1961. Besides reproducing the short and charming autobiography—written for his children, and not the world at large—this volume also contains reminiscences by his son, Sir Francis Darwin; notes and sketches by Charles Darwin during his research that led to the *Origin*; and letters relating to it, including those to Lyell and Hooker asking them what he should do about Wallace's essay, which he had just received and which so neatly duplicated all his own ideas about natural selection.

de Beer, G. 1964. *Charles Darwin: Evolution by Natural Selection*. Garden City, N.Y.: Doubleday. An interesting and insightful book explaining Darwin's geological and biological researches in the context of a scientific biography of the great naturalist.

Dobzhansky, T., F. J. Ayala, G. L. Stebbins, and J. W. Valentine. 1977. *Evolution*. San Francisco: W. H. Freeman and Company. An excellent and up-to-date treatment of evolution written for the advanced student as well as the more general reader, with many examples from the fossil record considered.

Kitts, D. B. 1974. "Paleontology and Evolutionary Theory," *Evolution*, vol. 28, pp. 458–472. Defends the idea that nothing *per se* can be learned about the physical objects we call "fossil." Rather, by our *a priori* knowledge and theory we attribute meaning and make interpretations with respect to fossils. Definitely challenges our assumptions about what fossils tell us.

Simpson, G. G. 1963. "Historical Science," in *The Fabric of Geology*, C. C. Albritton, Jr., ed. Reading, Mass.: Addison-Wesley, pp. 24–48. A thoughtful essay by a leading paleontologist about the historical perspective that is so much a crucial part of the geological sciences; clarifies the distinction between them and the nonhistorical sciences such as chemistry and physics.

Evolution

by Ernst Mayr
September 1978

Introducing a volume devoted to the history of life on the earth as it is understood in the light of the modern "synthetic" theory of evolution through natural selection, the organizing principle of biology today

The most consequential change in man's view of the world, of living nature and of himself came with the introduction, over a period of some 100 years beginning only in the 18th century, of the idea of change itself, of change over long periods of time: in a word, of evolution. Man's world view today is dominated by the knowledge that the universe, the stars, the earth and all living things have evolved through a long history that was not foreordained or programmed, a history of continual, gradual change shaped by more or less directional natural processes consistent with the laws of physics. Cosmic evolution and biological evolution have that much in common.

Yet biological evolution is fundamentally different from cosmic evolution in many ways. For one thing, it is more complicated than cosmic evolution, and the living systems that are its products are far more complex than any nonliving system; other differences will emerge in the course of this article. This *Scientific American* article deals with the origin, history and interrelations of living systems as they are understood in the light of the currently accepted general theory of life: the theory of evolution through natural selection, which was propounded more than 100 years ago by Charles Darwin, has since been modified and explicated by the science of genetics and stands today as the organizing principle of biology.

The creation myths of primitive peoples and of most religions had in common an essentially static concept of a world that, once it had been created, had not changed—and that indeed had not been in existence for very long. Bishop Ussher's 17th-century calculation that the world had been created in 4004 B.C. was noteworthy only for its misplaced precision in an age when the reach of history was still foreshortened by the limited arm's length of written records and tradition. It remained for the naturalists and philosophers of the 18th-century Enlightenment and the geologists and biologists of the 19th century to begin to extend the time dimension. In 1749 the French naturalist the Comte de Buffon first undertook to calculate the age of the earth. He reckoned it was at least 70,000 years (and suggested an age of as much as 500,000 years in his unpublished notes). Immanuel Kant was even more daring in his *Cosmogony* of 1755, in which he wrote in terms of millions or even hundreds of millions of years. Clearly both Buffon and Kant conceived of a physical universe that had evolved.

"Evolution" implies change with continuity, usually with a directional component. Biological evolution is best defined as change in the diversity and adaptation of populations of organisms. The first consistent theory of evolution was proposed in 1809 by the French naturalist and philosopher Jean Baptiste de Lamarck, who concentrated on the process of change over time: on what appeared to him to be a progression in nature from the smallest visible organisms to the most complex and most nearly perfect plants and animals and thence to man.

To explain the particular course of evolution Lamarck invoked four principles: the existence in organisms of a built-in drive toward perfection; the capacity of organisms to become adapted to "circumstances," that is, to the environment; the frequent occurrence of spontaneous generation, and the inheritance of acquired characters, or traits. The belief in the heritability of acquired characters, the error for which Lamarck is mainly remembered, was not new with him. It was a universal belief in his time, firmly grounded in folklore (one expression of which was the biblical story of Jacob and the division of the striped and speckled livestock). The belief persisted. Darwin, for example, assumed that the use or disuse of a structure by one generation would be reflected in the next generation, and so did many evolutionists until late in the century, when the German biologist August Weismann demonstrated the impossibility, or at least the improbability, of the inheritance of acquired characters. Lamarck's assumptions of a drive toward perfection and of frequent spontaneous generation were also not confirmed, but he was right in recognizing that much of evolution is what we now call adaptive. He understood, moreover, that one could explain the great diversity of living organisms only by postulating a great age for the earth, and that evolution was a gradual process.

Lamarck's main interest was evolution in the time dimension—in vertical

CHARLES DARWIN was 31 years old and had already published his journal of the round-the-world voyage of H.M.S. *Beagle* when he sat in 1840 for the watercolor portrait by George Richmond reproduced on the opposite page. By this time, judging from his notebooks, Darwin had already worked out the major features of his theory of evolution through natural selection. Recently married, he was living in London, writing a monograph on coral reefs and turning from time to time to the notes on species that were to lead in 1859 to *On the Origin of Species*.

evolution, so to speak. Darwin, in contrast, was initially intrigued by the problem of the origin of diversity, and more specifically by the origin of species through diversification in a geographical dimension—in horizontal evolution. His interest in diversification and speciation was aroused, as is well known, during his five-year voyage around the world, beginning in 1831, as naturalist on H.M.S. *Beagle*. In the Galápagos Islands, for example, he learned that each island had its own form of tortoise, of mocking bird and of finch; the various forms were closely related and yet distinctly different. Pondering his observations after his return to England, he came to the conclusion that each island population was an incipient species, and thus to the concept of the "transmutation," or evolution, of species. In 1838 he conceived of the mechanism that could account for evolution: natural selection. After more years of observation and experiment, informed by wide reading in geology, zoology and other fields, a preliminary statement of Darwin's theory of evolution through natural se-

lection was announced in 1858 in a report to the Linnean Society of London. Alfred Russel Wallace, a young English naturalist doing fieldwork in the East Indies, had come independently to the concept of natural selection and had set down his ideas in a manuscript he mailed to Darwin; his paper was read at the meeting along with Darwin's.

Darwin's full theory, buttressed with innumerable personal observations and carefully argued, was published on November 24, 1859, in *On the Origin of Species*. His broad explanatory scheme comprised a number of component subtheories, or postulates, of which I shall single out what I take to be the four principal ones. Two of them were consistent with Lamarck's thinking. The first was the postulate that the world is not static but is evolving. Species change continually, new ones originate and others become extinct. Biotas, as reflected in the fossil record, change over time, and the older they are the more they are seen to have differed from living organisms. Wherever one looks in living na-

ture one encounters phenomena that make no sense except in terms of evolution. Darwin's second Lamarckian concept was the postulate that the process of evolution is gradual and continuous; it does not consist of discontinuous saltations, or sudden changes.

Darwin's two other main postulates were essentially new concepts. One was the postulate of common descent. For Lamarck each organism or group of organisms represented an independent evolutionary line, having had a beginning in spontaneous generation and having constantly striven toward perfection. Darwin postulated instead that similar organisms were related, descended from a common ancestor. All mammals, he proposed, were derived from one ancestral species; all insects had a common ancestor, and so did all the organisms of any other group. He implied, in fact, that all living organisms might be traced back to a single origin of life.

Darwin's inclusion of man in the common descent of mammals was considered by many to be an unforgivable insult to the human race, and it aroused a storm of protest. The idea of common descent had such enormous explanatory power, however, that it was almost immediately adopted by most biologists. It explained both the Linnaean hierarchy of taxonomic categories and the finding by comparative anatomists that all organisms could be assigned to a limited number of morphological types.

Darwin's fourth subtheory was that of natural selection, and it was the key to his broad scheme. Evolutionary change, said Darwin, is not the result of any mysterious Lamarckian drive, nor is it a simple matter of chance; it is the result of selection. Selection is a two-step process. The first step is the production of variation. In every generation, according to Darwin, an enormous amount of variation is generated. Darwin did not know the source of this variation, which could not be understood until after the rise of the science of genetics. All he had was his empirical knowledge of a seemingly inexhaustible reservoir of large and small differences within species.

The second step is selection through survival in the struggle for existence. In most species of animals and plants a set of parents produces thousands if not millions of offspring. Darwin's reading of Thomas Malthus told him that very few of the offspring could survive. Which ones would have the best chance of surviving? They would be those individuals that have the most appropriate combination of characters for coping with the environment, including climate, competitors and enemies; they would have the greatest chance of surviving, of reproducing and of leaving survivors, and their characters would

IN ABOUT 1854, the year in which he published a large monograph on barnacles that had occupied him for some eight years, Darwin sat for this photograph. He continued meanwhile with what he called his "species work": reading, corresponding, collecting, experimenting and making notes on the subject of his major work but delaying the writing until 1856. The realization two years later that Alfred Russel Wallace had independently developed the concept of natural selection led Darwin to prepare the "abstract" we know as *On the Origin of Species*.

therefore be available for the next cycle of selection.

The concept of an evolving world rather than a static one was almost universally accepted by serious scientists even before Darwin's death in 1882, and those who accepted evolution also accepted the concept of common descent (although there were those who insisted on exempting man from the common lineage). The situation was very different, however, for Darwin's two other postulates, both of which were bitterly resisted by many learned and able men for the next 50 to 80 years.

One of the postulates was the concept of gradualism. Even T. H. Huxley, who was known as "Darwin's bulldog" for his vigorous championing of most aspects of the new theory, could not accept the gradual origin of higher types and new species; he proposed a saltational origin instead. Saltationism was also popular with such biologists as Hugo De Vries, one of the rediscoverers of Gregor Mendel's laws of inheritance. He proposed a theory in 1901 according to which new species originate by mutation. As late as 1940 the geneticist Richard B. G. Goldschmidt was defending "systemic mutations" as the source of new higher types.

Three developments eventually resulted in the abandonment of such saltational theories. One development was the gradual adoption of a new attitude toward the physical world and its variation. Since the time of Plato the dominant view had been what the philosopher Karl Popper has called "essentialism": the world consisted of a limited number of unvarying essences (Plato's *eide*), of which the visible world's variable manifestations are merely incomplete and imprecise reflections. In such a view genuine change could arise only through the origin of a new essence either by creation or through a spontaneous saltation (mutation). Classes of physical objects do consist of identical entities, and physical constants are unvarying under identical conditions, and so (in the 19th century) there was no conflict between mathematics or the physical sciences and the philosophy of essentialism.

Biology required a different philosophy. Living organisms are characterized by uniqueness; every population of organisms consists of uniquely distinct individuals. In "population thinking" the mean values are the abstractions; only the variant individual has reality. The importance of the population lies in its being a pool of variations (a gene pool, in the language of genetics). Population thinking makes gradual evolution possible, and it now dominates every aspect of evolutionary theory.

The second development that led to

IN ABOUT 1880 DARWIN was photographed at Down House in Kent, where he had lived and worked since 1842. When he died in 1882 at 73, he was buried in Westminster Abbey.

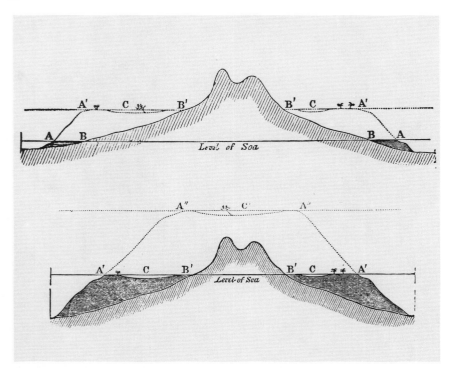

BIRTH OF AN ATOLL through subsidence of the ocean floor was illustrated by these woodcuts in Darwin's journal of the voyage of the *Beagle*. In the first stage (*top*) a fringing reef of coral (*A–B, B–A*) is built up at sea level around an island in the Pacific Ocean. As the island subsides, the coral polyps, which can survive only in shallow water, keep building the reef upward, forming a fringing reef (*A'–B', B'–A'*) that encloses a lagoon (*C*). The island continues to subside (*bottom*) until it is below sea level; the barrier reef, growing, becomes an atoll (*A"–A"*).

DARWIN'S FINCHES, which he observed in the Galápagos Islands and some of which were shown in this woodcut from his published journal, provided him with a major insight. Seeing the wide range of beak sizes and shapes in "one small, intimately related group of birds," he wrote, "one might really fancy that... one species had been taken and modified for different ends."

the rejection of saltation was the discovery of the immense variability of natural populations and the realization that a high variability of discontinuous genetic factors, provided there are enough of them and provided the gaps between them are sufficiently small, can manifest itself in continuous variation of the organism. The third development was the demonstration by naturalists that processes of gradual evolution are entirely capable of explaining the origin of discontinuities such as new species and new types and of evolutionary novelties such as the wings of birds and the lungs of vertebrates.

The other Darwinian concept that was long resisted by most biologists and philosophers was natural selection. At first many rejected it because it was not deterministic, and hence predictive, in the style of 19th-century science. How could a proposed "natural law" such as natural selection be entirely a matter of chance? Others attacked its "crass materialism." In the 19th century to attribute the harmony of the living world to the arbitrary workings of natural selection was to undermine the natural theologian's "argument from design," which held that the existence of a Creator could be inferred from the beautiful design of his works. Those who rejected natural selection on religious or philosophical grounds or simply because it seemed too random a process to explain evolution continued for many years to put forward alternative schemes with such names as orthogenesis, nomogenesis, aristogenesis or the "omega principle" of Teilhard de Chardin, each scheme relying on some built-in tendency or drive toward perfection or progress. All these theories were finalistic: they postulated some form of cosmic teleology, of purpose or program.

The proponents of teleological theories, for all their efforts, have been unable to find any mechanisms (except supernatural ones) that can account for their postulated finalism. The possibility that any such mechanism can exist has now been virtually ruled out by the findings of molecular biology. As the late Jacques Monod argued with particular force, the genetic material is constant; it can change only through mutation. Finalistic theories have also been refuted by the paleontological evidence, as George Gaylord Simpson has shown most clearly. When the evolutionary trend of any character—a trend toward larger body size or longer teeth, for example—is examined carefully, the trend is found not to be consistent but to change direction repeatedly and even to reverse itself occasionally. The frequency of extinction in every geological period is another powerful argument against any finalistic trend toward perfection.

As for the objection to the presumed random aspect of natural selection, it is not hard to deal with. The process is not at all a matter of pure chance. Although variations arise through random processes, those variations are sorted by the second step in the process: selection by survival, which is very much an anti-chance factor. And if it is nonetheless true that some evolution is the result of chance, it is now known that physical processes in general have a far larger probabilistic component than was recognized 100 years ago.

Even so, can natural selection explain the long evolutionary progression up to the "highest" plants and animals, including man, from the origin of life between three and four billion years ago [see "Chemical Evolution and the Origin of Life," by Richard E. Dickerson; SCIENTIFIC AMERICAN Offprint 1401]? How can natural selection account not only for differential survival and adaptive changes within a species but also for the rise of new and differently adapted species? Again it was Darwin who suggested the right answer. An organism competes not only with other individuals of the same species but also with individuals of other species. A new

adaptation or general physiological improvement will make an individual and its descendants stronger interspecific competitors and so contribute to diversification and specialization. Such specialization may often be a dead-end street, as it is in the case of adaptation to life in caves or hot springs. Many specializations, however, and particularly those that were acquired early in evolutionary history, opened up entirely new levels of adaptive radiation. These ranged from the invention of membranes and an organized cell nucleus [see "The Evolution of the Earliest Cells," by J. William Schopf, beginning on page 46] and the aggregation of cells to form multicellular organisms [see "The Evolution of Multicellular Plants and Animals," by James W. Valentine, beginning on page 28] to the advent of highly developed central nervous systems and the invention of long-continued parental care.

Evolution, as Simpson has emphasized, is recklessly opportunistic: it favors any variation that provides a competitive advantage over other members of an organism's own population or over individuals of different species. For billions of years this process has auto-

matically fueled what we call evolutionary progress. No program controlled or directed this progression; it was the result of the spur-of-the-moment decisions of natural selection.

Darwin's uncertainty concerning the source of the genetic variability that supplies raw material for natural selection left a major hole in his argument. That hole was plugged by the science of genetics. Mendel discovered in 1865 that the factors transmitting hereditary information are discrete units transmitted by each parent to the offspring, preserved uncontaminated and reassorted in each generation. Darwin never knew of Mendel's findings, which were largely ignored until they were rediscovered in 1900.

We now know that DNA in the cell nucleus is organized in numerous self-replicating genes (Mendel's hereditary units), which can mutate to form different alleles, or alternative forms. There are structural genes that encode the information for making a specific protein and there are regulatory genes that turn the structural genes on and off. A mutated structural gene can code for a variant

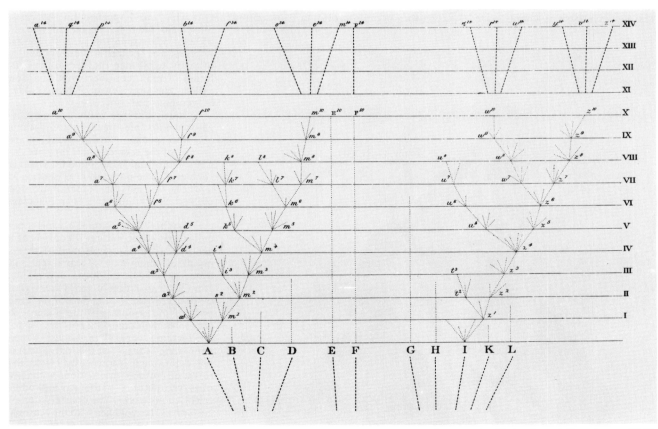

FORMATION OF NEW SPECIES through the divergence of characters and natural selection was illustrated in *On the Origin of Species*. The capital letters (*bottom*) represent species of the same genus. Horizontal lines marked by Roman numerals (*right*) represent, say, a 1,000-generation gap. Branching, diverging dotted lines represent varying offspring, the "profitable" ones of which are "preserved or naturally selected." Some species (*B, C* and so on) die out; some (*E, F*) remain essentially unchanged. Some (*A, I*) diverge widely, giving rise after many generations to new varieties (a^1, m^1, z^1) that diverge in turn, giving rise to increasingly divergent varieties that eventually become distinct new species (a^{14}, q^{14}, p^{14} and so on). After longer intervals these may become new genera or even higher categories.

protein, leading to a variant character. The genes are arrayed on chromosomes and may recombine with one another during meiosis, the cellular process that precedes the formation of germ cells in sexually reproducing species. The diversity of genotypes (full sets of genes) that can be produced during meiosis is almost unimaginably great, and much of that diversity is preserved in populations in spite of natural selection [see "The Mechanisms of Evolution," by Francisco J. Ayala; Scientific American Offprint 1407].

Strangely, the early Mendelians did not accept the theory of natural selection. They were essentialists and saltationists, and they looked on mutation as the probable driving force in evolution. That began to change with the development of population genetics in the 1920's. Eventually, during the 1930's and 1940's, a synthesis was achieved, expressed in and largely brought about by books written by Theodosius Dobzhansky, Julian Huxley, Bernhard Rensch, Simpson, G. Ledyard Stebbins and me. The new "synthetic theory" of evolution amplified Darwin's theory in the light of the chromosome theory of heredity, population genetics, the biological concept of a species and many other concepts of biology and paleontology. The new synthesis is characterized by the complete rejection of the inheritance of acquired characters, an emphasis on the gradualness of evolution, the realization that evolutionary phenomena are population phenomena and a reaffirmation of the overwhelming importance of natural selection.

The understanding of the evolutionary process achieved by the synthetic theory has had a profound effect on all biology. It led to the realization that every biological problem poses an evolutionary question, that it is legitimate to ask with respect to any biological structure, function or process: Why is it there? What was its selective advantage when it was acquired? Such questions have had an enormous impact on every area of biology, notably molecular biology, behavioral studies and ecology [see "The Evolution of Ecological Systems," by Robert M. May; Scientific American Offprint 1404].

Philosophers and physical scientists as well as lay people continue to have trouble understanding the modern theory of organic evolution through natural selection. At the risk of repeating some points I have already made in a historical context, let me outline the special features of the current theory, in particular drawing attention to what distinguishes organic evolution from cosmic evolution and other processes dealt with by physical scientists.

Evolution through natural selection is (I repeat!) a two-step process. The first step is the production (through recombination, mutation and chance events) of genetic variability; the second is the ordering of that variability by selection. Most of the variation produced by the first step is random in that it is not caused by, and is unrelated to, the current needs of the organism or the nature of its environment.

Natural selection can operate successfully because of the inexhaustible supply of variation made available to it owing to the high degree of individuality of biological systems. No two cells within an organism are precisely identical; each individual is unique, each species is unique and each ecosystem is unique. Many nonbiologists find the extent of organic variability incomprehensible. It is totally incompatible with traditional essentialist thinking and calls for a very different conceptual framework: population thinking. (The individuality of biological systems and the fact that there are multiple solutions for almost any environmental problem combine to make organic evolution nonrepeatable. Deterministically inclined astronomers are convinced by statistical reasoning that what has happened on the earth must also have happened on planets of stars other than the sun. Biologists, impressed by the inherent improbability of every single step that led to the evolution of man, consider what Simpson called "the prevalence of humanoids" exceedingly improbable.)

Uniquely different individuals are organized into interbreeding populations and into species. All the members are "parts" of the species, since they are derived from and contribute to a single gene pool. The population or species as a whole is itself the "individual" that undergoes evolution; it is not a class with members.

Every biological individual has a peculiarly dualistic nature. It consists of a genotype (its full complement of genes, not all of which may be expressed) and a phenotype (the organism that results from the translation of genes in the genotype). The genotype is part of the gene pool of the population; the phenotype competes with other phenotypes for reproductive success. This success (which defines the "fitness" of the individual) is not determined intrinsically but is the result of multiple interactions with enemies, competitors, pathogens and other selection pressures. The constellation of such pressures changes with the seasons, through the years and geographically.

The second step of natural selection, selection itself, is an extrinsic ordering principle. In a population of thousands or millions of unique individuals

JEAN BAPTISTE DE LAMARCK, the French naturalist and philosopher who was the first consistent evolutionist, understood that the earth is very old, that evolution is gradual and that organisms adapt. Lamarck also believed, however, in the inheritance of acquired characters.

some will have sets of genes that are better suited to the currently prevailing assortment of ecological pressures. Such individuals will have a statistically greater probability of surviving and of leaving survivors than other members of the population. It is this second step in natural selection that determines evolutionary direction, increasing the frequency of genes and constellations of genes that are adaptive at a given time and place, increasing fitness, promoting specialization and giving rise to adaptive radiation and to what may be loosely described as evolutionary progress [see "Adaptation," by Richard C. Lewontin, *page 16*].

Selectionist evolution, in other words, is neither a chance phenomenon nor a deterministic phenomenon but a two-step tandem process combining the advantages of both. As the pioneering population geneticist Sewall Wright wrote: "The Darwinian process of continued interplay of a random and a selective process is not intermediate between pure chance and pure determinism, but in its consequences qualitatively utterly different from either."

No Darwinian I know questions the fact that the processes of organic evolution are consistent with the laws of the physical sciences, but it makes no sense to say that biological evolution has been "reduced" to physical laws. Biological evolution is the result of specific processes that impinge on specific systems, the explanation of which is meaningful only at the level of complexity of those processes and those systems. And the classical theory of evolution has not been reduced to a "molecular theory of evolution," an assertion based on such reductionist definitions of evolution as "a change in gene frequencies in natural populations." This reductionist definition omits the crucial aspects of evolution: changes in diversity and adaptation. (Once I gave a lump of sugar to a raccoon in a zoo. He ran with it to his water basin and washed it vigorously until there was nothing left of it. No complex system should be taken apart to the extent that nothing of significance is left.)

After the new synthesis of the 1930's and 1940's was achieved a few non-evolutionists asked whether it did not mark the end of research in evolution, whether all the questions had not been answered. The answer to both questions is decidedly no, as is made clear by the exponential increase in the number of publications in evolutionary biology. Let me mention some problems that currently interest workers in the field.

One major subject of inquiry is the role of chance. As far back as 1871 it was proposed that perhaps only some evolutionary change is due to selection, with much or even most change being

ALFRED RUSSEL WALLACE, as a young naturalist working in the East Indies, independently developed a theory of natural selection; his paper on the subject was read along with Darwin's in 1858. Later he differed with Darwin about the mechanisms of human evolution: Wallace believed that natural selection alone could not account for man's higher capacities.

T. H. HUXLEY, distinguished for brilliant work in many areas of biology, took on himself the role of Darwin's "general agent" and "bulldog," explicating and praising *On the Origin of Species* in a book review published in *The Times* of London and in many articles and lectures.

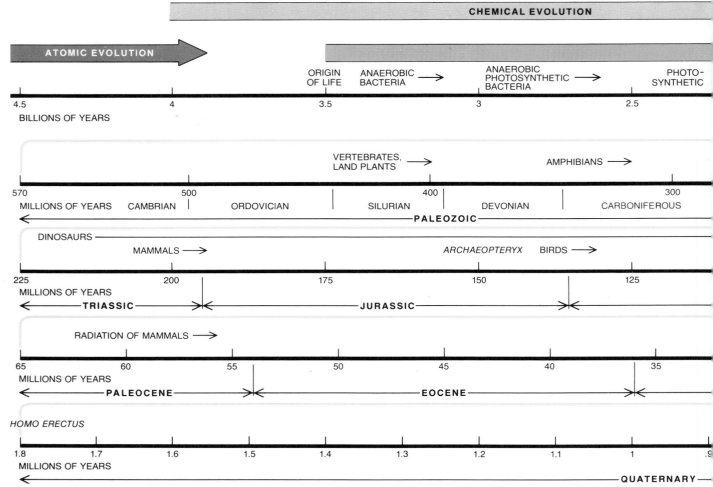

GEOLOGIC TIME is charted on these two pages. The top line shows the full sweep from the origin of the earth some 4.6 billion years ago to the present day. The relatively short span of Phanerozoic time (*col-* *or*), **the time during which the fossils of living things have been abundant in the geological record, is enlarged in the second line from the top, and successively shorter periods (***color***) are enlarged in the next**

due to accidental variation, or to what are now called "neutral" mutations; the suggestion has been repeated many times since then. The problem acquired a new dimension when the technique of electrophoresis made it possible to detect small differences in the composition of a particular enzyme in a large random sample of individuals, thereby revealing the enormous extent of allelic variability. What part of that variability is evolutionary "noise" and what part is due to selection? How can one partition the variability into neutral and into relatively significant alleles?

The discovery of molecular biology that there are regulatory genes as well as structural ones poses new evolutionary questions. Is the rate of evolution of the two kinds of genes the same? Are they equally susceptible to natural selection? Is one kind of gene more important than the other in speciation or in the origin of higher taxa? (For example, the structural genes of the chimpanzee and of man appear to be remarkably similar. Is it perhaps the regulatory genes that make for most of the difference between us

and them?) Are there still other kinds of genes?

Darwin's favorite problem, that of the multiplication of species, has again become a focus of research. In certain groups of organisms, such as birds, new species seem to originate exclusively by geographical speciation: through the genetic restructuring of populations isolated from the remainder of a species' range, as on an island. In plants and in a few groups of animals, however, a different form of speciation can be effected through polyploidy, the doubling of the set of chromosomes, because polyploid individuals are immediately isolated reproductively from their parents. Another mode of speciation is "sympatric" speciation in parasites or in insects that are adapted to life on a specific host plant. Occasionally a new host species is colonized accidentally, and the descendants of the immigrant, perhaps aided by having favorable genes, come to constitute a well-established colony. In such a case there will be strong selection of genes that favor reproduction with other individuals living on the new host spe-

cies, so that conditions may favor the development of a new race adapted to the new host, and eventually of a new host-specific species. The frequency of sympatric speciation is still a matter of controversy. The respective role of genes and chromosomes in speciation is yet another controversial area.

In few areas of biology has the introduction of evolutionary thinking been as productive as it has in behavioral biology. The classical ethologists showed that such behavior patterns as the signaling displays of courtship can be as indicative of taxonomic relations as structural characters are. Classifications based on behavior have been worked out that agree remarkably well with systems based on structure, and the behavioral data have often provided decisive clues where the morphological evidence was ambiguous. More important has been the demonstration that behavior often—perhaps invariably—serves as a pacemaker in evolution. A change in behavior, such as the selection of a new habitat or food source, sets up new selective pressures and may lead to important

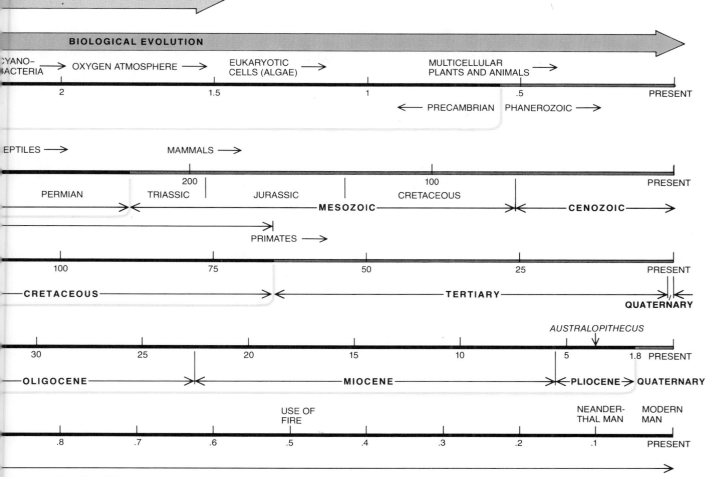

three lines. Three stages of evolution are shown at the top, with biological evolution beginning some 3.5 billion years ago with the appearance of the first living cells known. The three eras of Phanerozoic time (Paleozoic, Mesozoic and Cenozoic) are divided in turn into 11 periods; the Tertiary period is divided into five epochs, and the Quaternary period comprises the Pleistocene epoch and recent time.

adaptive shifts. There is little doubt that some of the most important events in the history of life, such as the conquest of land or of the air, were initiated by shifts in behavior. The selection pressures that potentiate such evolutionary progress are now receiving special attention [see "The Evolution of Behavior"; SCIENTIFIC AMERICAN Offprint 1405].

The perception that the world is not static but forever changing and that our own species is the product of evolution has inevitably had a fundamental impact on human understanding. We now know that the evolutionary line to which we belong arose from apelike ancestors over the course of millions of years, with the crucial steps having taken place during the past million years or so [see "The Evolution of Man," by Sherwood L. Washburn, beginning on page 182]. We know that natural selection must have been responsible for this advance. What do past events enable one to predict with regard to the future of mankind? Since there is no finalistic element in organic evolution and no inher-

itance of acquired characters, selection is obviously the only mechanism potentially capable of influencing human biological evolution.

That conclusion poses a dilemma. Eugenics, or deliberate selection, would be in conflict with cherished human values. Even if there were no moral objections, the necessary information on which to base such selection is simply not yet available. We know next to nothing about the genetic component of nonphysical human traits. There are innumerable and very different kinds of "good," "useful" or adapted human beings. Even if we could select a set of momentarily ideal characteristics, the changes generated in society by technological advances come so rapidly that no one could predict what particular blend of talents would lead in the future to the most harmonious human society. "Mankind is still evolving," Dobzhansky said, but we cannot know where it is headed biologically.

There is another kind of evolution, however: cultural evolution. It is a uniquely human process by which man

to some extent shapes and adapts to his environment. (Whereas birds, bats and insects became fliers by evolving genetically for millions of years, Dobzhansky pointed out, "man has become the most powerful flier of all, by constructing flying machines, not by reconstructing his genotype.") Cultural evolution is a much more rapid process than biological evolution. One of its aspects is the fundamental (and oddly Lamarckian) ability of human beings to evolve culturally through the transmission from generation to generation of learned information, including moral—and immoral—values. Surely in this area great advances can still be made, considering the modest level of moral values in mankind today. Even though we have no way of influencing our own biological evolution, we can surely influence our cultural and moral evolution. To do so in directions that are adaptive for all mankind would be a realistic evolutionary objective, but the fact remains that there are limits to cultural and moral evolution in a genetically unmanaged human species.

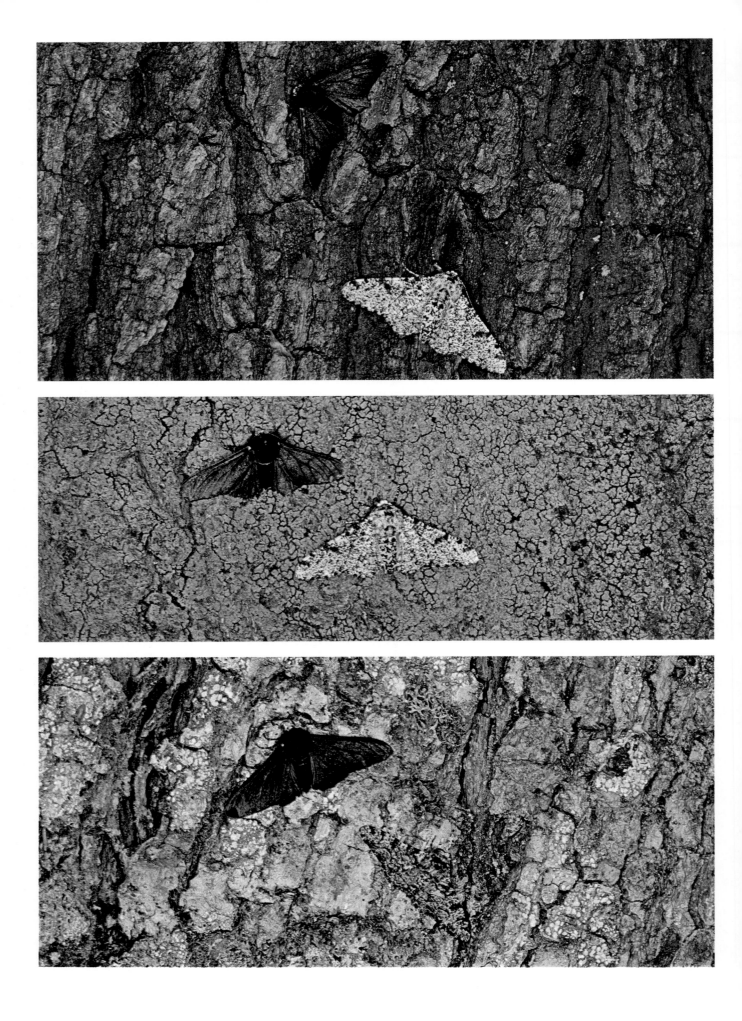

Adaptation

by Richard C. Lewontin
September 1978

The manifest fit between organisms and their environment is a major outcome of evolution. Yet natural selection does not lead inevitably to adaptation; indeed, it is sometimes hard to define an adaptation

The theory about the history of life that is now generally accepted, the Darwinian theory of evolution by natural selection, is meant to explain two different aspects of the appearance of the living world: diversity and fitness. There are on the order of two million species now living, and since at least 99.9 percent of the species that have ever lived are now extinct, the most conservative guess would be that two billion species have made their appearance on the earth since the beginning of the Cambrian period 600 million years ago. Where did they all come from? By the time Darwin published *On the Origin of Species* in 1859 it was widely (if not universally) held that species had evolved from one another, but no plausible mechanism for such evolution had been proposed. Darwin's solution to the problem was that small heritable variations among individuals within a species become the basis of large differences between species. Different forms survive and reproduce at different rates depending on their environment, and such differential reproduction results in the slow change of a population over a period of time and the eventual replacement of one common form by another. Different populations of the same species then diverge from one another if they occupy different habitats, and eventually they may become distinct species.

Life forms are more than simply multiple and diverse, however. Organisms fit remarkably well into the external world in which they live. They have morphologies, physiologies and behaviors that appear to have been carefully and artfully designed to enable each organism to appropriate the world around it for its own life.

It was the marvelous fit of organisms to the environment, much more than the great diversity of forms, that was the chief evidence of a Supreme Designer. Darwin realized that if a naturalistic theory of evolution was to be successful, it would have to explain the apparent perfection of organisms and not simply their variation. At the very beginning of the *Origin of Species* he wrote: "In considering the Origin of Species, it is quite conceivable that a naturalist ... might come to the conclusion that each species ... had descended, like varieties, from other species. Nevertheless, such a conclusion, even if well founded, would be unsatisfactory, until it could be shown how the innumerable species inhabiting this world have been modified, so as to acquire that perfection of structure and coadaptation which most justly excites our admiration." Moreover, Darwin knew that "organs of extreme perfection and complication" were a critical test case for his theory, and he took them up in a section of the chapter on "Difficulties of the Theory." He wrote: "To suppose that the eye, with all its inimitable contrivances for adjusting the focus to different distances, for admitting different amounts of light, and for the correction of spherical and chromatic aberration, could have been formed by natural selection, seems, I freely confess, absurd in the highest degree."

These "organs of extreme perfection" were only the most extreme case of a more general phenomenon: adaptation. Darwin's theory of evolution by natural selection was meant to solve both the problem of the origin of diversity and the problem of the origin of adaptation at one stroke. Perfect organs were a difficulty of the theory not in that natural selection could not account for them but rather in that they were its most rigorous test, since on the face of it they seemed the best intuitive demonstration that a divine artificer was at work.

The modern view of adaptation is that the external world sets certain "problems" that organisms need to "solve," and that evolution by means of natural selection is the mechanism for creating these solutions. Adaptation is the process of evolutionary change by which the organism provides a better and better "solution" to the "problem," and the end result is the state of being adapted. In the course of the evolution of birds from reptiles there was a successive alteration of the bones, the muscles and the skin of the forelimb to give rise to a wing; an increase in the size of the breastbone to provide an anchor for the wing muscles; a general restructuring of bones to make them very light but strong, and the development of feathers to provide both aerodynamic elements and lightweight insulation. This wholesale reconstruction of a reptile to make a bird is considered a process of major adaptation by which birds solved the problem of flight. Yet there is no end to adaptation. Having adapted to flight, some birds reversed the process: the penguins adapted to marine life by changing their wings into flippers and their feathers into a waterproof covering, thus solving the problem of aquatic existence.

The concept of adaptation implies a preexisting world that poses a problem to which an adaptation is the solution. A key is adapted to a lock by cutting and filing it; an electrical appliance is adapted to a different voltage by a transform-

ADAPTATION is exemplified by "industrial melanism" in the peppered moth (*Biston betularia*). Air pollution kills the lichens that would normally colonize the bark of tree trunks. On the dark, lichenless bark of an oak tree near Liverpool in England the melanic (*black*) form is better adapted: it is better camouflaged against predation by birds than the light, peppered wild type (*top photograph on opposite page*), which it largely replaced through natural selection in industrial areas of England in the late 19th century. Now air quality is improving. On a nearby beech tree colonized by algae and the lichen *Lecanora conizaeoides*, which is itself particularly well adapted to low levels of pollution, the two forms of the moth are equally conspicuous (*middle*). On the lichened bark of an oak tree in rural Wales the wild type is almost invisible (*bottom*), and in such areas it predominates. The photographs were made by J. A. Bishop of the University of Liverpool and Laurence M. Cook of the University of Manchester.

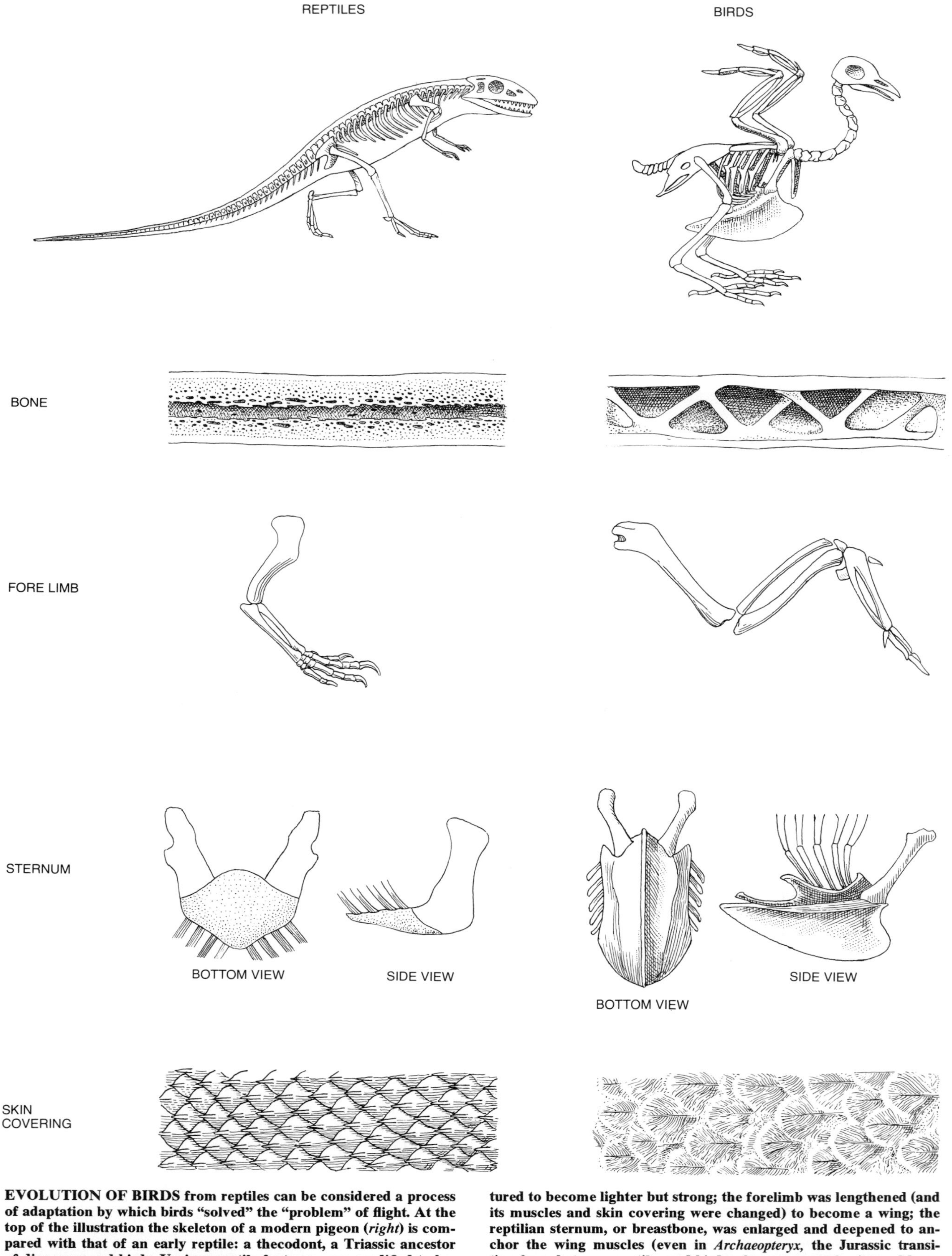

REPTILES

BIRDS

BONE

FORE LIMB

STERNUM

BOTTOM VIEW   SIDE VIEW

BOTTOM VIEW

SIDE VIEW

SKIN
COVERING

EVOLUTION OF BIRDS from reptiles can be considered a process of adaptation by which birds "solved" the "problem" of flight. At the top of the illustration the skeleton of a modern pigeon (*right*) is compared with that of an early reptile: a thecodont, a Triassic ancestor of dinosaurs and birds. Various reptile features were modified to become structures specialized for flight. Heavy, dense bone was restruc- tured to become lighter but strong; the forelimb was lengthened (and its muscles and skin covering were changed) to become a wing; the reptilian sternum, or breastbone, was enlarged and deepened to an- chor the wing muscles (even in *Archaeopteryx*, the Jurassic transi- tion form between reptiles and birds whose sternum is pictured here, the sternum was small and shallow); scales developed into feathers.

er. Although the physical world certain-ly'predated the biological one, there are certain grave difficulties for evolution-ary theory in defining that world for the process of adaptation. It is the difficulty of defining the "ecological niche." The ecological niche is a multidimensional description of the total environment and way of life of an organism. Its descrip-tion includes physical factors, such as temperature and moisture; biological factors, such as the nature and quantity of food sources and of predators, and factors of the behavior of the organism itself, such as its social organization, its pattern of movement and its daily and seasonal activity cycles.

The first difficulty is that if evolution is described as the process of adaptation of organisms to niches, then the niches must exist before the species that are to fit them. That is, there must be empty niches waiting to be filled by the evolu-tion of new species. In the absence of organisms in actual relation to the envi-ronment, however, there is an infinity of ways the world can be broken up into arbitrary niches. It is trivially easy to describe "niches" that are unoccupied. For example, no organism makes a liv-ing by laying eggs, crawling along the surface of the ground, eating grass and living for several years. That is, there are no grass-eating snakes, even though snakes live in the grass. Nor are there any warm-blooded, egg-laying animals that eat the mature leaves of trees, even though birds inhabit trees. Given any description of an ecological niche occu-pied by an actual organism, one can cre-ate an infinity of descriptions of unoccu-pied niches simply by adding another arbitrary specification. Unless there is some preferred or natural way to sub-divide the world into niches the con-cept loses all predictive and explanatory value.

A second difficulty with the specifica-tion of empty niches to which organisms adapt is that it leaves out of account the role of the organism itself in creating the niche. Organisms do not experience en-vironments passively; they create and define the environment in which they live. Trees remake the soil in which they grow by dropping leaves and putting down roots. Grazing animals change the species composition of herbs on which they feed by cropping, by dropping ma-nure and by physically disturbing the ground. There is a constant interplay of the organism and the environment, so that although natural selection may be adapting the organism to a particular set of environmental circumstances, the ev-olution of the organism itself changes those circumstances. Finally, organisms themselves determine which external factors will be part of their niche by their own activities. By building a nest the phoebe makes the availability of dried grass an important part of its

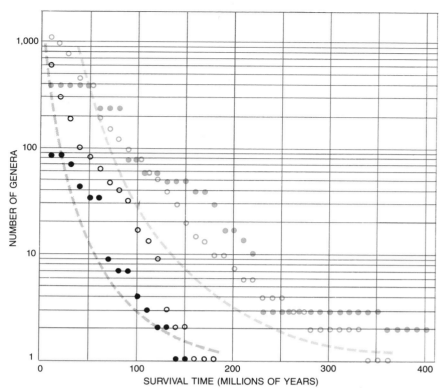

EXTINCTION RATES in many evolutionary lines suggest that natural selection does not nec-essarily improve adaptation. The data, from Leigh Van Valen of the University of Chicago, show the duration of survival of a number of living (*solid dots*) and extinct (*open circles*) gen-era of Echinoidea (*black*) and Pelecypoda (*color*), two classes of marine invertebrates. If natu-ral selection truly fitted organisms to environments, the points should fall along concave curves (*broken-line curves*) indicating a lower probability of extinction for long-lived genera. Actual-ly, points fall along rather straight lines, indicating constant rate of extinction for each group.

niche, at the same time making the nest itself a component of the niche.

If ecological niches can be specified only by the organisms that occupy them, evolution cannot be described as a process of adaptation because all organ-isms are already adapted. Then what is happening in evolution? One solution to this paradox is the Red Queen hypothe-sis, named by Leigh Van Valen of the University of Chicago for the character in *Through the Looking Glass* who had to keep running just to stay in the same place. Van Valen's theory is that the en-vironment is constantly decaying with respect to existing organisms, so that natural selection operates essentially to enable the organisms to maintain their state of adaptation rather than to im-prove it. Evidence for the Red Queen hypothesis comes from an examination of extinction rates in a large number of evolutionary lines. If natural selection were actually improving the fit of organ-isms to their environments, then we might expect the probability that a spe-cies will become extinct in the next time period to be less for species that have already been in existence for a long time, since the long-lived species are presumably the ones that have been im-

proved by natural selection. The data show, however, that the probability of extinction of a species appears to be a constant, characteristic of the group to which it belongs but independent of whether the species has been in exis-tence for a long time or a short one. In other words, natural selection over the long run does not seem to improve a species' chance of survival but simply enables it to "track," or keep up with, the constantly changing environment.

The Red Queen hypothesis also ac-counts for extinction (and for the occa-sional dramatic increases in the abun-dance and range of species). For a spe-cies to remain in existence in the face of a constantly changing environment it must have sufficient heritable variation of the right kind to change adaptively. For example, as a region becomes drier because of progressive changes in rain-fall patterns, plants may respond by evolving a deeper root system or a thick-er cuticle on the leaves, but only if their gene pool contains genetic variation for root length or cuticle thickness, and suc-cessfully only if there is enough genetic variation so that the species can change as fast as the environment. If the genetic variation is inadequate, the species will become extinct. The genetic resources of a species are finite, and eventually the environment will change so rapidly that the species is sure to become extinct.

The theory of environmental tracking seems at first to solve the problem of adaptation and the ecological niche. Whereas in a barren world there is no clear way to divide the environment into preexisting niches, in a world already occupied by many organisms the terms of the problem change. Niches are already defined by organisms. Small changes in the environment mean small changes in the conditions of life of those organisms, so that the new niches to which they must evolve are in a sense very close to the old ones in the multidimensional niche space. Moreover, the organisms that will occupy these slightly changed niches must themselves come from the previously existing niches, so that the kinds of species that can evolve are stringently limited to ones that are extremely similar to their immediate ancestors. This in turn guarantees that the changes induced in the environment by the changed organism will also be small and continuous in niche space. The picture of adaptation that emerges is the very slow movement of the niche through niche space, accompanied by a slowly changing species, always slightly behind, slightly ill-adapted, eventually becoming extinct as it fails to keep up with the changing environment because it runs out of genetic variation on which natural selection can operate. In this view species form when two populations of the same species track environments that diverge from each other over a period of time.

The problem with the theory of environmental tracking is that it does not predict or explain what is most dramatic in evolution: the immense diversification of organisms that has accompanied, for example, the occupation of the land from the water or of the air from the land. Why did warm-blooded animals arise at a time when cold-blooded animals were still plentiful and come to coexist with them? The appearance of entirely new life forms, of ways of making a living, is equivalent to the occupation of a previously barren world and brings us back to the preexistent empty niche waiting to be filled. Clearly there have been in the past ways of making a living that were unexploited and were then "discovered" or "created" by existing organisms. There is no way to explain and predict such evolutionary adaptations unless a priori niches can be described on the basis of some physical principles before organisms come to occupy them.

That is not easy to do, as is indicated by an experiment in just such a priori predictions that has been carried out by probes to Mars and Venus designed to detect life. The instruments are designed to detect life by detecting growth in nutrient solutions, and the solutions are prepared in accordance with knowledge of terrestrial microorganisms, so that the probes will detect only organisms whose ecological niches are like those on the earth. If Martian and Venusian life partition the environment in totally unexpected ways, they will remain unrecorded. What the designers of those instruments never dreamed of was that the reverse might happen: that the nature of the physical environment on Mars might be such that when it was provided with a terrestrial ecological niche, inorganic reactions might have a lifelike appearance. Yet that may be exactly what happened. When the Martian soil was dropped into the nutrient broth on the lander, there was a rapid production of carbon dioxide and then—nothing. Either an extraordinary kind of life began to grow much more rapidly than any terrestrial microorganism and then was poisoned by its own activity in a strange environment, or else the Martian soil is such that its contact with nutrient broths results in totally unexpected catalytic processes. In either case the Mars life-detection experiment has foundered on the problem of defining ecological niches without organisms.

Much of evolutionary biology is the working out of an adaptationist program. Evolutionary biologists assume that each aspect of an organism's morphology, physiology and behavior has been molded by natural selection as a solution to a problem posed by the

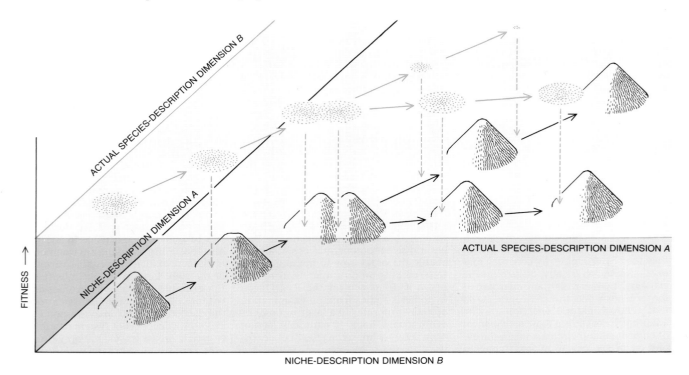

SPECIES TRACK ENVIRONMENT through niche space, according to one view of adaptation. The niche, visualized as an "adaptive peak," keeps changing (moving to the right); a slowly changing species population (*colored dots*) just manages to keep up with the niche, always a bit short of the peak. As the environment changes, the sin- **gle peak becomes two distinct peaks, and two populations diverge to form distinct species. One species cannot keep up with its rapidly changing environment, becomes less fit (lags farther behind changing peak) and extinct. Here niche space and actual-species space have only two dimensions; both of them are actually multidimensional.**

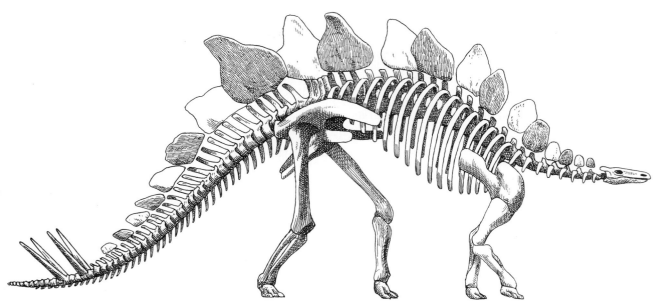

STEGOSAURUS, a large herbivorous dinosaur of the Jurassic period, had an array of bony plates along its back. Were they solutions to the problem of defense, courtship recognition or heat regulation? An engineering analysis reveals features characteristic of heat regulators: porous structure (suggesting a rich blood supply), particularly large plates over the massive part of the body, staggered arrangement along the midline, a constriction near the base and so on. This skeleton in the American Museum of Natural History is 18 feet long.

environment. The role of the evolutionary biologist is then to construct a plausible argument about how each part functions as an adaptive device. For example, functional anatomists study the structure of animal limbs and analyze their motions by time-lapse photography, comparing the action and the structure of the locomotor apparatus in different animals. Their interest is not, however, merely descriptive. Their work is informed by the adaptationist program, and their aim is to explain particular anatomical features by showing that they are well suited to the function they perform. Evolutionary ethologists and sociobiologists carry the adaptationist program into the realm of animal behavior, providing an adaptive explanation for differences among species in courting pattern, group size, aggressiveness, feeding behavior and so on. In each case they assume, like the functional anatomist, that the behavior is adaptive and that the goal of their analysis is to reveal the particular adaptation.

The dissection of an organism into parts, each of which is regarded as a specific adaptation, requires two sets of a priori decisions. First one must decide on the appropriate way to divide the organism and then one must describe what problem each part solves. This amounts to creating descriptions of the organism and of the environment and then relating the descriptions by functional statements; one can either start with the problems and try to infer which aspect of the organism is the solution or start with the organism and then ascribe adaptive functions to each part.

For example, for individuals of the same species to recognize each other at mating time is a problem, since mistakes about species mean time, energy and gametes wasted in courtship and mating without the production of viable offspring; species traits such as distinctive color markings, special courtship behavior, unique mating calls, odors and restricted time and place of activity can be considered specific adaptations for the proper recognition of potential mates. On the other hand, the large, leaf-shaped bony plates along the back of the dinosaur Stegosaurus constitute a specific characteristic for which an adaptive function needs to be inferred. They have been variously explained as solutions to the problem of defense (by making the animal appear to be larger or by interfering directly with the predator's attack), the problem of recognition in courtship and the problem of temperature regulation (by serving as cooling fins).

The same problems that arose in deciding on a proper description of the ecological niche without the organism arise when one tries to describe the organism itself. Is the leg a unit in evolution, so that the adaptive function of the leg can be inferred? If so, what about a part of the leg, say the foot, or a single toe, or one bone of a toe? The evolution of the human chin is an instructive example. Human morphological evolution can be generally described as a "neotenic" progression. That is, human infants and adults resemble the fetal and young forms of apes more than they resemble adult apes; it is as if human beings are born at an earlier stage of physical development than apes and do not

mature as far along the apes' development path. For example, the relative proportion of skull size to body size is about the same in newborn apes and human beings, whereas adult apes have much larger bodies in relation to their heads than we do; in effect their bodies "go further."

The exception to the rule of human neoteny is the chin, which grows relatively larger in human beings, whereas both infant and adult apes are chinless. Attempts to explain the human chin as a specific adaptation selected to grow larger failed to be convincing. Finally it was realized that in an evolutionary sense the chin does not exist! There are two growth fields in the lower jaw: the dentary field, which is the bony structure of the jaw, and the alveolar field, in which the teeth are set. Both the dentary and the alveolar fields do show neoteny. They have both become smaller in the human evolutionary line. The alveolar field has shrunk somewhat faster than the dentary one, however, with the result that a "chin" appears as a pure consequence of the relative regression rates of the two growth fields. With the recognition that the chin is a mental construct rather than a unit in evolution the problem of its adaptive explanation disappears. (Of course, we may go on to ask why the dentary and alveolar growth fields have regressed at different rates in evolution, and then provide an adaptive explanation for that phenomenon.)

Sometimes even the correct topology of description is unknown. The brain is divided into anatomical divisions corresponding to certain separable nervous functions that can be localized, but memory is not one of those functions. The memory of specific events seems to be stored diffusely over large regions of the cerebrum rather than being localized microscopically. As one moves from anatomy to behavior the problem of a correct description becomes more acute and the opportunities to introduce arbitrary constructs as if they were evolutionary traits multiply. Animal behavior is described in terms of aggression, division of labor, warfare, dominance, slave-making, cooperation—and yet each of these is a category that is taken directly from human social experience and is transferred to animals.

The decision as to which problem is solved by each trait of an organism is equally difficult. Every trait is involved in a variety of functions, and yet one would not want to say that the character is an adaptation for all of them. The green turtle *Chelonia mydas* is a large marine turtle of the tropical Pacific. Once a year the females drag themselves up the beach with their front flippers to the dry sand above the high-water mark. There they spend many hours laboriously digging a deep hole for their eggs, using their hind flippers as trowels. No one who has watched this painful process would describe the turtles' flippers as adaptations for land locomotion and digging; the animals move on land and dig with their flippers because nothing better is available. Conversely, even if a trait seems clearly adaptive, it cannot be assumed that the species would suffer in its absence. The fur of a polar bear is an adaptation for temperature regulation, and a hairless polar bear would certainly freeze to death. The color of a polar bear's fur is another matter. Although it may be an adaptation for camouflage, it is by no means certain that the polar bear would become extinct or even less numerous if it were brown. Adaptations are not necessary conditions of the existence of the species.

For extinct species the problem of judging the adaptive status of a trait is made more difficult because both the trait and its function must be reconstructed. In principle there is no way to be sure whether the dorsal plates of *Stegosaurus* were heat-regulation devices, a defense mechanism, a sexual recognition sign or all these things. Even in living species where experiments can be carried out a doubt remains. Some modern lizards have a brightly colored dewlap under the jaw. The dewlap may be a warning sign, a sexual attractant or a species-recognition signal. Experiments removing or altering the dewlap could decide, in principle, how it functions. That is a different question from its status as an adaptation, however, since the assertion of adaptation implies a historical argument about natural selection as the cause of its establishment. The large dorsal plates of *Stegosaurus* may have evolved because individuals with slightly larger plates were better able to gather food in the heat of the day than other individuals. If, when the plates reached a certain size, they incidentally frightened off predators, they would be a "preadaptation" for defense. The distinction between the primary adaptation for which a trait evolved and incidental functions it may have come to have cannot be made without the reconstruction of the forces of natural selection during the actual evolution of the species.

The current procedure for judging the adaptation of traits is an engineering analysis of the organism and its environment. The biologist is in the position of an archaeologist who uncovers a machine without any written record and attempts to reconstruct not only its operation but also its purpose. The hypothesis that the dorsal plates of *Stegosaurus* were a heat-regulation device is based on the fact that the plates were porous and probably had a large supply of blood vessels, on their alternate placement to the left and right of the midline (suggesting cooling fins), on their large size over the most massive part of the body and on the constriction near their base, where they are closest to the heat source and would be inefficient heat radiators.

Ideally the engineering analysis can be quantitative as well as qualitative and so provide a more rigorous test of the

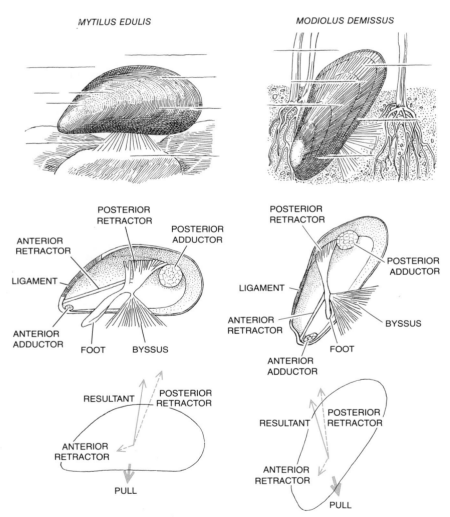

MYTILUS EDULIS

MODIOLUS DEMISSUS

POSTERIOR RETRACTOR

POSTERIOR ADDUCTOR

ANTERIOR RETRACTOR

LIGAMENT

ANTERIOR ADDUCTOR

FOOT   BYSSUS

POSTERIOR RETRACTOR

POSTERIOR ADDUCTOR

LIGAMENT

ANTERIOR RETRACTOR

BYSSUS

FOOT

ANTERIOR ADDUCTOR

RESULTANT   POSTERIOR RETRACTOR

ANTERIOR RETRACTOR

PULL

RESULTANT   POSTERIOR RETRACTOR

ANTERIOR RETRACTOR

PULL

FUNCTIONAL ANALYSIS indicates how the shape and musculature of two species of mussels are adapted to their particular environments. *Mytilus edulis* (*left*) attaches itself to rocks by means of its byssus, a beardlike group of threads (*top*). Its ventral, or lower, edge is flattened; the anterior and posterior retractor muscles are positioned (*middle*) so that their resultant force pulls the bottom of the shell squarely down to the substratum (*bottom*). *Modiolus demissus* (*right*) attaches itself to debris in marshes. Its ventral edge is sharply angled to facilitate penetration of the substratum; its retractor muscles are positioned to pull its anterior end down into the marsh. The analysis was done by Steven M. Stanley of Johns Hopkins University.

adaptive hypothesis. Egbert G. Leigh, Jr., of the Smithsonian Tropical Research Institute posed the question of the ideal shape of a sponge on the assumption that feeding efficiency is the problem to be solved. A sponge's food is suspended in water and the organism feeds by passing water along its cell surfaces. Once water is processed by the sponge it should be ejected as far as possible from the organism so that the new water taken in is rich in food particles. By an application of simple hydrodynamic principles Leigh was able to show that the actual shape of sponges is maximally efficient. Of course, sponges differ from one another in the details of their shape, so that a finer adjustment of the argument would be needed to explain the differences among species. Moreover, one cannot be sure that feeding efficiency is the only problem to be solved by shape. If the optimal shape for feeding had turned out to be one with many finely divided branches and protuberances rather than the compact shape observed, it might have been argued that the shape was a compromise between the optimal adaptation for feeding and the greatest resistance to predation by small browsing fishes.

Just such a compromise has been suggested for understanding the feeding behavior of some birds. Gordon H. Orians of the University of Washington studied the feeding behavior of birds that fly out from a nest, gather food and bring it back to the nest for consumption ("central-place foraging"). If the bird were to take food items indiscriminately as it came on them, the energy cost of the round trip from the nest and back might be greater than the energy gained from the food. On the other hand, if the bird chose only the largest food items, it might have to search so long that again the energy it consumed would be too great. For any actual distribution of food-particle sizes in nature there is some optimal foraging behavior for the bird that will maximize its net energy gain from feeding. Orians found that birds indeed do not take food particles at random but are biased in the direction of an optimal particle size. They do not, however, choose the optimal solution either. Orians' explanation was that the foraging behavior is a compromise between maximum energy efficiency and not staying away from the nest too long, because the young are exposed to predation when they are unattended.

The example of central-place foraging illustrates a basic assumption of all such engineering analyses, that of ceteris paribus, or all other things being equal. In order to make an argument that a trait is an optimal solution to a particular problem, it must be possible to view the trait and the problem in isolation, all other things being equal. If all

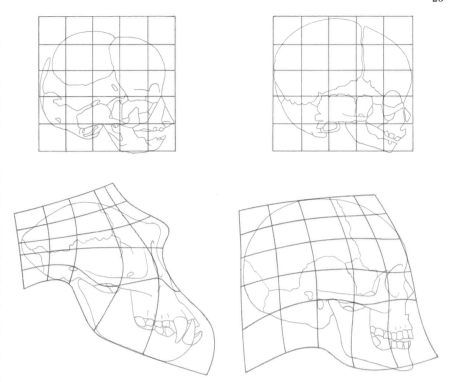

NEOTENY OF HUMAN SKULL is evident when the growth of the chimpanzee skull (*left*) and of the human skull (*right*) is plotted on transformed coordinates, which show the relative displacement of each part. The chimpanzee and the human skulls are much more similar at the fetal stage (*top*) than they are at the adult stage (*bottom*). The adult human skull also departs less from the fetal form than the adult chimpanzee skull departs from its fetal form, except in the case of the chin, which becomes relatively larger in human beings. The chin is a mental construct, however: the result of allometry, or differential growth, of different parts of human jaw.

other things are not equal, if a change in a trait as a solution to one problem changes the organism's relation to other problems of the environment, it becomes impossible to carry out the analysis part by part, and we are left in the hopeless position of seeing the whole organism as being adapted to the whole environment.

The mechanism by which organisms are said to adapt to the environment is that of natural selection. The theory of evolution by natural selection rests on three necessary principles: Different individuals within a species differ from one another in physiology, morphology and behavior (the principle of variation); the variation is in some way heritable, so that on the average offspring resemble their parents more than they resemble other individuals (the principle of heredity); different variants leave different numbers of offspring either immediately or in remote generations (the principle of natural selection).

These three principles are necessary and sufficient to account for evolutionary change by natural selection. There must be variation to select from; that variation must be heritable, or else there will be no progressive change from gen-

eration to generation, since there would be a random distribution of offspring even if some types leave more offspring than others. The three principles say nothing, however, about adaptation. In themselves they simply predict change caused by differential reproductive success without making any prediction about the fit of organisms to an ecological niche or the solution of ecological problems.

Adaptation was introduced by Darwin into evolutionary theory by a fourth principle: Variations that favor an individual's survival in competition with other organisms and in the face of environmental stress tend to increase reproductive success and so tend to be preserved (the principle of the struggle for existence). Darwin made it clear that the struggle for existence, which he derived from Thomas Malthus' *An Essay on the Principle of Population,* included more than the actual competition of two organisms for the same resource in short supply. He wrote: "I should premise that I use the term Struggle for Existence in a large and metaphorical sense.... Two canine animals in a time of dearth, may be truly said to struggle with each other which shall get food and live. But a plant on the edge of the desert is said to struggle for life against the drought."

The diversity that is generated by various mechanisms of reproduction and mutation is in principle random, but the diversity that is observed in the real world is nodal: organisms have a finite number of morphologies, physiologies and behaviors and occupy a finite number of niches. It is natural selection, operating under the pressures of the struggle for existence, that creates the nodes. The nodes are "adaptive peaks," and the species or other form occupying a peak is said to be adapted.

More specifically, the struggle for existence provides a device for predicting which of two organisms will leave more offspring. An engineering analysis can determine which of two forms of zebra can run faster and so can more easily escape predators; that form will leave more offspring. An analysis might predict the eventual evolution of zebra locomotion even in the absence of existing differences among individuals, since a careful engineer might think of small improvements in design that would give a zebra greater speed.

When adaptation is considered to be the result of natural selection under the pressure of the struggle for existence, it is seen to be a relative condition rather than an absolute one. Even though a species may be surviving and numerous, and therefore may be adapted in an absolute sense, a new form may arise that has a greater reproductive rate on the same resources, and it may cause the extinction of the older form. The concept of relative adaptation removes the apparent tautology in the theory of natural selection. Without it the theory of natural selection states that fitter individuals have more offspring and then defines the fitter as being those that leave more offspring; since some individuals will always have more offspring than others by sheer chance, nothing is explained. An analysis in which problems of design are posed and characters are understood as being design solutions breaks through this tautology by predicting in advance which individuals will be fitter.

The relation between adaptation and natural selection does not go both ways. Whereas greater relative adaptation leads to natural selection, natural selection does not necessarily lead to greater adaptation. Let us contrast two evolutionary scenarios. We begin with a resource-limited population of 100 insects of type A requiring one unit of food resource per individual. A muta-

tion to a new type a arises that doubles the fecundity of its bearers but does absolutely nothing to the efficiency of the utilization of resources. We can calculate what happens to the composition, size and growth rate of the population over a period of time [see illustration below]. In a second scenario we again begin with the population of 100 individuals of type A, but now there arises a different mutation a, which does nothing to the fecundity of its bearers but doubles their efficiency of resource utilization. Again we can calculate the population history.

In both cases the new type a replaces the old type A. In the case of the first mutation nothing changes but the fecundity; the adult population size and the growth rate are the same throughout the process and the only effect is that twice as many immature stages are being produced to die before adulthood. In the second case, on the other hand, the population eventually doubles its adult members as well as its immature members, but not its fecundity. In the course of its evolution the second population has a growth rate greater than 1 for a while but eventually attains a constant size and stops growing.

In which of these populations, if in either, would the individuals be better

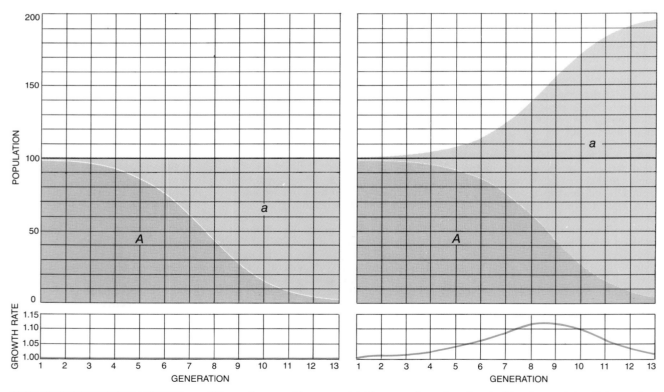

TWO DIFFERENT MUTATIONS have different demographic results for a resource-limited population of 100 insects. In one case (left) a mutation arises that doubles the fecundity of its bearers. The new type (a) replaces the old type (A), but the total population does not increase: the growth rate (bottom) remains 1.00. In the other case (right) a mutation arises that doubles the carrier's efficiency of resource utilization. Now the new population grows more rapidly, but only for a short time: eventually the growth rate falls back to 1.00 and the total population is stabilized at 200. The question is: Has either mutation given rise to a population that is better adapted?

adapted than those in the old population? Those with higher fecundity would be better buffered against accidents such as sudden changes in temperature since there would be a greater chance that some of their eggs would survive. On the other hand, their offspring would be more susceptible to the epidemic diseases of immature forms and to predators that concentrate on the more numerous immature forms. Individuals in the second population would be better adapted to temporary resource shortages, but also more susceptible to predators or epidemics that attack adults in a density-dependent manner. Hence there is no way we can predict whether a change due to natural selection will increase or decrease the adaptation in general. Nor can we argue that the population as a whole is better off in one case than in another. Neither population continues to grow or is necessarily less subject to extinction, since the larger number of immature or adult stages presents the same risks for the population as a whole as it does for individual families.

Unfortunately the concept of relative adaptation also requires the ceteris paribus assumption, so that in practice it is not easy to predict which of two forms will leave more offspring. A zebra having longer leg bones that enable it to run faster than other zebras will leave more offspring only if escape from predators is really the problem to be solved, if a slightly greater speed will really decrease the chance of being taken and if longer leg bones do not interfere with some other limiting physiological process. Lions may prey chiefly on old or injured zebras likely in any case to die soon, and it is not even clear that it is speed that limits the ability of lions to catch zebras. Greater speed may cost the zebra something in feeding efficiency, and if food rather than predation is limiting, a net selective disadvantage might result from solving the wrong problem. Finally, a longer bone might break more easily, or require greater developmental resources and metabolic energy to produce and maintain, or change the efficiency of the contraction of the attached muscles. In práctice relative-adaptation analysis is a tricky game unless a great deal is known about the total life history of an organism.

Not all evolutionary change can be understood in terms of adaptation. First, some changes will occur directly by natural selection that are not adaptive, as for example the changes in fecundity and feeding efficiency in the hypothetical example I cited above.

Second, many changes occur indirectly as the result of allometry, or differential growth. The rates of growth of different parts of an organism are different,

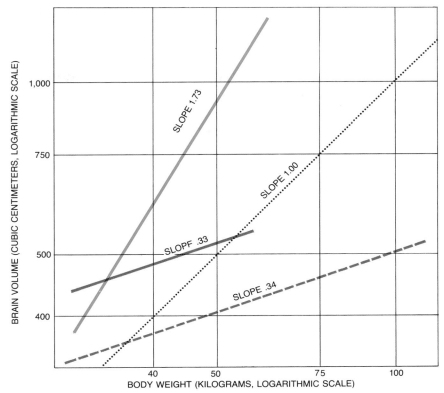

ALLOMETRY, or differential growth rates for different parts, is responsible for many evolutionary changes. Allometry is illustrated by this comparison of the ratio of brain size to body weight in a number of species of the pongids, or great apes (*broken black curve*), of *Australopithecus*, an extinct hominid line (*solid black*), and of hominids leading to modern man (*color*). A slope of less than 1.00 means the brain has grown more slowly than the body. The slope of more than 1.00 for the human lineage indicates a clear change in the evolution of brain size.

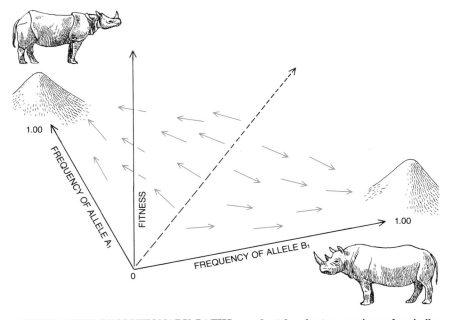

ALTERNATIVE EVOLUTIONARY PATHS may be taken by two species under similar selection pressures. The Indian rhinoceros has one horn and the African rhinoceros has two horns. The horns are adaptations for protection in both cases, but the number of horns does not necessarily constitute a specifically adaptive difference. There are simply two adaptive peaks in a field of gene frequencies, or two solutions to the same problem; some variation in the initial conditions led two rhinoceros populations to respond to similar pressures in different ways. For each of two hypothetical genes there are two alleles: A_1 and A_2, B_1 and B_2. A population of genotype A_1B_2 has one horn and a population of genotype A_2B_1 has two horns.

so that large organisms do not have all their parts in the same proportion. This allometry shows up both between individuals of the same species and between species. Among primate species the brain increases in size more slowly than the body; small apes have a proportionately larger brain than large apes. Since the differential growth is constant for all apes, it is useless to seek an adaptive reason for gorillas' having a relatively smaller brain than, say, chimpanzees.

Third, there is the phenomenon of pleiotropy. Changes in a gene have many different effects on the physiology and development of an organism. Natural selection may operate to increase the frequency of the gene because of one of the effects, with pleiotropic, or unrelated, effects being simply carried along. For example, an enzyme that helps to detoxify poisonous substances by converting them into an insoluble pigment will be selected for its detoxification properties. As a result the color of the organism will change, but no adaptive explanation of the color per se is either required or correct.

Fourth, many evolutionary changes may be adaptive and yet the resulting differences among species in the character may not be adaptive; they may simply be alternative solutions to the same problem. The theory of population genetics predicts that if more than one gene influences a character, there may often be several alternative stable equilibriums of genetic composition even when the force of natural selection remains the same. Which of these adaptive peaks in the space of genetic composition is eventually reached by a population depends entirely on chance events at the beginning of the selective process. (An exact analogy is a pinball game. Which hole the ball will fall into under the fixed force of gravitation depends on small variations in the initial conditions as the ball enters the game.) For example, the Indian rhinoceros has one horn and the African rhinoceros has two. Horns are an adaptation for protection against predators, but it is not true that one horn is specifically adaptive under Indian conditions as opposed to two horns on the African plains. Beginning with two somewhat different developmental systems, the two species responded to the same selective forces in slightly different ways.

Finally, many changes in evolution are likely to be purely random. At the present time population geneticists are sharply divided over how much of the evolution of enzymes and other molecules has been in response to natural selection and how much has resulted from the chance accumulation of mutations. It has proved remarkably difficult to get compelling evidence for changes in enzymes brought about by selection, not to speak of evidence for adaptive changes; the weight of evidence at present is that a good deal of amino acid substitution in evolution has been the result of the random fixation of mutations in small populations. Such random fixations may in fact be accelerated by natural selection if the unselected gene is genetically linked with a gene that is undergoing selection. The unselected gene will then be carried to high frequency in the population as a "hitchhiker."

If the adaptationist program is so fraught with difficulties and if there are so many alternative explanations of evolutionary change, why do biologists not abandon the program altogether?

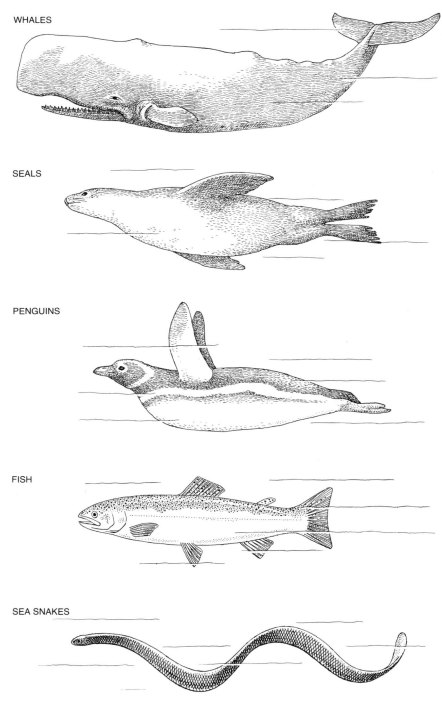

WHALES

SEALS

PENGUINS

FISH

SEA SNAKES

REALITY OF ADAPTATION is demonstrated by the indisputable fact that unrelated groups of animals do respond to similar selective pressures with similar adaptations. Locomotion in water calls for a particular kind of structure. And the fact is that whales and seals have flippers and flukes, penguins have paddles, fish have fins and sea snakes have a flat cross section.

There are two compelling reasons. On the one hand, even if the assertion of universal adaptation is difficult to test because simplifying assumptions and ingenious explanations can almost always result in an ad hoc adaptive explanation, at least in principle some of the assumptions can be tested in some cases. A weaker form of evolutionary explanation that explained some proportion of the cases by adaptation and left the rest to allometry, pleiotropy, random gene fixations, linkage and indirect selection would be utterly impervious to test. It would leave the biologist free to pursue the adaptationist program in the easy cases and leave the difficult ones on the scrap heap of chance. In a sense, then, biologists are forced to the extreme adaptationist program because the alternatives, although they are undoubtedly operative in many cases, are untestable in particular cases.

On the other hand, to abandon the notion of adaptation entirely, to simply observe historical change and describe its mechanisms wholly in terms of the different reproductive success of different types, with no functional explanation, would be to throw out the baby with the bathwater. Adaptation is a real phenomenon. It is no accident that fish have fins, that seals and whales have flippers and flukes, that penguins have paddles and that even sea snakes have become laterally flattened. The problem of locomotion in an aquatic environment is a real problem that has been solved by many totally unrelated evolutionary lines in much the same way. Therefore it must be feasible to make adaptive arguments about swimming appendages. And this in turn means that in nature the ceteris paribus assumption must be workable.

It can only be workable if both the selection between character states and reproductive fitness have two characteristics: continuity and quasi-independence. Continuity means that small changes in a characteristic must result in only small changes in ecological relations; a very slight change in fin shape cannot cause a dramatic change in sexual recognition or make the organism suddenly attractive to new predators. Quasi-independence means that there is a great variety of alternative paths by which a given characteristic may change, so that some of them will allow selection to act on the characteristic without altering other characteristics of the organism in a countervailing fashion; pleiotropic and allometric relations must be changeable. Continuity and quasi-independence are the most fundamental characteristics of the evolutionary process. Without them organisms as we know them could not exist because adaptive evolution would have been impossible.

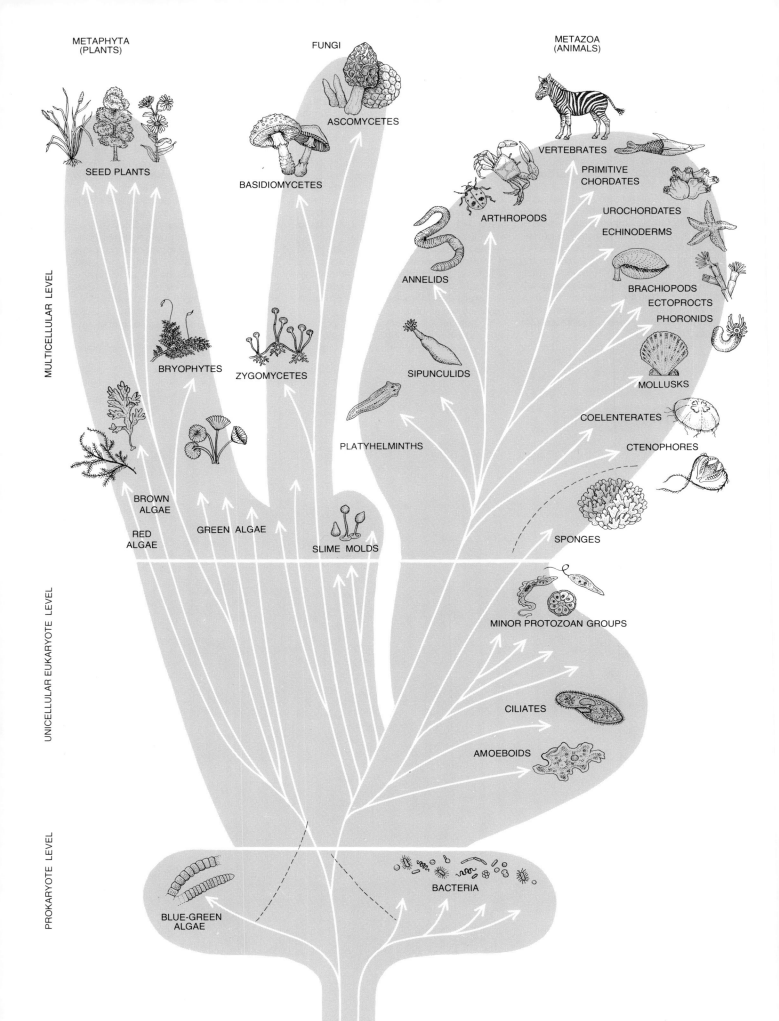

METAPHYTA
(PLANTS)

FUNGI

METAZOA
(ANIMALS)

SEED PLANTS

ASCOMYCETES

VERTEBRATES

BASIDIOMYCETES

PRIMITIVE
CHORDATES

ARTHROPODS

UROCHORDATES

ECHINODERMS

ANNELIDS

BRACHIOPODS
ECTOPROCTS
PHORONIDS

BRYOPHYTES

ZYGOMYCETES

SIPUNCULIDS

MOLLUSKS

MULTICELLULAR LEVEL

COELENTERATES

CTENOPHORES

PLATYHELMINTHS

BROWN
ALGAE

RED
ALGAE

GREEN ALGAE

SLIME MOLDS

SPONGES

MINOR PROTOZOAN GROUPS

UNICELLULAR EUKARYOTE LEVEL

CILIATES

AMOEBOIDS

PROKARYOTE LEVEL

BACTERIA

BLUE-GREEN
ALGAE

3

The Evolution of Multicellular Plants and Animals

by James W. Valentine
December 1978

It has been only during the last fifth of the history of life on the earth that multicellular organisms have existed. They appear to have arisen from unicellular organisms on numerous occasions

The animals and plants one sees on the land, in the air and on the water are all multicellular, made up of millions and in some cases billions of individual cells. Even the simplest multicellular organisms include several different kinds of cells, and the more complicated ones have as many as 200 different kinds. All the multicellular plants and animals have evolved from unicellular eukaryotes of the kind described by J. William Schopf in the preceding article. G. Ledyard Stebbins of the University of California at Davis estimates that multicellular organisms have evolved independently from unicellular ancestors at least 17 times. At least two million multicellular species exist today, and many others have come and gone over the ages.

Clearly the multicellular grade of construction is advantageous and successful. The chief advantages of multicellularity stem from the repetition of cellular machinery it entails. From this feature flows the ability to live longer (since individual cells can be replaced), to produce more offspring (since many cells can be devoted to reproduction), to be larger and so to have a greater internal physiological stability, and to construct bodies with a variety of architectures. Moreover, cells can become differentiated (specialized for a particular function, as nerve and muscle cells are), with a resulting increase in functional efficiency. The particular advantages that were the keys to the evolution of multicellularity probably varied from case to case.

The largest categories employed for the classification of organisms are the kingdoms. Multicellular organisms are placed in one or another of three kingdoms on the basis of their broad modes of life and particularly of their modes of obtaining energy. Plants, which are autotrophs (meaning that they require only inorganic compounds as nutrients), utilize the energy of the sun to create living matter through photosynthesis; they make up the kingdom Metaphyta. Fungi (such as mushrooms, which are plantlike but feed by ingesting organic substances) make up the kingdom Fungi. Animals, which are also ingesters, comprise the kingdom Metazoa. Each kingdom includes more than one lineage that evolved independently from the kingdom Protista, consisting of eukaryotic unicellular organisms.

Much of what is known about the evolution of multicellular organisms comes from the fossil record. The Fungi are so poorly represented as fossils that their evolution is obscure, but the other kingdoms have a rich fossil history.

The patterns of adaptation that one can observe today amply demonstrate the effectiveness of evolution in shaping organisms to cope with their environment. Each environment contains animals particularly suited to exploit the conditions in it; each kind of organism has been developed by selection to perform a role in the biosphere. When an environment changes, natural selection acts to change the adaptations; sometimes new environmental roles are evolved. The history of life reflects an interaction between environmental change on the one hand and the evolutionary potential of organisms on the other. It is therefore of interest to briefly examine the major causes of the more biologically important environmental changes.

Notable among them are the processes of plate tectonics. Continents, riding on huge plates of the lithosphere (the outer layer of the solid earth) that move at a rate of a few centimeters per year, break up or collide and can become welded together following a collision. Thus a continent can fragment or grow, the number of continents can increase or decrease and the geographic patterns of a continent can change radically. Ocean basins too alter their sizes, numbers, positions and patterns. The consequences for living organisms can be profound.

Consider only one of the possible events resulting from plate tectonics: the collision of two continents, once widely separated, to form a single larger continent. The accompanying changes in the biological environment are far-reaching. The most obvious change is that the barriers to migration are broken down and the biotas of two continents now compete for existence on one continent. For many land animals the continental interior is now farther from the sea, and the moderating effects of marine air and temperature are diminished. Mountains rising along the suture will diversify the environment further, creating rain "shadows" and perhaps deserts if they happen to interrupt major flows of moisture-laden wind.

Entirely new environmental conditions may thus appear. They create opportunities for new modes of life, as does the general diversification of conditions following the collision. The biotas of the two former continents are subjected to competition and to environmental conditions for which they have not been adapted, and at the same time they are presented with novel opportunities. Evolution can be expected to produce a considerable change in the flora

KINGDOMS OF ORGANISMS are charted according to a concept originated by Robert H. Whittaker of Cornell University. The relatively simple unicellular Monera, which are prokaryotic (lacking a nucleus), gave rise to the more complex unicellular Protista, from which all three multicellular kingdoms have arisen. Multicellular organisms are placed in the kingdoms Metaphyta, Fungi and Metazoa mainly on the basis of the process by which they obtain their energy.

and fauna. The marine organisms of the shallow continental shelves, where 90 percent of marine species live today, will also be affected. For many of them the increased continental area leads to a lowering of environmental stability, requiring new adaptive strategies. In general it is expected that fewer marine species could be supported around one large continent than could be supported around two separate smaller ones.

As oceans widen or narrow, as continents drift into cooler or warmer zones, as winds and ocean currents are channeled in new directions, the pattern and the quality of the environment change. The pace of change may often be slow, because of the low rate of sea-floor spreading, the phenomenon that drives the drifting of the continents and the opening of the oceans. At other times more rapid and dramatic changes can be expected, as for example when continents finally collide after millions of years of approaching each other or when an ocean current is finally deflected from an ancient path.

Just as changes in the environment can affect organisms, so can the activities of organisms affect the environment to create new conditions. A fateful example is the rise of free oxygen in the atmosphere, owing chiefly to photosynthesis. Early organisms could not have existed with free oxygen, whereas most contemporary organisms cannot exist without it. Another point to bear in mind is that organisms form part of one another's environment and interact in numerous ways: as predators, competitors, hosts and habitats. As populations of organisms increase, decrease or change, the environment changes too.

In the course of the diversification of the multicellular kingdoms over the past 700 million years major new types of organisms have appeared and several revolutions have taken place within established groups. In many instances it is possible to identify the kind of environmental opportunity (or, with extinctions, the environmental foreclosure) to which the biota is responding. The history of animals is known best. They arose at least twice: sponges from one protistan ancestor and the rest of the metazoans from another. The major categories into which animals are divided are phyla; over the aeons at least 35 phyla have evolved, of which 26 are living and nine are extinct.

The fossil record that reveals something of the circumstances of the early members of the phyla varies in quality according to the type of fossil. Certain trace fossils are the most easily preserved, particularly burrows and trails left behind in sediments by the activity of organisms. Next come durable skeletal remains, such as seashells and the bones of vertebrates. Finally, the fossils of entirely soft-bodied animals turn up on rare occasions, usually as impressions or films in ancient sea-floor sediments.

The earliest animal fossils are burrows that begin to appear in rocks younger than 700 million years, late in the Precambrian era. Both long horizontal burrows and short vertical ones are found, comparable in size to the burrows of many modern marine organisms. The ability to burrow implies that the animals had evolved hydrostatic skeletons, that is, fluid-filled body spaces that work against muscles, so that the animal could dig in the sea bed. Although some simple animals such as sea anemones manage to employ their water-filled gut as a hydrostatic skeleton and so to burrow weakly, long horizontal burrows suggest a more active animal, probably one with a coelom, or true body cavity. This is quite an advanced grade of organization to find near the base of the fossil record of multicellular forms. Trace fossils are rather rare until about 570 million years ago, when they increase remarkably in kind and number.

The next animal fossils, surprisingly, are soft-bodied remains from between 680 and 580 million years ago, called the Ediacaran fauna from the region in southern Australia where they are best known. The ones that are clearly identifiable with modern phyla are all jellyfishes and their allies, which are at a simple grade of construction. The other fossils are more enigmatic; a few may be allied with living phyla (one resembles annelid worms) and some may not be.

MILLIONS OF YEARS AGO	ERA	PERIOD	EPOCH	EVENTS
0	CENOZOIC	QUATERNARY	PLEISTOCENE	EVOLUTION OF MAN
50	CENOZOIC	TERTIARY	PLIOCENE MIOCENE OLIGOCENE EOCENE PALEOCENE	MAMMALIAN RADIATION
100	MESOZOIC	CRETACEOUS		LAST DINOSAURS FIRST PRIMATES FIRST FLOWERING PLANTS
150	MESOZOIC	JURASSIC		DINOSAURS FIRST BIRDS
200	MESOZOIC	TRIASSIC		FIRST MAMMALS THERAPSIDS DOMINANT
250	PALEOZOIC	PERMIAN		MAJOR MARINE EXTINCTION PELYCOSAURS DOMINANT
300	PALEOZOIC	CARBONIFEROUS / PENNSYLVANIAN		FIRST REPTILES
	PALEOZOIC	CARBONIFEROUS / MISSISSIPPIAN		SCALE TREES, SEED FERNS
350	PALEOZOIC	DEVONIAN		FIRST AMPHIBIANS JAWED FISHES DIVERSIFY
400	PALEOZOIC	SILURIAN		FIRST VASCULAR LAND PLANTS
450	PALEOZOIC	ORDOVICIAN		BURST OF DIVERSIFICATION IN METAZOAN FAMILIES
500–550	PALEOZOIC	CAMBRIAN		FIRST FISH FIRST CHORDATES
600	PRECAMBRIAN	EDIACARAN		FIRST SKELETAL ELEMENTS
650–700	PRECAMBRIAN	EDIACARAN		FIRST SOFT-BODIED METAZOANS FIRST ANIMAL TRACES (COELOMATES)

MAJOR EVENTS in the evolution of multicellular organisms over the past 700 million years are depicted chronologically. The data are based principally on what is revealed by fossils.

SOFT-BODIED ANIMAL of middle Cambrian times left a record in the Burgess Shale of British Columbia. The creature was a polychaete worm with a number of setae, or bristlelike parts, that appear clearly here. The photograph of the fossil, which is enlarged about five diameters, was made in ultraviolet radiation by S. Conway Morris of the University of Cambridge. Lamp was set at a high angle to specimen.

PLANT FOSSILS include a tree fern from Jurassic times (*left*) and a birch leaf from the Miocene epoch (*right*). The ferns were among the earliest land plants, part of the dominant land flora in Devonian times. Birch leaf represents a more advanced group, vascular land plants.

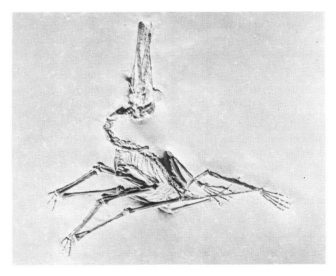

SKELETAL ANIMALS represented in the fossil record include a bed of crinoids (*left*) from the upper Cretaceous period and a pterodactyl (*right*) of late Jurassic times. The crinoids were marine echinoderms, a group represented today by such organisms as starfishes and sea urchins. The pterodactyl was a flying reptile that had a membranous, featherless wing. This specimen was found in Germany.

It is probable that some are coelomates.

Durable skeletal remains finally appear in the fossil record in rocks that are about 580 million years old. The earliest ones are minute scraps, denticles and plates of unknown affinity that were parts of larger animals. Then, starting some 570 million years ago and continuing over the next 50 million years or so, nearly all the coelomate phyla that possess durable skeletons appear in what is in evolutionary terms a quick succession. The exceptions are the phylum Chordata (which nonetheless does appear as a soft-bodied fossil) and the phylum Bryozoa, which finally appears less than 500 million years ago. These durable skeletonized invertebrates seem to have one thing in common: they all originally lived on the sea floor rather than burrowing in it, although one group (the extinct Trilobita) probably grubbed extensively in the sea floor, digging shallow pits or perhaps burrows in search of food.

A highly valuable fossil assemblage from near the middle of the Cambrian period is found in British Columbia in a rock unit named the Burgess Shale. Much of the fauna from this shale is soft-bodied, trapped by rapidly deposited muds and preserved as mineral films by a process that has yet to be determined. Here is an array of more or less normal invertebrate skeletons associated with such soft-bodied phyla as the Annelida (the group that includes the living earthworms), the Priapulidae (possibly pseudocoelomate worms) and the Chordata. The Burgess Shale also contains several animals that represent phyla not previously known. Only one of the phyla could be ancestral to a living phylum; the others are entirely distinct lineages that arose from unknown late Precambrian ancestors and have all become extinct.

From these facts and from the wealth of accumulated evidence on the comparative embryology and morphology of the living representatives of these fossil groups it is possible to build a picture of the rise of the major animal groups. The animals for which the fewest clues exist are the first ones, the founding metazoans, although there is no shortage of speculation on what they may have been like.

Since the evolution of novel organisms involves adaptation to new or previously unexploited conditions, one can imagine the adaptive pressures the earliest multicellular forms would have faced and can thereby derive a variety of plausible animals radiating from a trunk lineage into the available modes of life. Bottom-dwelling animals can eat deposited or suspended food. Suspension feeders must maximize the volume of water on which they can draw. Body shapes such as domes that jut up into the water and create turbulence in the ambient currents would increase the volume of water washing the animal and would be advantageous. In quiet waters taller cylinders or extensions of the body, perhaps in the form of filaments or tentacles, would increase the feeding potential. For deposit feeders flattened shapes would maximize the area of contact with the bottom.

A differentiation of function among cells would be an early trend among such animals: In a deposit feeder, for example, the cells on the bottom would be ingesting food and so could most easily specialize for digestion, perhaps becoming internalized to increase their number and stabilize the digestive processes. Covering cells would be supportive and protective. Marginal cells could specialize for locomotion. Well-nourished cells surrounding the digestive area might specialize in reproduction.

Patterns of differentiation would vary in organisms of other shapes. In quiet-water suspension feeders the digestive cells might lie uppermost. Floating animals would have other shapes, perhaps radial or globose. Any such simple animal type could have been ancestral to the remaining phyla. Such animals are essentially at the same grade as sponges, but not one of them survives today. (Sponges are so distinctive in their developmental pattern that they probably arose from unicellular organisms independently of all other animal phyla.)

It is impossible to estimate how long before 700 million years ago the first animals evolved; it could easily have been as little as 50 million years or as much as 500 million. At any rate the evolutionary trends of differentiation in organs and tissues led to the rather complex invertebrates, most or all of which arose as burrowers in the sea floor. From the body plan of living phyla that first appeared during the Cambrian period one can infer the types of body plan that must have been present among late Precambrian animals. For example, two of the most important types of coelomates in Cambrian times were metamerous and oligomerous forms.

In the metamerous type the coelom is divided by transverse septa into a large number of compartments, as it is in earthworms. The muscular activity associated with burrowing affects only the segments in the immediate vicinity of the contractions, so that the efficiency of burrowing is enhanced. The oligomerous body plan has only three (sometimes only two) coelomic compartments separated by transverse septa. Each coelomic region functions differently. In a particularly primitive body type, represented by living phoronid worms, the regions correspond to a long trunk in which the coelom is used in burrowing and to a tentacular crown that is employed in feeding.

It seems that between the development (perhaps more than once) of the coelom about 700 million years ago and the appearance of animals with durable skeletons about 570 million years ago coelomate worms diversified considerably within their adaptive zones. Most of these groups lived in burrows within soft sediments of the ancient sea floor. Then, beginning some 570 million years ago, a rich fauna with durable skeletons appeared. The animals lived on the sea floor, where conditions are considerably different from those within it. Most of the lineages were modified in adapting to this new zone. Animals that descended from oligomerous suspension feeders tended to evolve skeletons that served to protect and support their feeding apparatus. Some of the more mobile burrowers, such as metamerous worms, developed a jointed external skeleton with jointed appendages, which operated largely as a system of levers. Their living descendants include insects, crabs and shrimps (phylum Arthropoda).

In these and other instances the evolution of a durable skeleton was associated with a large number of coadapted changes in the anatomy of the soft parts. For example, the arthropod coelom was reduced, since the exoskeleton took over locomotion and limbs replaced the wavelike action of peristalsis in achieving movement. Septa between segments were no longer required, and they disappeared. The original metamerous architecture is still evidenced by a regular series of paired internal organs, inherited from a segmented ancestry.

All the evidence points to a major phylum-level diversification near the beginning of the Paleozoic era. Phyla that still exist appeared, along with the extinct soft-bodied phyla represented in the Burgess Shale. Some early skeletal types that are now extinct may also have been distinctive phyla. One phylum resembled sponges, and another may have been allied to mollusks. At least a third again as many phyla as are alive are known from Paleozoic fossils, and there must have been even more.

The environmental context of this radiation is not understood. It has been

LIVING ANIMAL PHYLA are grouped on the opposite page according to their body architecture, with a representative of each phylum portrayed. The groups are (*a*) simple multicellular forms in which a single tissuelike layer surrounds a central cavity; (*b*) forms with two tissue layers, one of which surrounds a well-defined gut; (*c*) wormlike forms with three tissue layers, the middle one forming the core of the body surrounding the gut; (*d*) small, usually parasitic forms with three tissue layers and a primitive body cavity, and (*e–h*) four groups with "true" body cavities that are insulated from the external environment; the form of the cavity differs somewhat from group to group. There are 26 living phyla and at least nine extinct ones.

PORIFERA (SPONGES)

a

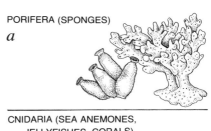

CNIDARIA (SEA ANEMONES, JELLYFISHES, CORALS)

b

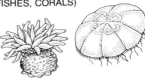

CTENOPHORA (COMB JELLIES)

PLATYHELMINTHES (FLATWORMS)

c

NEMERTINA (PROBOSCIS OR RIBBON WORMS)

ACANTHOCEPHALA (SPINY-HEADED WORMS)

d

ROTIFERA (WHEEL ANIMALS)

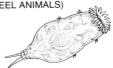

GASTROTRICHA (SCALED WORMS)

KINORHYNCHA (SPINY-SKINNED WORMS)

NEMATODA (ROUNDWORMS)

PRIAPULIDA (PRIAPUS WORMS)

d

ENTOPROCTA (TENTACLED, STALKED ANIMALS)

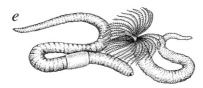

ANNELIDA (EARTHWORMS, FAN WORMS)

e

ARTHROPODA (INSECTS, CRABS, SHRIMPS, BARNACLES)

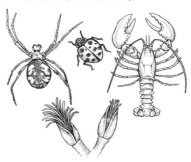

SIPUNCULIDA (PEANUT WORMS)

f

ECHIUROIDEA (SAUSAGE-SHAPED MARINE WORMS)

MOLLUSCA (CLAMS, SNAILS, OCTOPUS, SQUID)

g

PHORONIDA (HORSESHOE WORMS)

h

BRACHIOPODA (LAMPSHELLS)

h

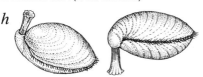

ECTOPROCTA (BRYOZOA OR MOSS ANIMALS)

CHAETOGNATHA (ARROWWORMS)

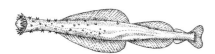

POGONOPHORA (DEEP-SEA WORMS)

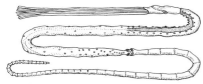

ECHINODERMATA (STARFISHES, SEA URCHINS, SAND DOLLARS)

HEMICHORDATA (ACORN WORMS)

UROCHORDATA (SEA SQUIRTS)

CHORDATA (AMPHIOXUS, FISHES, AMPHIBIANS, REPTILES, BIRDS, MAMMALS)

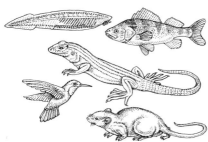

suggested that oxygen may finally have reached a level sufficiently high to support active animals, stimulating an evolutionary burst. Another idea is that the environment became stabler.

Among the many lineages evolving novel modes of life at the time was one that developed a swimming specializa-tion and diverged significantly from other animal groups. This was a suspension-feeding oligomerous animal that evolved a stiffening but flexible dorsal cord and characteristic chevron-shaped blocks of muscle that could flex the cord from side to side for swimming. Eventually the animals developed durable parts: an outer armor of plates for protection and an axial skeleton of articulated vertebrae with elongated lateral flanges to support the body walls. These were the earliest fishes. They were jawless and had unpaired fins. They fed by taking in water through an anterior mouth, ingesting suspended material as

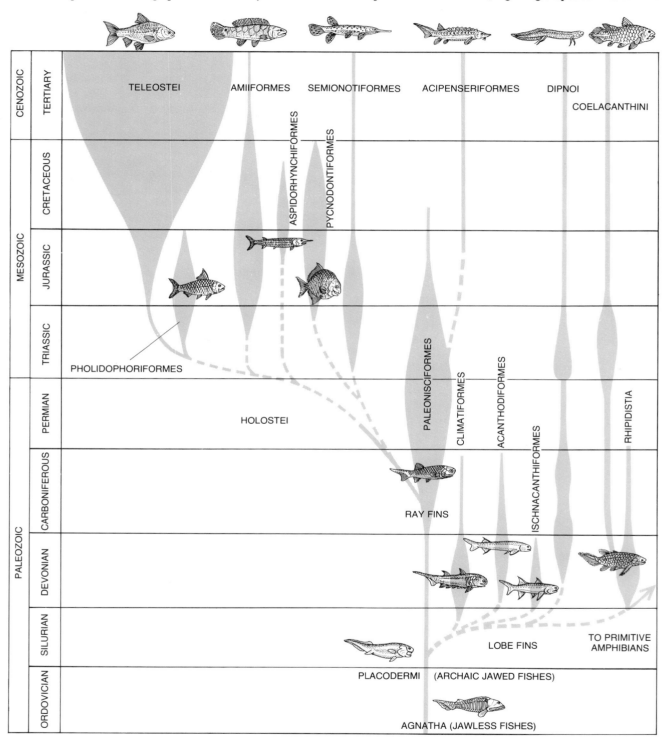

EVOLUTION OF FISHES began in the Cambrian period with the appearance of jawless species that had evolved from a simpler multicellular marine organism. In the Devonian period jaws evolved from a pair of gill slits, and fins became paired. One major line of jawed fishes, the ray fins, was ancestral to most of the fish species now living. The other line, the lobe fins, did not fare as well but led eventually to the earliest amphibians, the ancestors of all four-footed vertebrates. In large part the ability of the early amphibians to venture onto the land was due to the evolution there of multicellular plants that sustained them, giving rise later to animals that lived entirely on land.

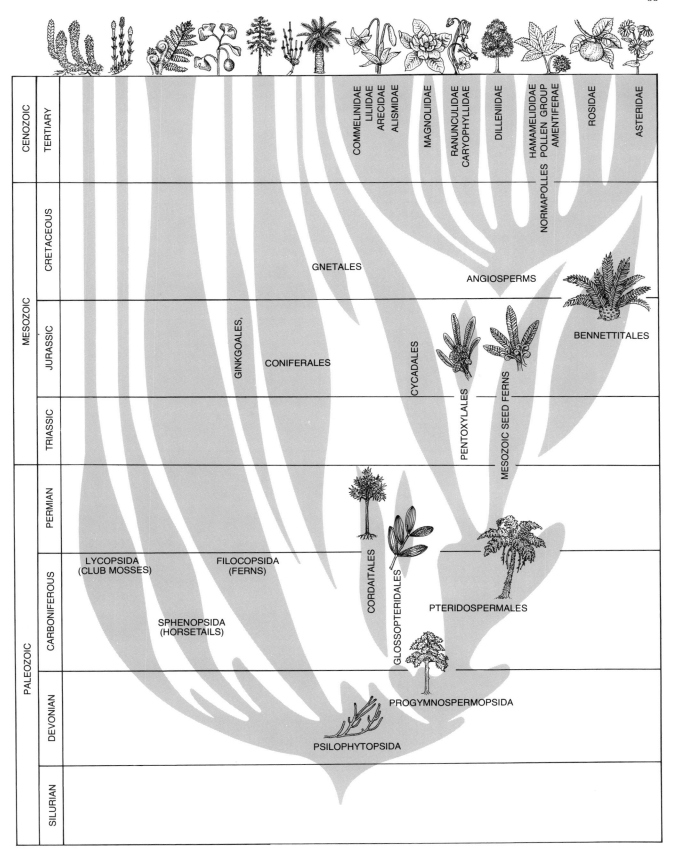

VASCULAR LAND PLANTS probably evolved according to the pattern indicated in this diagram. In the Devonian period the primitive assemblage of horsetails, club mosses and ferns dominated the flora. Those forms reproduce by spores and prefer humid conditions. Seed- and pollen-bearing plants developed by Devonian times, and by the Permian the conifers had begun an expansion that made them dominant in the Mesozoic era. Late in the Cretaceous period the angiosperms (flowering plants) spread explosively and became dominant.

food and expelling the filtered water through their gills.

The early jawless fishes, the agnathans, are first known from late Cambrian fossils. They continued as a successful group into the middle of the Paleozoic era, but they diversified only modestly. At some time in the Devonian period two developments revolutionized the fishes: jaws evolved from the anterior pair of gill slits and fins became paired. With a greatly expanded choice of food items and an improved swimming balance, the jawed fishes diversified spectacularly. (The agnathans declined, although they may still be represented by such forms as the lampreys.)

Two major types of jawed fishes were the ray fins and the lobe fins. The great majority of living fishes have descended from the ray fins. The lobe fins were far less successful as fishes (they survive only as lungfishes and a

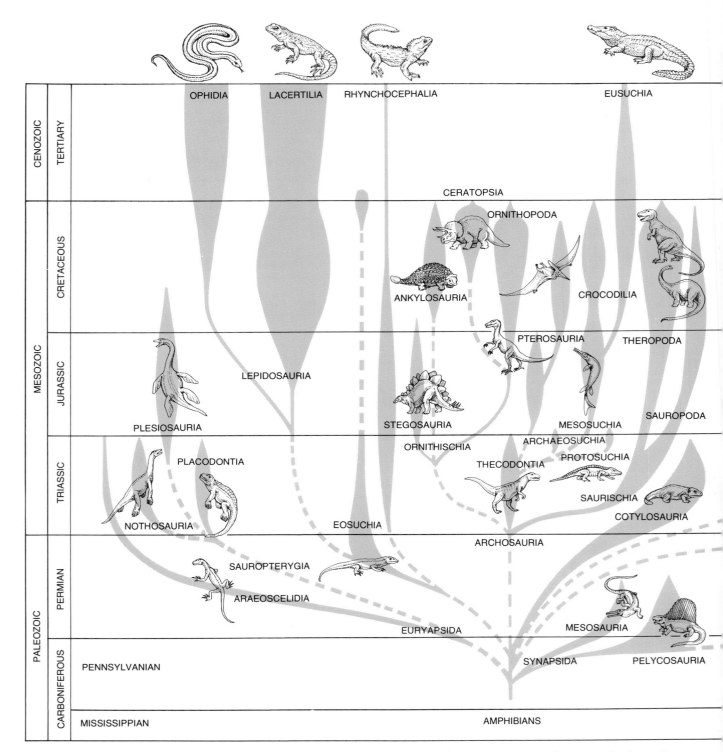

REPTILES EVOLVED from a primitive amphibian stock, which became free from ties to the water by evolving an egg that could develop in a terrestrial setting. Diversifying rapidly, the reptiles eventually came to dominate life on the land. The pelycosaurs and the therapsids were two particularly successful reptile groups, the therapsids gradually becoming the more diversified. In the Triassic period the dinosaurs evolved in the reptile line and began their 150 million years of dominance of the terrestrial environment. In this chart and in the

few relict marine forms), but they had bony supports within their fins from which limbs evolved. The earliest amphibians arose from a primitive group of lobe fins (the rhipidistians), and so all four-footed vertebrates (tetrapods) and their descendants also evolved from this vanished fish stock.

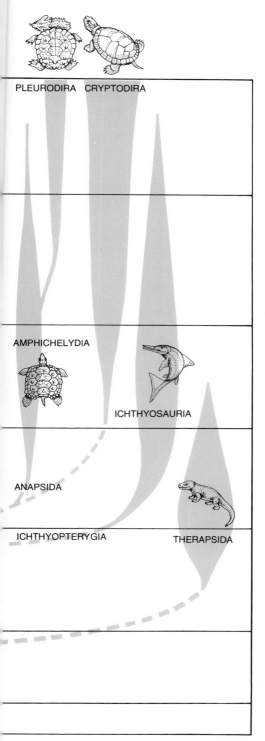

PLEURODIRA CRYPTODIRA

AMPHICHELYDIA

ICHTHYOSAURIA

ANAPSIDA

ICHTHYOPTERYGIA THERAPSIDA

other accompanying charts tracing the evolution of animal and plant groups the width of a line or a band indicates relative size of that group at the corresponding geological period.

The energy to support the early marine animals must have been supplied by photosynthesizing unicellular organisms at first (and it still is supplied chiefly by such forms), but at some time in the late Precambrian era multicellular marine algae evolved. Little is known of the diversity or abundance of these plants; they must have contributed detritus to invertebrate communities and may have been grazed directly. It is possible that soil bacteria, fungi and lower plant forms had colonized the land by the Cambrian period, and perhaps the fringes of swamps and embayments supported hardy, semiaquatic plant types. The first nonaquatic plant lines whose descendants form the major elements in terrestrial flora arose, however, in the Silurian period. The early plants spread from marshes and swamps into drier upland habitats. As the green belt expanded, animals followed it ashore: arthropods and probably worms, feeding on plant debris and eventually on the plants themselves. Thus prey items existed on land to support the populations of larger tetrapods that appeared during the Devonian period.

Many of the early amphibian lineages developed rather large body sizes and radiated into the available habitats, becoming herbivores and predators on many food items in aquatic, semiaquatic and terrestrial settings. Although they must have been quite hardy, perhaps rivaling reptiles in this respect, they were still bound to the water for reproduction, as frogs and salamanders are today. Modern amphibians are quite unlike the large forms that ruled the terrestrial domain for some 75 million years. The modern antecedents probably arose during the late Paleozoic era: small-bodied forms adapted to marginal habitats and utilizing resources not utilized by their larger relatives. Perhaps in this way they escaped competition with later vertebrates.

Reptiles arose from a primitive amphibian lineage, freed from ties to the water by the evolution of an egg that could develop in a terrestrial environment. The reptiles diversified rapidly and spread into all the environments occupied by their large amphibian cousins, becoming successful predators and competitors. The large amphibians declined to extinction late in the Triassic period. Even by late Permian times reptiles were well on the way to dominance. Two groups were particularly successful: the pelycosaurs (known by their large dorsal fins) and the therapsids, which may have been more active and aggressive in view of the fact that they eclipsed the pelycosaurs in diversity early in the Triassic.

The therapsids were replaced in turn as dominant reptiles by the dinosaurs, which evolved in Triassic times and did not become extinct until 150 million years later. During this interval they underwent several severe waves of extinction, which carried off the larger species disproportionately, but each time the dinosaurs reradiated from the surviving smaller stocks to maintain their ascendancy. Finally they disappeared at the close of the Cretaceous period about 65 million years ago.

During most of their tenure the dinosaurs shared the terrestrial world with a group of small, active, hairy animals that evolved from a predatory therapsid lineage: the mammals. The evolution of the mammals is particularly well recorded by fossils, which display the gradual appearance of mammalian skeletal features from reptilian ones. Unfortunately many mammalian characteristics cannot ordinarily be determined from fossils. They include warm-bloodedness, hair, a respiratory diaphragm, increased agility and facial muscles that allow suckling. The mammals must have shared at least some of these features with their therapsid ancestors. In mammals they form a unique adaptive assemblage.

Once the dinosaurs became extinct, mammals radiated into the vacant habitats to dominate in their turn the terrestrial environment. Since mammals are alert and active compared with living reptiles, it is puzzling that they did not radiate sooner to challenge the dinosaurs. A possible reason is that the dinosaurs themselves were active, alert and warm-blooded. The posture of dinosaurs suggests agility; their bones, compared with the bones of warm-blooded and cold-blooded tetrapods alive today, contain channels that suggest warm-bloodedness, and the fossil record of predator-prey ratios suggests that dinosaurs required a large amount of food, like warm-blooded mammals and unlike cold-blooded reptiles.

If dinosaurs were warm-blooded, it would help to explain their long dominance. The dinosaurs were larger than their mammalian contemporaries; even dinosaur hatchlings were larger than most mammals. Clearly the mammals played the smaller-animal roles and the dinosaurs the larger-animal ones. Even when repeated extinctions reduced the diversity of the dinosaurs, the survivors were still larger than the mammals and were able to hold their own and to furnish the basic lineages from which evolution re-created large animals.

The cause of the final extinction of the dinosaurs remains a mystery, but it appears to be related to the carrying capacity of the Cretaceous environment. Only small animals weighing no more than about 20 pounds survived the wave of terrestrial extinctions that closed the Cretaceous period; the mammals were smaller still.

During their long coexistence with the dinosaurs the mammals developed improvements that have stood them in

38

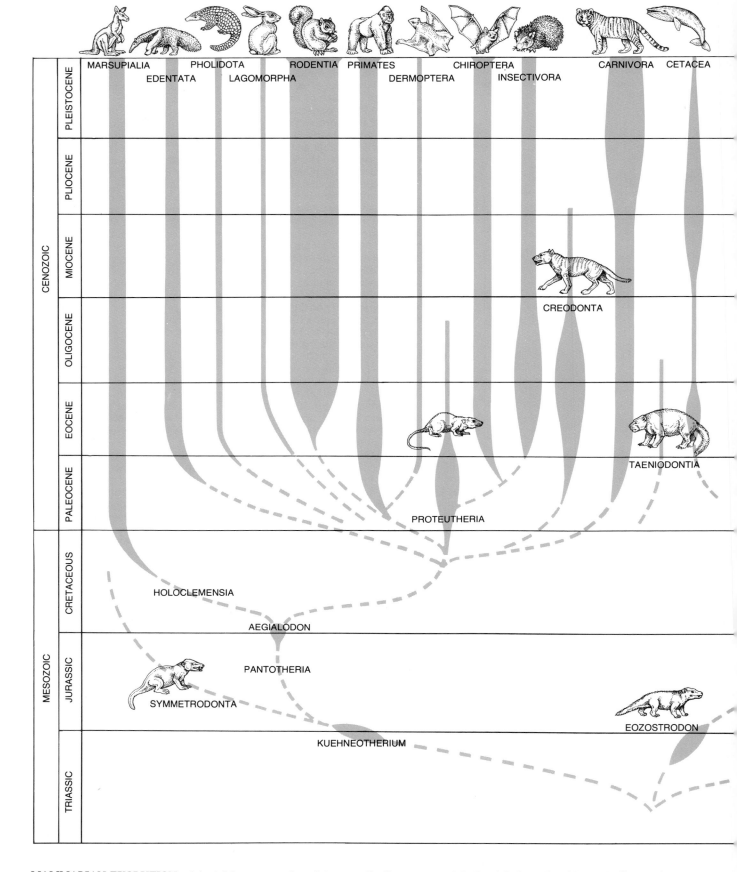

MARSUPIALIA   PHOLIDOTA   RODENTIA PRIMATES   CHIROPTERA   CARNIVORA CETACEA
EDENTATA   LAGOMORPHA   DERMOPTERA   INSECTIVORA

CREODONTA

TAENIODONTIA

PROTEUTHERIA

HOLOCLEMENSIA

AEGIALODON

PANTOTHERIA

SYMMETRODONTA

EOZOSTRODON

KUEHNEOTHERIUM

MAMMALIAN EVOLUTION originated in a group of predatory therapsids. The first mammals were quite small; the ones that survived the heavy extinction of land animals that came at the end of the Cretaceous period all weighed less than 20 pounds. Several lines of mammals survived the dinosaurs, and during the Cenozoic era the mammal group diversified widely. The first primates appeared while

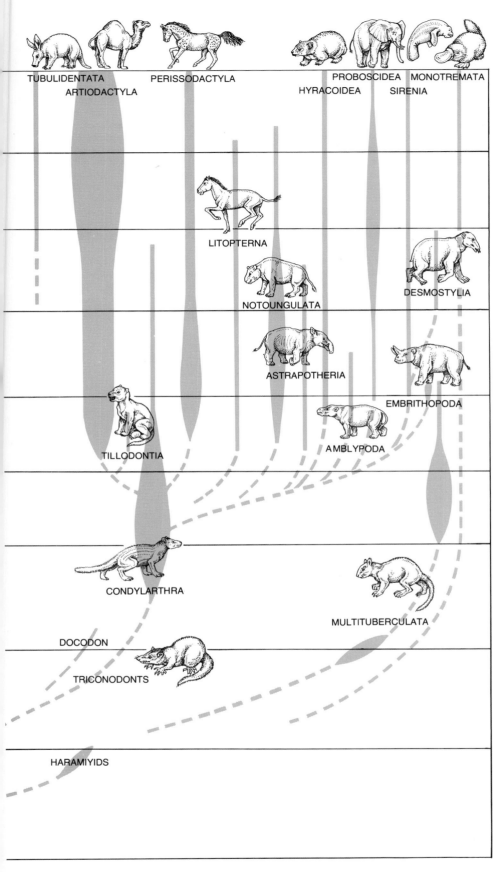

TUBULIDENTATA
ARTIODACTYLA
PERISSODACTYLA
PROBOSCIDEA MONOTREMATA
HYRACOIDEA SIRENIA

LITOPTERNA

NOTOUNGULATA

DESMOSTYLIA

ASTRAPOTHERIA

EMBRITHOPODA

AMBLYPODA

TILLODONTIA

CONDYLARTHRA

MULTITUBERCULATA

DOCODON

TRICONODONTS

HARAMIYIDS

the dinosaurs were still dominant; those primates were small, tree-dwelling creatures with habits probably resembling those of squirrels. The mammal line that led eventually to man took up life on the ground, probably only as foragers at first but later as highly successful hunters.

good stead. Mammalian forms with a placenta arose during the Cretaceous period and diversified moderately. Primates, the order including man, appeared while the dinosaurs still dominated. Hence several mammal lineages survived the dinosaurs. During the Cenozoic era they diversified impressively. A number of lineages that originated then have become extinct, so that fewer orders exist today than were living at times during the Cenozoic era.

The evolutionary activity of the mammals reached its peak within the past two million years, perhaps because of the great climatic diversity associated with the late Cenozoic glaciation. The latest mammalian episode has been a wave of extinction that was particularly severe for large mammals, including manlike species.

The primates of early Cenozoic times were small and probably squirrel-like in their habits. Many characteristic primate features, such as overlapping binocular visual fields, a short face, grasping forepaws and increased brain size and alertness, are probably adaptations to an arboreal existence. The lineage that led eventually to man descended from the trees to the forest floor to forage and eventually to hunt, perhaps coming to live at the margins of forests with access to moderately open country. A continuing adaptation to a terrestrial habitat led to an erect posture. Hunter-gatherer bands developed; perhaps the distinctive tooth arrangement of human beings, with its reduced canine teeth, was associated with dietary shifts as this kind of social evolution proceeded. It has been suggested that the final rise of the human species was associated with a further shift to big-game hunting, increasing the value of cunning, intelligence and cooperation.

The pattern of the history of land plants is similar to that of land vertebrates, with waves of extinction and replacement and the episodic rise of new forms to dominance. In the Devonian period, when the early forests appeared, the primitive assemblage of horsetails, club mosses and ferns spread and came to dominate the land flora. Plants of this form reproduce by spores and prefer humid conditions.

Plants bearing seeds and pollen developed as early as Devonian times. They diversified during the Carboniferous period; by the Permian one lineage, the conifers, began an expansion that led them to dominate the Mesozoic floras. The shift to conifers was associated with the appearance of drier climates.

Still another shift came late in the Cretaceous period, when flowering plants (angiosperms) spread explosively to conquer the terrestrial realm. (About a quarter of a million species of angiosperms are living today.) The earliest flowering plants seem to have been weedy, opportunistic species adapted for rapid reproduction. The reproductive specializations, including the development of flowers and the appearance of insect pollinating systems, were transformed into a general advantage over the more slowly growing conifers.

The details of the diversity and abundance of plant species through the Paleozoic and Mesozoic eras are largely unknown. The major transitions in dominant floral elements resemble what was happening to land animals, but as far as one can tell they do not correspond to the events that were affecting the animals. For example, the angiosperms were well established long before the dinosaurs were extinguished. Moreover, the several waves of extinction of tetrapods during the Mesozoic era are not reflected in the history of the land plants as it is now known.

The evolutionary history of animals and plants, from the rise of the kingdoms and their principal subdivisions to the origin of recent species, appears as a series of biological responses to environmental opportunities. Early diversifications near the beginning of the Cambrian period produced a series of animal body plans, each adapted to a particular mode of life. Many of these plans proved to be preadapted to further life modes and were extensively diversified; today they are what are called phyla. Others became extinct sooner or later. Extinctions are in some ways a measure of the success of evolution in adapting organisms to particular environmental conditions. When those conditions vanish, the organisms vanish. Opportunities are thus created for selection to develop novel kinds of organisms from among the survivors. The new forms are sometimes spectacularly successful, particularly when their adaptations provide entry into a relatively empty niche. The groups that disappear are not replaced by totally new groups but by branches from the remaining lineages. Therefore as time goes by the number of distinctive kinds of organisms declines, whereas the remaining groups become on the average more diverse.

The groups that disappeared were not necessarily less well adapted than or structurally inferior to the survivors, except in relation to the sequence of environmental changes that happened to occur. This sequence, ultimately controlled by physical changes within a dynamic earth, is unrelated to the life forms clinging to adaptation as the means of survival on the continents and in the oceans. If the environmental changes had been different, different survivors would have emerged. One cannot say what life forms would now inhabit the earth if the physical history of the planet had been different. One can only be certain they would differ from the ones that are here today.

II

EARLIEST TRACES
OF LIFE

EARLIEST TRACES OF LIFE II

INTRODUCTION

Given the phenomenon of evolution, we naturally next ask: When and how did it all begin? Paleontologists and biologists postulate that the process of speciation and diversification has gone on for a very long time, and can be projected backward through the Cenozoic, Mesozoic, and into the dawn of the Paleozoic. But until fairly recently, scientists have been puzzled by this abrupt and relatively late boundary between the rocks of the lower Paleozoic, with its abundant and diverse, although primitive, fossils, and the rocks of the Precambrian, which seemed virtually to lack fossils. The apparent puzzle was that the Precambrian spanned about seven-eighths of earth history, whereas the Phanerozoic (Paleozoic, Mesozoic, and Cenozoic eras) included but one-eighth. Moreover, the shelly invertebrates found in basal Cambrian strata included trilobites, who were members of the highly complex and advanced phylum of the arthropods.

It was argued that surely the Precambrian must have been a time of gradual evolution and diversification of life, leading up to the Cambrian fossil faunas. But where were these Precambrian fossils? Except for fossilized algal mounds and scattered tracks and trails, Precambrian rocks failed to reveal suitable presumptive ancestors of early Paleozoic life. Various ideas were offered to explain away this problem: Perhaps Precambrian rocks are too metamorphosed to allow preservation of fossils; maybe the environments in which these primitive organisms lived were not recorded in the rocks of the Precambrian; if the fossils and environments were initially preserved, perhaps there was a subsequent period of widespread erosion that removed the crucial evidence; or maybe Precambrian organisms just lacked hard parts that could be readily fossilized.

Increasingly, however, attempts to rationalize the absence of a good Precambrian fossil record seemed weaker and weaker. Among other things, examples of relatively unmetamorphosed rocks of many different sedimentary environments became known. In addition, a number of places in the world showed that there was no significant erosion separating upper Precambrian and lower Cambrian rocks. As for the lack of hard parts, one would have expected that somewhere within the Precambrian impressions or traces ought to be found, just as they are, even if rarely, in younger rocks. Moreover, to some paleontologists it seemed biologically unlikely that soft-bodied trilobites or brachiopods existed, because their very anatomy would require mineralized supporting structures.

Gradually, it dawned on people that maybe we should accept the Precambrian fossil record, such as it is, as a true reflection of what in fact occurred during this period of evolution. One immediate implication of this new way of viewing the fossil record was the need to explain the sharp biological discontinuity between late Precambrian (without Cambrian precursors) and the abundant and diverse shelly invertebrates of the Cambrian. A related consequence of this new approach was the need to explain why this discontinuity happened so relatively late in the history of the planet.

Just about this time, during the 1950s, geologists began discovering a diverse, but *microscopic*, assemblage of primitive algae and bacterium-like organisms in various parts of the Precambrian on several different continents, as well as evidence of soft-bodied, primitive invertebrates near the end of the Precambrian. These discoveries illustrate an interesting sidelight about the nature of scientific inquiry. As long as paleontologists expected to find macroscopic examples of prototrilobites, protobrachiopods, or protomolluscs in the Precambrian—as suggested by earlier ideas about how Precambrian evolution must have taken place—they not only would not find these fossils (because they didn't exist), but also would fail to observe what was really there. However, as opinions changed about what the record might look like, important discoveries were quickly made and, more significantly, were appreciated as such. This episode demonstrates that what we see or do not see in our scientific investigations is often influenced by what we think we should be seeing. So much for scientific "objectivity."

"The Evolution of the Earliest Cells" by J. William Schopf traces the evolution of life from its origin through the single-celled prokaryotes (blue-green algae and bacteria) to the more advanced single-celled eukaryotes. He indicates some of the important chemical pathways that were developed and describes the fossil and sedimentary rock evidence that supports these interpretations. The evolution described here spans an interval of earth history more than four times that during which multicellular plants and animals evolved.

For people unfamiliar with rocks and fossils, it usually comes as a surprise to discover that the initial atmosphere lacked any free oxygen. Only with the subsequent evolution of photosynthesis around three billion years ago did our present-day oxidizing atmosphere, so rich in free oxygen, begin to slowly accumulate. The presence of thick accumulations of banded iron ores, ranging in age from 1.8 to 3.2 billion years, in all the Precambrian continental shields suggests that the early formed photosynthetic oxygen swept the seas of their soluble ferrous iron into the virtually insoluble ferric oxides. (Most of the world's steel comes from this early biological activity.) Free oxygen in the atmosphere also resulted in the ozone layer, which screens out much of the biologically damaging, solar ultraviolet radiation. Shallow sea life was restricted and terrestrial life prevented until this protective layer became established.

Another important milestone was passed about one billion years ago, when the eukaryote cell appeared, differentiated in part from the earlier prokaryote cells by being able to undergo sexual reproduction. Fossils documenting this have been found in rocks from Australia that strongly suggest various steps in meiosis, the process whereby a nucleated cell undergoes nuclear duplication before subsequent cell division. Sexual reproduction creates much genetic variability and so may have provided the trigger for the rapid evolution of life, particularly animals, soon after.

Martin F. Glaessner's "Pre-Cambrian Animals" picks up the story where Schopf leaves off. Australian rocks formed some 650 million years ago contain excellent fossil impressions along clay partings in shallow marine sandstones. These impressions are of soft-bodied animals reminiscent of modern jellyfish, sea pens (cousins to corals), flatworms, and segmented worms. Since this first discovery, these distinctive fossils have been found in many other parts of the world in rocks of about the same age and are now judged to be representatives of an Ediacaran fauna (named after the Australian locality where first described) that was ancestral to the diverse early Cambrian faunas about one hundred million years later.

The Precambrian fossil record thus supports present-day ideas about evolution during this interval of earth history, which are outlined as follows (refer to the chart near the end of Schopf's article):

1. Formation of the earth; about 4.5 billion years ago.

2. Outgassing creates a reducing atmosphere; chemical evolution leading to earliest life forms; 3.5 to 4.5 billion years ago.

3. Heterotrophs appear and "feed" on inorganically synthesized compounds. As this reservoir becomes depleted, photosynthesis—the process by which organisms make their own food, using carbon dioxide and water—evolves the so-called autotrophs; about 3.3 million years ago.

4. Diverse autotrophic prokaryote cells; increasing free oxygen in the atmosphere and consequent development of ozone layer; about 2 billion years ago.

5. Rise of eukaryote cell and sexual reproduction with greatly increased genetic variation; almost 1 billion years ago, perhaps 1.5 billion.

6. Appearance of soft-bodied multicellular animals into basic ground plans followed by the higher invertebrates; about 700 million years ago.

7. Diversification of invertebrates and beginning of skeletonization for protection and support; almost 600 million years ago, the beginning of the Cambrian Period.

Recall that in Valentine's article in the preceding section, the appearance of the hydrostatic skeleton led to the rapid radiation of diverse marine invertebrates followed by hard-part secretion by many of these different organisms. The fine record of life we have from fossils is, of course, largely due to the ability of many animals and some plants to make mechanically robust and chemically resistant skeletons that can survive transportation, burial, and lithification within sediments for eons. Because the post-Precambrian, or Phanerozoic, rock record is so rich in fossils, we sometimes forget how much of the ancient world is not preserved. The final article in this section, "The Animals of the Burgess Shale" by Simon Conway Morris and H. B. Whittington, describes one of those few instances whereby soft-bodied organisms are abundantly preserved. Rare as it is, the middle Cambrian Burgess Shale fauna not only gives us some insight about how much is missing in the fossil record, but also provides us with a sense of how much "experimentation" went on among the various invertebrate phyla before successful, long-lived ground plans were hit upon. Similar experimentation occurred during the initial stages of the Cenozoic radiation of mammals. (For example, by one count there were as many as 28 orders of placental mammals during the early Cenozoic, with but 16 surviving today.)

Besides being intrinsically interesting as an expanded window on the Cambrian marine world, the Burgess Shale provides a cautionary lesson about not overinterpreting the fossil record. It is easy to forget that the fossil record is the result of a *destructive* process, however grateful we are for whatever it does finally preserve for our scrutiny and wonder.

SUGGESTED FURTHER READING

Cloud, P. 1976. "Beginnings of Biosphere Evolution and Their Biogeochemical Consequences," *Paleobiology*, vol. 2, pp. 351–387. An excellent review of the state of knowledge about the major events leading to the evolution of higher forms of life, with special emphasis on the interactions between the biosphere and the atmosphere, hydrosphere, and the rock record. Cloud carefully points out the different degrees of confidence to be placed on various facts and their interpretations regarding early life on earth. The bibliography of 172 articles provides a comprehensive entry into the subject.

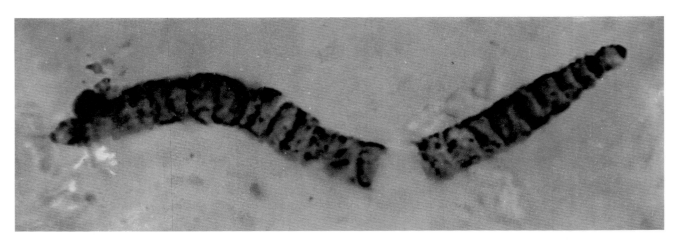

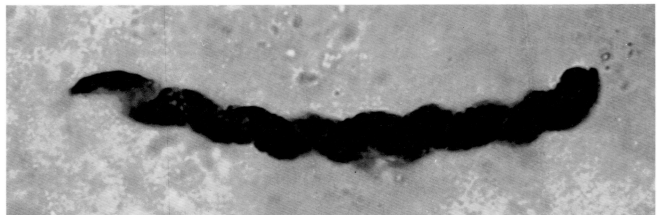

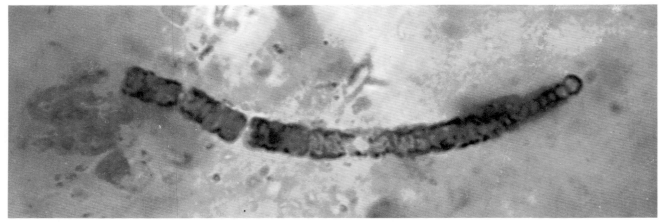

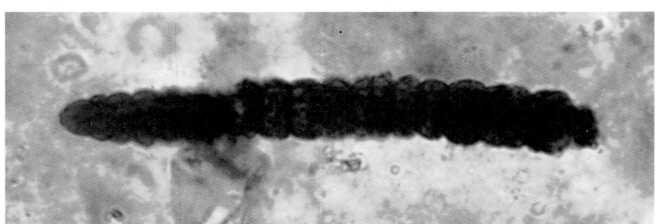

4

The Evolution of the Earliest Cells

by J. William Schopf
September 1978

*For some three billion years the only living things were
primitive microorganisms. These early cells gave rise
to biochemical systems and the oxygen-enriched
atmosphere on which modern life depends*

When *On the Origin of Species* appeared in 1859, the history of life could be traced back to the beginning of the Cambrian period of geologic time, to the earliest recognized fossils, forms that are now known to have lived more than 500 million years ago. A far longer prehistory of life has since been discovered: it extends back through geologic time almost three billion years more. During most of that long Precambrian interval the only inhabitants of the earth were simple microscopic organisms, many of them comparable in size and complexity to modern bacteria. The conditions under which these organisms lived differed greatly from those prevailing today, but the mechanisms of evolution were the same. Genetic variations made some individuals better fitted than others to survive and to reproduce in a given environment, and so the heritable traits of the better-adapted organisms were more often represented in succeeding generations. The emergence of new forms of life through this principle of natural selection worked great changes in turn on the physical environment, thereby altering the conditions of evolution.

One momentous event in Precambrian evolution was the development of the biochemical apparatus of oxygen-generating photosynthesis. Oxygen released as a by-product of photosynthesis accumulated in the atmosphere and effected a new cycle of biological adaptation. The first organisms to evolve in response to this environmental change could merely tolerate oxygen; later cells could actively employ oxygen in metabolism and were thereby enabled to extract more energy from foodstuff.

A second important episode in Precambrian history led to the emergence of a new kind of cell, in which the genetic material is aggregated in a distinct nucleus and is bounded by a membrane. Such nucleated cells are more highly organized than those without nuclei. What is most important, only nucleated cells are capable of advanced sexual reproduction, the process whereby the genetic variations of the parents can be passed on to the offspring in new combinations. Because sexual reproduction allows novel adaptations to spread quickly through a population its development accelerated the pace of evolutionary change. The large, complex, multicellular forms of life that have appeared and quickly diversified since the beginning of the Cambrian period are without exception made up of nucleated cells.

The history of life in its later phases, since the start of the Cambrian period, has been reconstructed mainly from the study of fossils preserved in sedimentary rocks. In the 18th and 19th centuries it gradually became apparent that the fossil record has appreciable chronological and geographical continuity. The fossil deposits form recognizable layers, which can be identified in widely separated geological formations. Boundaries between such layers, where one characteristic suite of fossils gives way to another, provide the basis for dividing geologic time into eras, periods and epochs.

One of the most dramatic boundaries in the rock record is the one that separates the Cambrian period from all that came before. The 11 periods of geologic time since the start of the Cambrian are referred to collectively as the Phanerozoic era, which might be translated from the Greek as the era of manifest life. The preceding era is called simply the Precambrian.

By itself the geologic time scale cannot provide dates for fossil deposits; it only lists their sequence. Ages can be calculated, however, from the constant rate of decay of radioactive isotopes in the earth's crust. By determining how much of an isotope has decayed since the minerals in a rock unit crystallized, a date can be assigned to that unit and to nearby strata containing fossils. Radioactive-isotope studies of this kind, carried out on rocks from many parts of the world, have established a rather well-defined date for the start of the Phanerozoic era: it began about 570 million years ago. The same method indicates that the earth itself and the rest of the solar system are 4.6 billion years old. Thus the Precambrian era encompasses some seven-eighths of the earth's entire history.

The boundary between the Precambrian era and the Cambrian period has traditionally been viewed as a sharp discontinuity. In Cambrian strata there are abundant fossils of marine plants and animals: seaweeds, worms, sponges, mollusks, lampshells and, what are perhaps most characteristic of the period, the early arthropods called trilobites. It was thought for many years that fossils were entirely absent in the underlying Precambrian strata. The Cambrian fauna seemed to come into existence abruptly and without known predecessors.

Life could not have begun with organisms as complex as trilobites. In *On the Origin of Species* Darwin wrote: "To the question why we do not find rich fossiliferous deposits belonging to...periods prior to the Cambrian system, I can give no satisfactory answer.... The case at present must remain inexplicable; and

MICROSCOPIC FOSSILS on the opposite page are the remains of organisms that were once the dominant form of life on the earth. The fossils are from silica-rich rocks in the Bitter Springs formation of central Australia, deposited about 850 million years ago, or late in the Precambrian era. The rocks have the layered structure of stromatolites, sedimentary deposits that were formed by matlike communities of microorganisms. Among Precambrian fossils these specimens are exceptionally well preserved; their petrified cell walls are composed of organic matter and have retained their three-dimensional form. In size, structure and ecological setting they resemble living cyanobacteria, or blue-green algae. Like their modern counterparts, the fossil forms were presumably capable of photosynthesis, and similar cyanobacteria some billion years earlier were evidently responsible for the first rapid release of oxygen into the earth's atmosphere. Organisms in these photomicrographs are about 60 micrometers long.

may be truly urged as a valid argument against the views here entertained." The argument is no longer valid, but it is only in the past 20 years or so that a definitive answer to it has been found.

One part of the answer lies in the discovery of primitive fossil animals in rocks below the earliest Cambrian strata. The fossils include the remains of jellyfishes, various kinds of worms and possibly sponges, and they make up a fauna quite distinct from that of the predominantly shelled animals of the Cambrian period. These discoveries, however, extend the fossil record by only about 100 million years, less than four percent of the Precambrian era. It can still be asked: What came before?

Since the 1950's a far-reaching explanation has emerged. It has come to be recognized that not only are many Precambrian rocks fossil-bearing but also Precambrian fossils can be found even in some of the most ancient sedimentary deposits known. These fossils had escaped notice earlier largely because they are the remains only of microscopic forms of life.

An important clue in the search for Precambrian life was discovered in the early years of the 20th century, but its significance was not fully appreciated until much later. The clue came in the form of masses of thinly layered limestone rock discovered by Charles Doo-

little Walcott in Precambrian strata from western North America. Walcott found numerous moundlike or pillarlike structures made up of many draped horizontal layers, like tall stacks of pancakes. These structures are now called stromatolites, from the Greek *stroma*, meaning bed or coverlet, and *lithos*, meaning stone.

Walcott interpreted the stromatolites as being fossilized reefs that had probably been formed by various types of algae. Other workers were skeptical, and for many years the stromatolites were widely attributed to some nonbiological origin. The first convincing evidence substantiating Walcott's hypothesis came in 1954, when Stanley A. Tyler of the University of Wisconsin and Elso S. Barghoorn of Harvard University reported the discovery of fossil microscopic plants in an outcropping of Precambrian rocks called the Gunflint Iron formation near Lake Superior in Ontario. Most of the Gunflint fossils, which form the layers of dome-shaped and pillarlike stromatolites, resemble modern blue-green algae and bacteria. More recently, living stromatolites have been identified in several coastal habitats, most notably in a lagoon at Shark Bay on the western coast of Australia. They are indeed built up by communities of blue-green algae and bacteria, and they are strikingly similar in form to the fossilized Precambrian structures.

Today microfossils have been identified in some 45 stromatolitic deposits. (All but three of these fossilized communities have been found in the past 10 years.) The fossils are often well preserved, the cell walls being petrified in three-dimensional form, and they have become a prime source of documentation for the early history of life. In recent years the search for Precambrian microfossils in other kinds of sediments, such as shales deposited in offshore environments, has also been rewarded. These fossils are generally not as well preserved as the ones in stromatolites, most of them having been flattened by pressure; on the other hand, they supply information about Precambrian life in a habitat quite different from that of the shallow-water stromatolites.

A surprising amount of information can be derived from the fossil remains of a microorganism. Size, shape and degree of morphological complexity are among the most easily recognized features, but under favorable circumstances even details of the internal structure of cells can be discerned. In retracing the course of Precambrian evolution, however, there is no need to rely exclusively on the fossil record. An entirely independent archive has been preserved in the metabolism and the biochemical pathways of modern, living cells. No living organism is biochemically identical with its Precambri-

FOSSIL STROMATOLITES typically exhibit the appearance of mounds or pillars made up of many thin layers piled one on top of another. The stromatolites were formed by communities of cyanobacteria and other prokaryotes (cells without a nucleus) in shallow water; each layer represents a stage in the growth of the community. Stromatolites formed throughout much of the Precambrian era. They are an important source of Precambrian fossils. These specimens are in limestone about 1,300 million years old in Glacier National Park.

LIVING STROMATOLITES were photographed at Shark Bay in Australia. Elsewhere stromatolites are rare because of grazing by invertebrates. Here the invertebrates are excluded because the water is too salty for them; in the Precambrian era they had not yet evolved. In size and form the modern stromatolites are much like the fossil structures, and they are produced by the growth of cyanobacteria and other prokaryotes in matlike communities. The discovery of such living stromatolites has confirmed the biological origin of the fossil ones.

an antecedents, but vestiges of earlier biochemistries have been retained. By studying their distribution in modern forms of life it is sometimes possible to deduce when certain biochemical capabilities first appeared in the evolutionary sequence.

Still another independent source of information about the early evolutionary progression is based neither on living nor on fossil organisms but on the inorganic geological record. The nature of the minerals found there reflects physical conditions at the time the minerals were deposited, conditions that may have been influenced by biological innovations. In order to understand the introduction of oxygen into the early atmosphere, for example, all three fields of study must be called on to testify. The mineral record tells when the change took place, the fossil record reveals the organisms responsible and the distribution of biochemical capabilities among modern organisms puts the development in its proper evolutionary context.

Since the 1960's it has become apparent that the greatest division among living organisms is not between plants and animals but between organisms whose cells have nuclei and those that lack nuclei. In terms of biochemistry, metabolism, genetics and intracellular organization, plants and animals are very similar; all such higher organisms, however, are quite different in these features from bacteria and blue-green algae, the principal types of non-nucleated life. Recognition of this discontinuity has been important for understanding the early stages of biological history.

Organisms whose cells have nuclei are called eukaryotes, from the Greek roots *eu-*, meaning well or true, and *karyon*, meaning kernel or nut. Cells without nuclei are prokaryotes, the prefix *pro-* meaning before. All green plants and all animals are eukaryotes. So are the fungi, including the molds and the yeasts, and protists such as *Paramecium* and *Euglena*. The prokaryotes include only two groups of organisms, the bacteria and the blue-green algae. The latter produce oxygen through photosynthesis like other algae and higher plants, but they have much stronger affinities with the bacteria than they do with eukaryotic forms of life. I shall therefore refer to blue-green algae by an alternative and more descriptive name, the cyanobacteria.

Several important traits distinguish eukaryotes from prokaryotes. In the nucleus of a eukaryotic cell the DNA is organized in chromosomes and is enclosed by an intracellular membrane; many prokaryotes have only a single loop of DNA, which is loose in the cytoplasm of the cell. Prokaryotes reproduce asexually by the comparatively simple process of binary fission. In contrast, asexual reproduction in eukaryot-

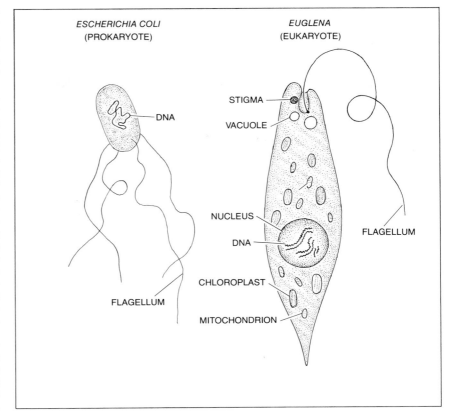

	PROKARYOTES	EUKARYOTES
ORGANISMS REPRESENTED	BACTERIA AND CYANOBACTERIA	PROTISTS, FUNGI, PLANTS AND ANIMALS
CELL SIZE	SMALL, GENERALLY 1 TO 10 MICROMETERS	LARGE, GENERALLY 10 TO 100 MICROMETERS
METABOLISM AND PHOTOSYNTHESIS	ANAEROBIC OR AEROBIC	AEROBIC
MOTILITY	NONMOTILE OR WITH FLAGELLA MADE OF THE PROTEIN FLAGELLIN	USUALLY MOTILE, CILIA OR FLAGELLA CONSTRUCTED OF MICROTUBULES
CELL WALLS	OF CHARACTERISTIC SUGARS AND PEPTIDES	OF CELLULOSE OR CHITIN, BUT LACKING IN ANIMALS
ORGANELLES	NO MEMBRANE-BOUNDED ORGANELLES	MITOCHONDRIA AND CHLOROPLASTS
GENETIC ORGANIZATION	LOOP OF DNA IN CYTOPLASM	DNA ORGANIZED IN CHROMOSOMES AND BOUNDED BY NUCLEAR MEMBRANE
REPRODUCTION	BY BINARY FISSION	BY MITOSIS OR MEIOSIS
CELLULAR ORGANIZATION	MAINLY UNICELLULAR	MAINLY MULTICELLULAR, WITH DIFFERENTIATION OF CELLS

GREATEST DIVISION among organisms is the one separating cells with nuclei (eukaryotes) from those without nuclei (prokaryotes). The only prokaryotes are bacteria and cyanobacteria, and here they are represented by the bacterium *Escherichia coli* (*top left*). All other organisms are eukaryotes, including higher plants and animals, fungi and protists such as *Euglena* (*top right*). Eukaryotic cells are by far the more complex ones, and some of the organelles they contain, such as mitochondria and chloroplasts, may be derived from prokaryotes that established a symbiotic relationship with the host cell. Prokaryotes vary widely in their tolerance of or requirement for free oxygen, and they are thought to have evolved during a period of fluctuating oxygen. All eukaryotes require oxygen for metabolism and for the synthesis of various substances, and they must have emerged after an atmosphere rich in oxygen became established.

ic cells takes place through the complicated process of mitosis, and most eukaryotes can also reproduce sexually through meiosis and the subsequent fusion of sex cells. (The "parasexual" reproduction of some prokaryotes differs markedly from advanced eukaryotic sexuality.) Eukaryotic cells are generally larger than prokaryotic ones, although the range of sizes overlaps, and almost all prokaryotes are unicellular organisms whereas the majority of eukaryotes are large, complex and many-celled. A mammalian animal, for example, can be made up of billions of cells,

which are highly differentiated in both structure and function.

An intriguing feature of eukaryotic cells is that they have within them smaller membrane-bounded subunits, or organelles, the most notable being mitochondria and chloroplasts. Mitochondria are present in all eukaryotes, where they play a central role in the energy economy of the cell. Chloroplasts are present in some protists and in all green plants and are responsible for the photosynthetic activities of those organisms. It has been suggested that both mitochondria and chloroplasts may be evo-

lutionary derivatives of what were once free-living microorganisms, an idea discussed in particular by Lynn Margulis of Boston University. The modern chloroplast, for example, may be derived from a cyanobacterium that was engulfed by another cell and that later established a symbiotic relationship with it. In support of this hypothesis it has been noted that both mitochondria and chloroplasts contain a small fragment of DNA whose organization is somewhat like that of prokaryotic DNA. In the past several years the testing of this hypothesis has generated a large body of

METABOLIC PATHWAYS by which cells extract energy from foodstuff apparently evolved in response to an increase in free oxygen. In all organisms the only usable energy derived from the breakdown of carbohydrates such as glucose is the fraction stored in high-energy phosphate bonds, denoted ~P; the rest is lost as heat. In anaerobic organisms (those that live without oxygen) glucose is broken down through fermentation: each molecule of glucose is split into two molecules of pyruvate, the process called glycolysis, with a net gain of two phosphate bonds. In bacterial fermentation the pyruvate is converted, in a step that provides no usable energy, into products such as lactic acid or ethyl alcohol and carbon dioxide, which are excreted as wastes. The metabolic system of aerobic organisms (those that require oxygen) is respiration. It begins with glycolysis, but the pyruvate is treated not as a waste but as a substrate for a further series of reactions that make up the citric acid cycle. In these reactions pyruvate is decomposed one carbon atom at a time and combined with oxygen, the ultimate products being carbon dioxide and water. Respiration releases far more energy than fermentation, and the proportion of the energy recovered in useful form is also greater; as a result 36 phosphate bonds are formed instead of two. Respiratory metab-

data on the comparative biochemistry of modern microorganisms, data that also provide clues to the evolution of life in the Precambrian.

One further difference between prokaryotes and eukaryotes is of particular importance in the study of their evolution: the extent to which the two types of organisms tolerate oxygen. Among the prokaryotes oxygen requirements are quite variable. Some bacteria cannot grow or reproduce in the presence of oxygen; they are classified as obligate anaerobes. Others can tolerate oxygen but can also survive in its absence; they are facultative anaerobes. There are also prokaryotes that grow best in the presence of oxygen but only at low concentrations, far below that of the present atmosphere. Finally, there are fully aerobic prokaryotes, forms that cannot survive without oxygen.

In contrast to this variety of adaptations the eukaryotes present a pattern of great consistency: with very few exceptions they have an absolute requirement for oxygen, and even the exceptions seem to be evolutionary derivatives of oxygen-dependent organisms. This observation leads to a simple hypothesis: the prokaryotes evolved during a period when environmental oxygen concentrations were changing, but by the time the eukaryotes arose the oxygen content was stable and relatively high.

One indication that eukaryotic cells have always been aerobic is provided by mitotic cell division, a process that can be considered a definitive characteristic of the group. Many eukaryotic cells can survive temporary deprivation of oxygen and can even carry on some metabolic functions; it appears that no cell, however, can undergo mitosis unless oxygen is available at least in low concentration.

The pathways of metabolism itself—the biochemical mechanisms by which an organism extracts energy from foodstuff—provide more detailed evidence. In eukaryotes the central metabolic process is respiration, which in overall terms can be described as the burning of the sugar glucose with oxygen to yield carbon dioxide, water and energy. Some prokaryotes (the aerobic or facultative ones) are also capable of respiration, but many derive their energy solely from the simpler process of fermentation. In bacterial fermentation glucose is not combined with oxygen (or with any other substance from outside the cell) but is simply broken down into smaller molecules. In both respiration and fermentation part of the energy released through the decomposition of glucose is captured in the form of high-energy phosphate bonds, usually in molecules of adenosine triphosphate (ATP). The rest of the energy is lost from the cell as heat.

Respiratory metabolism has two main components: a short series of chemical reactions, collectively called glycolysis, and a longer series called the citric acid cycle. In glycolysis a glucose molecule, with six carbon atoms, is broken down into two molecules of pyruvate, each having three carbon atoms. No oxygen is required for glycolysis, but on the other hand it releases only a little energy with a net gain of only two molecules of ATP.

The fuel for the citric acid cycle is the pyruvate formed by glycolysis. Through a series of enzyme-controlled reactions the carbon atoms of the pyruvate are oxidized and the oxidations are coupled to other reactions that result in the synthesis of ATP. For each two molecules of pyruvate (and hence for each molecule of glucose entering the sequence) 34 additional molecules of ATP are formed. The complete respiratory pathway is thus far more effective than glycolysis alone. In respiration the proportion of energy released that can be recovered in useful form (as ATP) is higher than it is in fermentation, about 38 percent instead of only some 30 percent, and in respiration the net energy yield to the cell is some 18 times greater. By breaking down the glucose to simple inorganic molecules (carbon dioxide and water) respiration liberates virtually all the biologically usable energy stored in the chemical bonds of the sugar.

The metabolism of the prokaryotes immediately suggests an evolutionary relationship between them and the eukaryotes: up to a point fermentation is indistinguishable from glycolysis. In bacterial fermentation a molecule of glucose is split into two molecules of pyruvate, with a net yield of two molecules of ATP. As in glycolysis, no oxygen is required for the process. In anaerobic prokaryotes, however, the metabolic pathway essentially ends at pyruvate. The only further reactions transform the pyruvate into such compounds as lactic acid, ethyl alcohol or carbon dioxide, which are excreted by the cell as wastes.

The similarity of fermentation in prokaryotes to glycolysis in eukaryotes seems too close to be a coincidence, and the assumption of an evolutionary relationship between the two groups provides a ready explanation. It seems likely that anaerobic fermentation became established as an energy-yielding process early in the history of life. When atmospheric oxygen became available for metabolism, it offered the potential for extracting 18 times as much useful energy from carbohydrate: a net yield of 36 molecules of ATP instead of only two molecules. The oxygen-dependent reactions did not, however, simply replace the anaerobic ones; they were appended to the existing anaerobic pathway.

Further evidence for this proposed evolutionary sequence can be found in the behavior of some eukaryotic cells under conditions of oxygen deprivation. In mammalian muscle cells, for example, prolonged exertion can demand more oxygen than the lungs and the blood can supply. The citric acid cycle is then disabled, but the cells continue to function, albeit at reduced efficiency, through glycolysis alone. Under such conditions of oxygen debt pyruvate is not consumed in the cell, but in the liver it can be converted back into glucose (at a cost in energy of six ATP molecules). Significantly the pyruvate itself is not transported to the liver but instead is

GAIN IN (P)	TOTAL ENERGY RELEASED (P) PLUS HEAT (KILOCALORIES)	AVAILABLE ENERGY (P) (KILOCALORIES)	CALCULATED EFFICIENCY (PERCENT)
2	57	14.6	26
2	47	14.6	31
36	686	262.8	38

olism could have evolved, however, only when free oxygen became readily available; it appears to have developed simply by appending the citric acid cycle to the glycolytic pathway. When aerobic cells are deprived of oxygen, many revert to fermentative metabolism, converting pyruvate into lactic acid. In vertebrates the lactic acid from muscle cells is exported to the liver, where it is returned to the form of pyruvate and converted into glucose.

converted into lactic acid, which in the liver must then be returned to the form of pyruvate. This use of lactic acid may represent a vestige of an earlier, bacterial pathway that under aerobic conditions has been suppressed. Indeed, the oxygen-starved muscle cell seems to revert to a more primitive, entirely anaerobic form of metabolism.

The development of an oxygen-dependent biochemistry can also be traced through a consideration of reaction sequences in the synthesis of various biological molecules. Once again stages in the synthetic pathway that emerged early in the Precambrian can be expected to proceed in the absence of oxygen. Reaction steps nearer the end product of the pathway, which were presumably added at a later age, might with

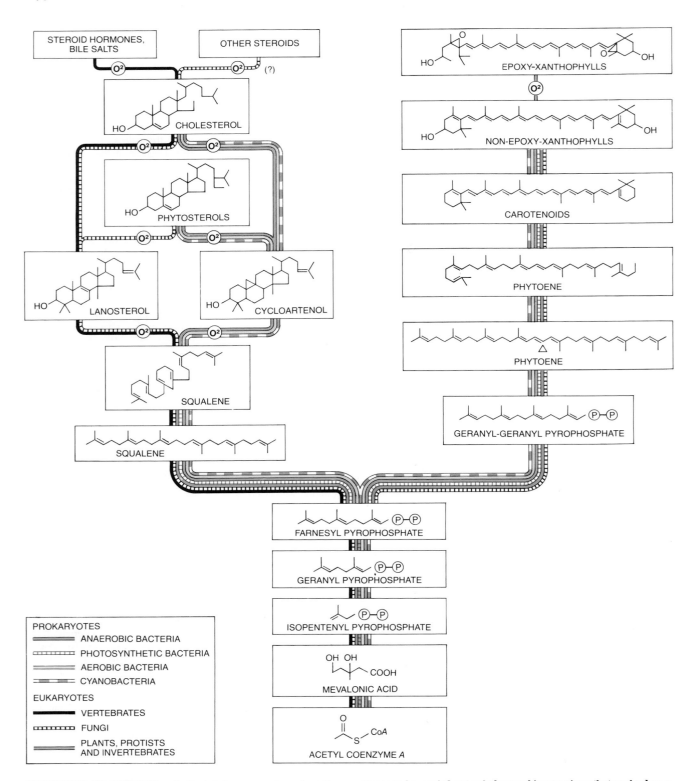

SYNTHESIS OF STEROLS and of related compounds such as the carotenoid pigments of plants requires molecular oxygen (O_2) only for steps near the end of the reaction sequence. Oxygen-dependent steps can be carried out only by aerobic organisms that evolved comparatively late in history of Precambrian life. Organic molecules are in schematic form with most carbon and hydrogen atoms omitted.

increasing frequency require oxygen. The distribution of the oxygen-demanding steps among various kinds of organisms could also have evolutionary significance. If only one pathway has evolved for the synthesis of a class of biochemical substances, then primitive forms of life might be expected to exhibit only the initial, anaerobic steps. Organisms that arose later might exhibit progressively longer, oxygen-dependent synthetic sequences.

In aerobic organisms it might seem at first that virtually all biochemical syntheses require oxygen; eukaryotic cells exhibit relatively little synthetic activity under anoxic conditions. For the most part, however, the oxygen requirement of such syntheses is simply for metabolism: the construction of biological molecules demands energy in the form

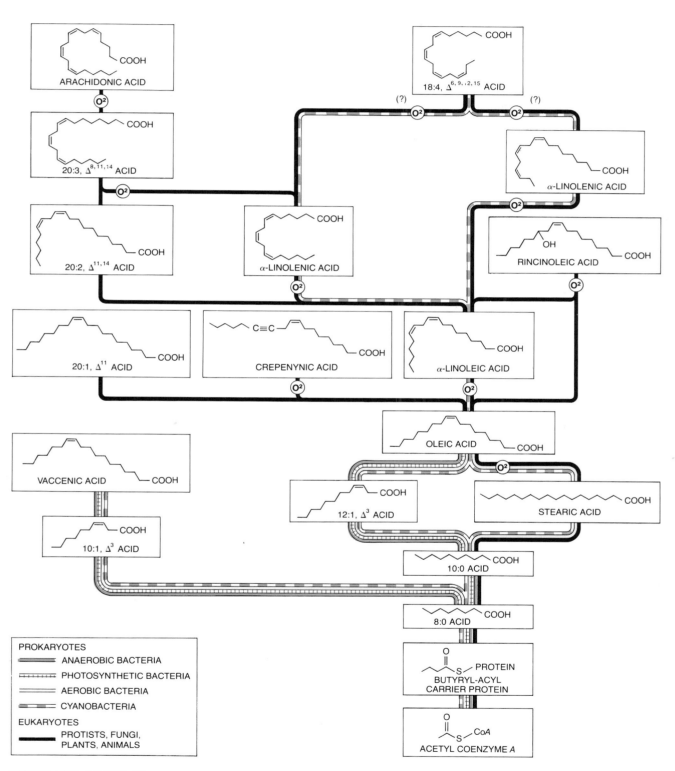

FATTY-ACID SYNTHESIS also follows a pattern suggesting the late addition of oxygen-dependent steps. Most prokaryotes can make mono-unsaturated fatty acids (those having one double bond) by inserting double bond during elongation of molecule. Eukaryotes and some prokaryotes first make a fully saturated molecule, stearic acid, then introduce double bonds by the process of oxidative desaturation.

of ATP, and most of the ATP is supplied through the oxygen-dependent citric acid cycle. If ATP is made available from some other source, many synthetic pathways can proceed unimpaired.

Some syntheses, however, have an intrinsic requirement for oxygen, quite apart from metabolic demands. Molecular oxygen is needed, for example, in the synthesis of bile pigments in vertebrates, of chlorophyll *a* in higher plants and of the amino acids hydroxyproline and, in animals, tyrosine. The oxygen dependence of two synthetic pathways in particular has been determined in detail. One of these pathways controls the manufacture of a class of compounds that includes the sterols and the carotenoids and the other is concerned with the synthesis of fatty acids.

Sterols, such as cholesterol and the steroid hormones, are flat, platelike molecules derived from the compound squalene, which has 30 carbon atoms. Carotenoids are derived from the 40-carbon compound phytoene; they are pigments, such as carotene, the orange-yellow compound in carrots, and they are found in virtually all photosynthetic organisms. A common starting point for the synthesis of both groups of compounds is isoprene, a five-carbon molecule that is also the repeating unit in synthetic rubber. In the biological synthesis two isoprene subunits are joined head to tail; then a third isoprene is added to form a 15-carbon polymer, farnesyl pyrophosphate. At this point there is a fork in the pathway. In one continuation of the synthesis two farnesyl chains are joined to form squalene, the 30-carbon precursor of the sterols. In the other continuation a fourth isoprene subunit is added, and only then are two of the chains joined. The product in this case is phytoene, the 40-carbon precursor of the carotenoids and of other pigments derived from them, such as the xanthophylls.

Up to this step in the synthetic pathway none of the reactions requires the participation of molecular oxygen. The next step in the synthesis of sterols, however, is the conversion of the linear squalene molecule to a 30-carbon ring, and this transformation does require oxygen; so do most of the subsequent steps in sterol synthesis. On the other branch of the pathway there are a few more anaerobic reactions, and indeed carotenoids can be made from phytoene without oxygen. Several further modifications of the carotenoids, however, such as the production of the pigments called epoxy-xanthophylls, are oxygen dependent.

Two observations about the evolution of these biosynthetic pathways are appropriate. Even in groups of organisms that have long been aerobic the first steps in the synthesis are independent of the oxygen supply; molecular oxygen enters the reaction sequence only at later stages. In a similar way the most primitive living organisms, the anaerobic bacteria, are capable only of the first segments of the pathway, the anaerobic segments. The more complex aerobic bacteria and the photosynthetic cyanobacteria have longer synthetic pathways, including some steps that require oxygen. Advanced eukaryotes, such as vertebrate animals and higher plants, have long, branched synthetic pathways, with many steps in which molecular oxygen is required.

A similar pattern can be discerned in the synthesis of fatty acids and their derivatives. The fatty acids are straight carbon-chain compounds that have a carboxyl group (COOH) at one end. A fatty acid is said to be saturated if there are no double bonds between carbon atoms in the chain; it is saturated with hydrogen, which fills all the available bonding positions. An unsaturated fatty acid has a double bond between two carbon atoms or it may have several such double bonds; for each double bond two hydrogen atoms must be removed from the molecule.

In the synthesis of fatty acids the molecule grows by the repeated addition of units two carbon atoms long. The first few steps in the synthesis are identical in all organisms, and they yield fully saturated fatty acids. The first branch in the pathway comes when the developing chain is eight carbons long. At that point many prokaryotes can introduce a double bond, which eukaryotes cannot. There is a second branch at the next step when the saturated chain is 10 carbons long; a double bond can similarly be introduced at that point by many prokaryotes but not by eukaryotes. No matter which branch is followed, elongation of the chain ends at 18 carbons. At that point the fatty acids produced by many prokaryotes contain a double bond, but in eukaryotes the product is always the fully saturated molecule, stearic acid. None of the steps in this sequence, whether in prokaryotes or eukaryotes, requires molecular oxygen.

If no subsequent transformations of fatty acids were possible, eukaryotic cells would be incapable of synthesizing any but the fully saturated forms. Actually extensive modifications can be accomplished through the process of oxidative desaturation, in which double bonds are formed by removing two hydrogen atoms and combining them with oxygen to form water. Oxidative desaturation can take place only in the presence of molecular oxygen (O_2). Through this mechanism cyanobacteria make unsaturated fatty acids with two, three and four double bonds, and eukaryotes form

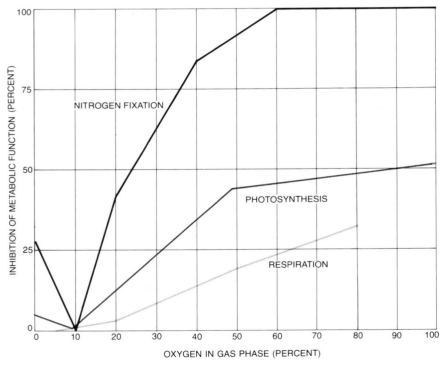

OXYGEN INHIBITION of metabolic functions in cyanobacteria suggests that these aerobic prokaryotes are adapted to an optimum oxygen concentration of about 10 percent, or roughly half the oxygen concentration of the earth's present atmosphere. Nitrogen fixation is completely halted by high oxygen levels, but even respiration, which requires oxygen, can be partially inhibited. Data are for the heterocyst-forming cyanobacterium *Anabaena flos-aquae.*

polyunsaturated fatty acids (with multiple double bonds).

As in the sterol-carotenoid synthesis, an analysis of the fatty-acid pathway argues for a pattern of biochemical evolution in which the increasing availability of atmospheric oxygen played a central role. The first steps in the synthetic sequence are common to all organisms capable of making fatty acids, and in the most primitive organisms those are the only steps. Hence the reactions that come first in the biochemical sequence apparently also developed early in the history of life; these first steps are all anaerobic. Organisms that presumably emerged somewhat later (such as aerobic bacteria and cyanobacteria) have longer pathways, including a few steps of oxidative desaturation. In advanced eukaryotes a substantial proportion of the steps are oxygen dependent.

Comparisons of the metabolism and biochemistry of prokaryotes and eukaryotes thus provide strong evidence that the latter group arose only after a substantial quantity of oxygen had accumulated in the atmosphere. Hence it is of interest to ask when eukaryotic cells first appeared. It seems apparent that an oxygen-rich atmosphere cannot have developed later than this signal evolutionary event.

The primary means of assigning a date to the origin of the eukaryotes is through the fossil record. Because this field of study is so new, however, the available information is scanty and often difficult to interpret. It is rarely a straightforward task to identify a microscopic, single-cell organism as being eukaryotic merely from an examination of its fossilized remains. And even when a fossil has been identified as unequivocally eukaryotic the available radioactive-isotope methods of dating can rarely assign it a precise age. At best such methods have an accuracy of only about plus or minus 5 percent. What is more, the age determinations are generally carried out on rocks that were once molten, such as volcanic lavas, whereas the fossils are found in sedimentary deposits. Consequently the stratum of the fossil itself usually cannot be dated; it is merely assigned an age somewhere between the ages of the nearest underlying and overlying datable rock units.

In spite of these difficulties there is now substantial evidence for the existence of eukaryotic fossils in rocks hundreds of millions of years older than the earliest Phanerozoic strata. The evidence is of two kinds: microfossils that display a morphological or organizational complexity judged to be of eukaryotic character, and the presence of fossil cells whose size is typical only of eukaryotes.

The evidence from relatively complex

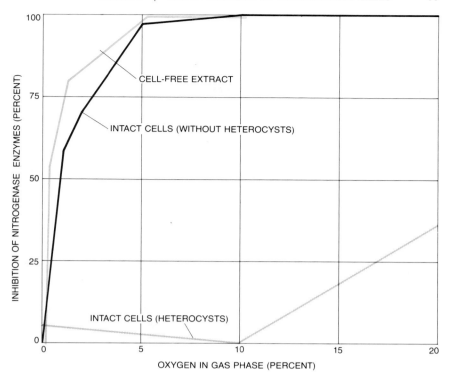

INHIBITION OF NITROGEN FIXATION in the presence of oxygen is caused by the deactivation of the nitrogenase enzymes. In cell-free extracts of the cyanobacterium *Plectonema boryanum* the nitrogenase enzymes are inhibited by even minute quantities of oxygen, and the intact cells of this species, which does not form heterocysts, offer little protection against the inhibition; such organisms can fix nitrogen only in an anoxic habitat. The thick cell walls and other special features of heterocyst cells, such as those formed by *Nostoc muscorum*, allow fixation to continue in a fully aerobic environment. Data suggest that the capability for nitrogen fixation evolved before significant quantities of oxygen had accumulated in the atmosphere.

microscopic fossils includes the following: (1) branched filaments, made up of cells with distinct cross walls and resembling modern fungi or green algae, from the Olkhin formation of Siberia, a deposit thought to be about 725 million years old (but with a known age of between 680 and 800 million years); (2) complex, flask-shaped microfossils from the Kwagunt formation in the eastern Grand Canyon, thought to be about 800 (or 650 to 1,150) million years old; (3) fossils of unicellular algae containing intracellular membranes and small, dense bodies that may represent preserved organelles, from the Bitter Springs formation of central Australia, dated at approximately 850 (or 740 to 950) million years; (4) a group of four sporelike cells in a tetrahedral configuration that may have been produced by mitosis or possibly meiosis, also from Bitter Springs rocks; (5) spiny cells or algal cysts several hundred micrometers in diameter and with unquestionable affinities to eukaryotic organisms, from Siberian shales that are reportedly 950 (or 750 to 1,050) million years old; (6) highly branched filaments of large diameter and with rare cross walls, similar in some respects to certain green or golden-green eukaryotic algae, from the

Beck Spring dolomite of southeastern California (1,300, or 1,200 to 1,400 million years old) and from the Skillogalee dolomite of South Australia (850, or 740 to 867 million years old); (7) spheroidal microfossils described as exhibiting two-layered walls and having "medial splits" on their surface, and which may represent an encystment stage of a eukaryotic alga, from shales 1,400 (or 1,280 to 1,450) million years old in the McMinn formation of northern Australia; (8) a tetrahedral group of four small cells, resembling spores produced by mitotic cell division of some green algae, from the Amelia dolomite of northern Australia, approaching 1,500 (or 1,390 to 1,575) million years in age; (9) unicellular fossils that appear to be exceptionally well preserved and that are reported to contain small membrane-bounded structures that could be remnants of organelles, from the Bungle-Bungle dolomite in the same region as the Amelia dolomite and of approximately the same age.

Thus the earliest of these eukaryote-like fossils are probably somewhat less than 1,500 million years old. Numerous types of microfossils have been discovered in older sediments, but none of them seems to be a strong candidate for

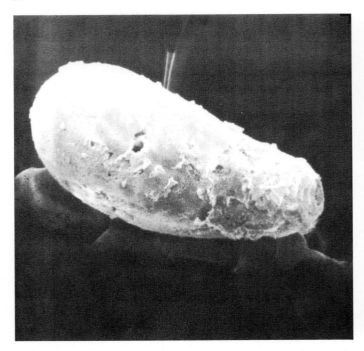

EARLY EUKARYOTIC CELLS may be represented among Precambrian microfossils. The gourd-shaped cell at the left, from shales in the Grand Canyon thought to be 800 million years old, is morphologically more complex than any known prokaryote; it is also larger, about 100 micrometers long. The second cell from the left is some two millimeters in diameter and hence is more than 30 times the size of

identification as eukaryotic. For example, the well-studied Canadian fossils of the Gunflint and Belcher Island iron formations, which are about two billion years old, have been interpreted as exclusively prokaryotic.

The testimony of these as yet rare and unusual specimens can be checked through statistical studies of the sizes of known Precambrian microfossils. The size ranges of prokaryotes and eukaryotes overlap, so that a particular fossil cannot always be classified unambiguously on the basis of size alone; by cataloguing the measured sizes in a large sample of fossils, however, it may be possible to determine whether or not eukaryotic cells are present. Among modern species of spheroidal cyanobacteria about 60 percent are very small, less than five micrometers in diameter; of the remaining species only a few are larger than 20 micrometers and none is larger than 60 micrometers. Unicellular eukaryotes, such as green or red algae, can be much larger. Typically they fall in the range between five and 60 micrometers, but several percent of living species are larger than 60 micrometers and a few are larger than 1,000 micrometers (one millimeter).

Systematic size measurements have been made on some 8,000 fossil cells from 18 widely dispersed Precambrian deposits. On the basis of those data certain tentative conclusions can be drawn. Cells larger than 100 micrometers, and hence of distinctly eukaryotic dimensions, are unknown in rocks older than

about 1,450 million years. Virtually all the unicellular fossils from rocks of that age, whether they grew in shallow-water stromatolites or were deposited in offshore shales, are of prokaryotic size.

Cells larger than modern prokaryotes (greater than 60 micrometers in diameter) first become abundant in rocks about 1,400 million years old. Algae of this type were apparently free-floating rather than mat-forming species, and they are therefore particularly common in shales, sediments deposited in deeper water. Such eukaryote-size fossils have been known for several years from shales of this age in China and in the U.S.S.R. Recently cells more than 100 micrometers in diameter have also been discovered in the Newland limestone of Montana, and cells more than 600 micrometers in size (10 times the size of the largest spheroidal prokaryote) have been found in the McMinn formation of Australia; the age of both of these fossil-bearing deposits is about 1,400 million years.

In somewhat younger Precambrian sediments there are still larger cells, fossils greater than one millimeter in diameter (with some as large as eight millimeters). They were first described in 1899 by Walcott, who discovered them in rocks from the Grand Canyon. They have since been found in nearly a dozen other rock units throughout the world. The oldest seem to be those from Utah and from Siberia, each about 950 million years old, and those from northern India, which could be even older (from 910 to 1,150 million years old).

Studies of both the morphology and the size of unicellular fossils therefore suggest that there is a break in the fossil record between 1,400 and 1,500 million years ago. Below this horizon cells with eukaryotelike traits are rare or absent; above it they become increasingly common. Moreover, the data suggest that the diversification of the eukaryotes began shortly after the cell type first appeared, apparently within the next few hundred million years. By a billion years ago there had been substantial increases in cell size, in morphological complexity and in the diversity of species. All these indicators also suggest, of course, that oxygen-dependent metabolism, which is highly developed even in the most primitive eukaryotes, had already become established by about 1.5 billion years ago.

The prokaryotes that must have held exclusive sway over the earth before the development of eukaryotic cells were less diverse in form, but they were probably more varied in metabolism and biochemistry than their eukaryotic descendants. Like modern prokaryotes, the ancient species presumably varied over a broad range in their tolerance of oxygen, all the way from complete intolerance to absolute need. In this regard one group of prokaryotes, the cyanobacteria, are of particular interest in that they were largely responsible for the development of an oxygen-rich atmosphere.

Like higher plants, cyanobacteria carry out aerobic photosynthesis, a process

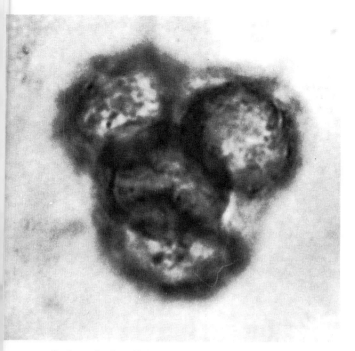
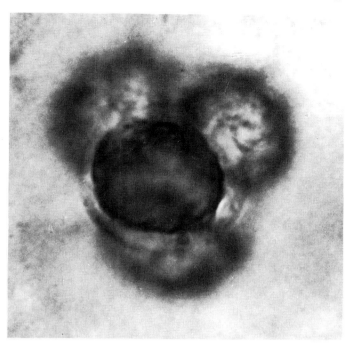

the largest spheroidal prokaryote; it was found in Utah shales 950 million years old. The cluster of cells shown in two views at the right is from sediments in central Australia thought to be

850 million years old. The cells are only 10 micrometers across, but their tetrahedral arrangement suggests they formed as a result of mitosis or possibly meiosis, mechanisms of cell division known only in eukaryotes.

that in overall effect (although not in mechanism) is the reverse of respiration. The energy of sunlight is employed to make carbohydrates from water and carbon dioxide, and molecular oxygen is released as a by-product. The cyanobacteria can tolerate the oxygen they produce and can make use of it both metabolically (in aerobic respiration) and in synthetic pathways that seem to be oxygen dependent (as in the synthesis of chlorophyll a). Nevertheless, the biochemistry of the cyanobacteria differs from that of green, eukaryotic plants and suggests that the group originated during a time of fluctuating oxygen concentration. For example, although many cyanobacteria can make unsaturated fatty acids by oxidative desaturation, some of them can also employ the anaerobic mechanism of adding a double bond during the elongation of the chain. In a similar manner oxygen-dependent syntheses of certain sterols can be carried out by some cyanobacteria, but the amounts of the sterols made in this way are minuscule compared with the amounts typical of eukaryotes. In other cyanobacteria those sterols are not found at all, the biosynthetic pathway being terminated after the last anaerobic step: the formation of squalene. Hence in their biochemistry the cyanobacteria seem to occupy a middle ground between the anaerobes and the eukaryotes.

In metabolism too the cyanobacteria occupy an intermediate position. They flourish today in fully oxygenated environments, but physiological experi-

ments indicate that for many species optimum growth is obtained at an oxygen concentration of about 10 percent, which is only half that of the present atmosphere. Both photosynthesis and respiration are increasingly inhibited when the oxygen concentration exceeds that optimum level. It has recently been discovered that some cyanobacteria can switch the cellular machinery of aerobic metabolism on and off according to the availability of oxygen. Under anoxic conditions these species not only halt respiration but also adopt an anaerobic mode of photosynthesis, employing hydrogen sulfide (H_2S) instead of water and releasing sulfur instead of oxygen. This capability for anaerobic metabolism is probably a relic of an earlier stage in the evolutionary development of the group.

Another activity of some cyanobacteria that seems to reflect an earlier adaptation to anoxic conditions is nitrogen fixation. Nitrogen is an essential element of life, but it is biologically useful only in "fixed" form, for example combined with hydrogen in ammonia (NH_3). Only prokaryotes are capable of fixing nitrogen (although they often do so in symbiotic relationships with higher plants). The crucial complex of enzymes for fixation, the nitrogenases, is highly sensitive to oxygen. In cell-free extracts nitrogenases are partially inhibited by as little as .1 percent of free oxygen, and they are irreversibly inactivated in minutes by exposure to oxygen concentrations of only about 5 percent.

Such a complex of enzymes could

have originated only under anoxic conditions, and it can operate today only if it is protected from exposure to the atmosphere. Many nitrogen-fixing bacteria provide that protection simply by adopting an anaerobic habitat, but among the cyanobacteria a different strategy has developed: the nitrogenase enzymes are protected in specialized cells, called heterocysts, whose internal milieu is anoxic. The heterocysts lack certain pigments essential for photosynthesis, and so they generate no oxygen of their own. They have thick cell walls and are surrounded by a mucilaginous envelope that retards the diffusion of oxygen into the cell. Finally, they are equipped with respiratory enzymes that quickly consume any uncombined oxygen that may leak in.

Because of the thick cell walls heterocysts should be comparatively easy to recognize in fossil material. Indeed, possible heterocysts have been reported from several Precambrian rock units, the oldest being about 2.2 billion years in age. If these cells are indeed heterocysts, they may be taken as a sign that free oxygen was present by then, at least in small concentrations.

Nitrogen fixation has a high cost in energy, and the capability for it would therefore seem to confer a selective advantage only when fixed nitrogen is a scarce resource. Today the main sources of fixed nitrogen are biological and industrial, but biologically usable nitrate (NO_3^-) is formed by the reaction of atmospheric nitrogen and oxygen. In the anoxic atmosphere of the early Precambrian the latter mechanism would obviously have been impossible. The lack of atmospheric oxygen would also have indirectly reduced the concentration of ammonia to very low levels. Ammonia is dissociated into nitrogen and hydrogen by ultraviolet radiation, most of which is filtered out today by a layer of ozone (O_3) high in the atmosphere; without free oxygen there would have been little ozone, and without this protective shield atmospheric ammonia would have been quickly destroyed.

It is likely that the capability for nitrogen fixation developed early in the Precambrian among primitive prokaryotic organisms and in an environment where fixed nitrogen was in short supply. The vulnerability of the nitrogenase enzymes to oxidation was of no consequence then, since the atmosphere had little oxygen. Later, as the photosynthetic activities of the cyanobacteria led to an increase in atmospheric oxygen, some nitrogen fixers adopted an anaerobic habitat and others developed heterocysts. By the time eukaryotes appeared, apparently more than half a billion years later, oxygen was abundant and fixed nitrogen (both NH_3 and NO_3^-) was probably less scarce, and so the eukaryotes never developed the enzymes needed for nitrogen fixation.

At present oxygen-releasing photosynthesis by green plants, cyanobacteria and some protists is responsible for the synthesis of most of the world's organic matter. It is not, however, the only mechanism of photosynthesis. The alternative systems are confined to a few groups of bacteria that on a global scale seem to be of minor importance today but that may have been far more significant in the geological past.

The several groups of photosynthetic bacteria differ from one another in their pigmentation, but they are alike in one important respect: unlike the photosynthesis of cyanobacteria and eukaryotes, all bacterial photosynthesis is a totally anaerobic process. Oxygen is not given off as a by-product of the reaction, and the photosynthesis cannot proceed in the presence of oxygen. Whereas oxygen appears to be required in green plants for the synthesis of chlorophyll *a*, oxygen inhibits the synthesis of bacteriochlorophylls.

The anaerobic nature of bacterial photosynthesis seems to present a paradox: photosynthetic organisms thrive where light is abundant, but such environments are also generally ones having high concentrations of oxygen, which poisons bacterial photosynthesis. These contradictory needs can be explained if it is assumed that anaerobic photosynthesis evolved among primitive bacteria early in the Precambrian, when the atmosphere was essentially anoxic. The photosynthesizers could thus have lived in matlike communities in shallow water and in full sunlight.

Somewhat later such bacteria gave rise to the first organisms capable of aerobic photosynthesis, the precursors of modern cyanobacteria. For the anaerobic photosynthetic bacteria the molecular oxygen released by this mutant strain was a toxin, and as a result the aerobic photosynthesizers were able to supplant the anaerobic ones in the upper portions of the mat communities. The anaerobic species became adapted to the lower parts of the mat, where there is less light but also a lower concentration of oxygen. Many photosynthetic bacteria occupy such habitats today.

Photosynthetic bacteria were surely not the first living organisms, but the history of life in the period that preceded their appearance is still obscure. What little information can be inferred about that early period, however, is consistent with the idea that the environment was then largely anoxic. One tentative line of evidence rests on the assumption that among organisms living today those that are simplest in structure and in biochemistry are probably the most closely related to the earliest forms of life. Those simplest organisms are bacteria of the clostridial and methanogenic types, and they are all obligate anaerobes.

There is even a basis for arguing that anoxic conditions must have prevailed during the time when life first emerged on the earth. The argument is based on the many laboratory experiments that have demonstrated the synthesis of organic compounds under conditions simulating those of the primitive planet. These syntheses are inhibited by even small concentrations of molecular oxygen. Hence it appears that life probably would not have developed at all if the early atmosphere had been oxygen-rich. It is also significant that the starting materials for such experiments often include hydrogen sulfide and carbon monoxide (CO), and that an intermediate in many of the reactions is hydrogen cyanide (HCN). All three compounds are poisonous gases, and it seems paradoxical that they should be forerunners of the earliest biochemistry. They are poisonous, however, only for aerobic forms of life; indeed, for many anaerobes hydrogen sulfide not only is harmless but also is an important metabolite.

It was argued above that oxygen must have been freely available by the time the first eukaryotic cells appeared, probably 1,400 to 1,500 million years ago. Hence the proliferation of cyanobacteria that released the oxygen must have taken place earlier in the Precambrian. How much earlier remains in question. The best available evidence bearing on the issue comes from the study of sedimentary minerals, some of which may have been influenced by the concentration of free oxygen at the time they were deposited. In recent years a number of workers have investigated this possibility, most notably Preston E. Cloud, Jr., of the University of California at Santa Barbara and the U.S. Geological Survey.

One mineral of significance in this argument is uraninite (UO_2), which is found in several deposits that were laid down in Precambrian streambeds. In the presence of oxygen, grains of uraninite are readily oxidized (to U_3O_8) and are thereby dissolved. David E. Grandstaff of Temple University has shown that streambed deposits of the mineral probably could not have accumulated if the concentration of atmospheric oxygen was greater than about 1 percent. Uraninite-bearing deposits of this type are found in sediments older than about two billion years but not in younger strata, suggesting that the transition in oxygen concentration may have come at about that time.

Another kind of mineral deposit, the iron-rich formations called red beds, exhibits the opposite temporal pattern: red beds are known in sedimentary sequences younger than about two billion years but not in older ones. The red beds are composed of particles coated with iron oxides (mostly the mineral hematite, Fe_2O_3), and many are thought to have formed by exposure to oxygen in the atmosphere rather than under water. It has been proposed that the oxygen may have been biologically generated. This hypothesis is consistent with several lines of evidence, but objections to it have also been raised. For example, most red beds are continental deposits rather than marine ones and are therefore susceptible to erosion; it is thus conceivable that red beds were formed earlier than two billion years ago as well as later but that the earlier beds have been destroyed. It is also possible that the oxygen in the red beds had a nonbiological origin; it may have come from the splitting of water by ultraviolet radiation. This has apparently happened on Mars to create a vast red bed across the surface of that planet, where there are only traces of free oxygen and there is no evidence of life.

Perhaps the most intriguing mineral evidence for the date of the oxygen transition comes from another kind of iron-rich deposit: the banded iron formation. These deposits include some tens of billions of tons of iron in the form of oxides embedded in a silica-rich matrix; they are the world's chief economic reserves of iron. A major fraction of them was deposited within a comparatively brief period of a few hundred million years beginning somewhat earlier than two billion years ago.

A transition in oxygen concentration could explain this major episode of iron sedimentation through the following hypothetical sequence of events. In a primitive, anoxic ocean, iron existed in the ferrous state (that is, with a valence of +2) and in that form was soluble in seawater. With the development of aerobic photosynthesis small concentrations of oxygen began diffusing into the upper portions of the ocean, where it reacted with the dissolved iron. The iron was thereby converted to the ferric form (with a valence of +3), and as a result hydrous ferric oxides were precipitated and accumulated with silica to form rusty layers on the ocean floor. As the process continued virtually all the dissolved iron in the ocean basins was precipitated: in a matter of a few hundred million years the world's oceans rusted.

As in the deposition of red beds, an inorganic origin could also be proposed for the oxygen in the banded iron formations; the oxygen in some formations laid down in the very early Precambrian might well have come from such a source. For the extensive iron formations of about two billion years ago, however, inorganic processes such as the photochemical splitting of water do not appear to be adequate; they probably could not have produced the necessary quantity of oxygen quickly enough to account for the enormous volume of iron ores deposited at about that time. Indeed, only one mechanism is known that could release oxygen at the required rate: aerobic photosynthesis, followed by the sedimentation and burial of the organic matter thereby produced. (The burial is a necessary condition, since aerobic decomposition of the organic remains would use up as much oxygen as had been generated.)

In relation to this hypothesis it is notable that fossil stromatolites first become abundant in sediments deposited about 2,300 million years ago, shortly before the major episode of iron-ore deposition. It is therefore possible that the first widespread appearance of stromatolites might mark the origin and the earliest diversification of oxygen-producing cyanobacteria. Even at that early date the cyanobacteria would probably have released oxygen at a high rate, but for several hundred million years the iron dissolved in the oceans would have served as a buffer for the oxygen concentration of the atmosphere, reacting with the gas and precipitating it as ferric oxides almost as quickly as it was generated. Only when the oceans had been swept free of unoxidized iron and similar materials would the concentration of oxygen in the atmosphere have begun to rise toward modern levels.

Although much remains uncertain, evidence from the fossil record, from modern biochemistry and from geology and mineralogy make possible a tentative outline for the history of Precam-

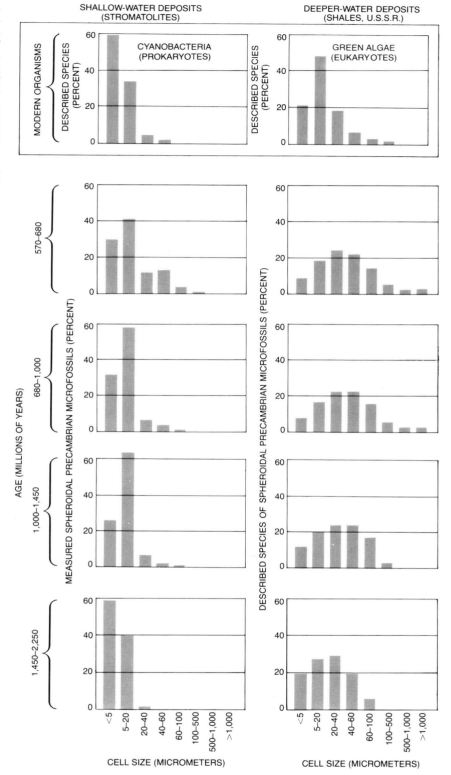

SIZE OF FOSSIL CELLS provides evidence on the origin of the eukaryotes. Spheroidal microfossils of various ages were measured and assigned to eight size categories; a similar procedure was followed with spheroidal members of two groups of modern microorganisms, the prokaryotic cyanobacteria and the eukaryotic green algae. The range of sizes for the modern species overlaps, but the largest cells are observed only among the eukaryotes. The oldest fossils examined have a distribution of sizes similar to that of prokaryotes, but Precambrian rock units younger than 1,450 million years include larger cells that are probably eukaryotic, and the proportion of larger cells increases in later periods. The larger cells tend to be more abundant in shales than in stromatolitic sediments. Because shales are deposited offshore that fact would be explained if early eukaryotes were predominantly free-floating rather than mat-forming.

brian life. The most primitive forms of life with recognizable affinities to modern organisms were presumably spheroidal prokaryotes, perhaps comparable to modern bacteria of the clostridial type. Initially at least they probably derived their energy from the fermentation of materials that were organic in nature but were of nonbiological origin. These materials were synthesized in the anoxic early atmosphere and were of the type that during the age of chemical evolution had led to the development of the first cells.

The first photosynthetic organisms apparently arose earlier than about three billion years ago. They were anaerobic prokaryotes, the precursors of modern photosynthetic bacteria. Most of them probably lived in matlike communities in shallow water, and they may have been responsible for building the earliest fossil stromatolites known, which are estimated to be about three billion years old.

The rise of aerobic photosynthesis in the mid-Precambrian introduced a change in the global environment that was to influence all subsequent evolution. The resulting increase in oxygen concentration probably led to the extinction of many anaerobic organisms, and others were forced to adopt mar-

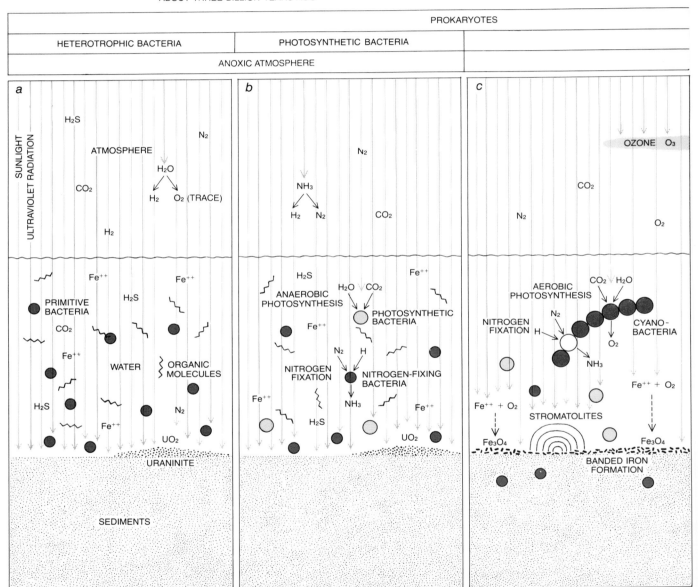

ORGANISM AND ENVIRONMENT evolved in counterpoint during the Precambrian. The first living cells (a) were presumably small, spheroidal anaerobes. Only traces of oxygen were present. They survived by fermenting organic molecules formed nonbiologically in the anoxic environment. The role of such ready-made nutrients was diminished, however, when the first photosynthetic organisms evolved (b). This earliest mode of photosynthesis was entirely anaerobic. Another early development was nitrogen fixation, required in part because ultraviolet radiation that could then freely penetrate the atmosphere would have quickly destroyed any ammonia (NH_3) present. A little more than two billion years ago (c) aerobic photosynthesis began in the precursors of modern cyanobacteria. Oxygen was generated by these stromatolite-building microorganisms, but for some 100 million years little of it accumulated in the atmosphere; instead it reacted with iron dissolved in the oceans, which was then precipitated to create massive banded iron formations. Only when the oceans had

ginal habitats, such as the lower reaches of bacterial mat communities. Nitrogen-fixing organisms also retreated to anaerobic habitats or developed heterocyst cells. With little competition for those regions having optimum light the cyanobacteria were able to spread rapidly and came to dominate virtually all accessible habitats. With the development of the citric acid cycle and its more efficient extraction of energy from food-

stuff, the dominance of the biological community by aerobic organisms was confirmed. When the major episode of deposition of banded iron formations ended some 1,800 million years ago, the trend toward increasing oxygen concentration became irreversible.

By the time eukaryotic cells arose 1,500 to 1,400 million years ago a stable, oxygen-rich atmosphere had long prevailed. Adaptive strategies needed

by earlier organisms to cope with fluctuations in the oxygen level were unnecessary for eukaryotes, which were from the start fully aerobic. The diversity of eukaryote cell types present by about a billion years ago suggests that some form of sexual reproduction may have evolved by then. Within the next 400 million years the rapid diversification of eukaryotic organisms had led to the emergence of multicellular forms of

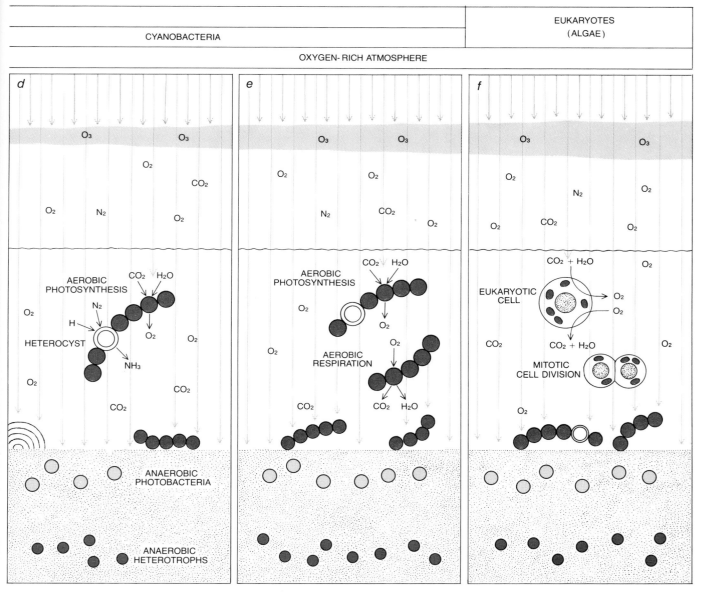

ABOUT TWO BILLION YEARS AGO

ABOUT 1.5 BILLION YEARS AGO

| CYANOBACTERIA | EUKARYOTES (ALGAE) |

OXYGEN-RICH ATMOSPHERE

been swept free of iron and similar materials (d) did the concentration of free oxygen begin to rise toward modern levels. This biologically induced change in the environment had several effects on biological development. Anaerobic organisms were forced to retreat to anoxic habitats, leaving the best spaces for photosynthesis to the cyanobacteria. In a similar manner nitrogen-fixing organisms had to adopt an anaerobic way of life or develop protective heterocyst cells. Atmospheric oxygen also created a layer of ozone (O_3) that filtered

out most ultraviolet radiation. Once the oxygen-rich atmosphere was fully established (e) cells evolved that not only could tolerate oxygen but also could employ it in respiration. The result was a great improvement in metabolic efficiency. Finally, about 1,450 million years ago, the first eukaryotic cells emerged (f). From the start they were adapted to a fully aerobic environment. The new modes of reproduction possible in eukaryotes, in particular the advanced sexual reproduction that evolved later, led to rapid diversification of the group.

62

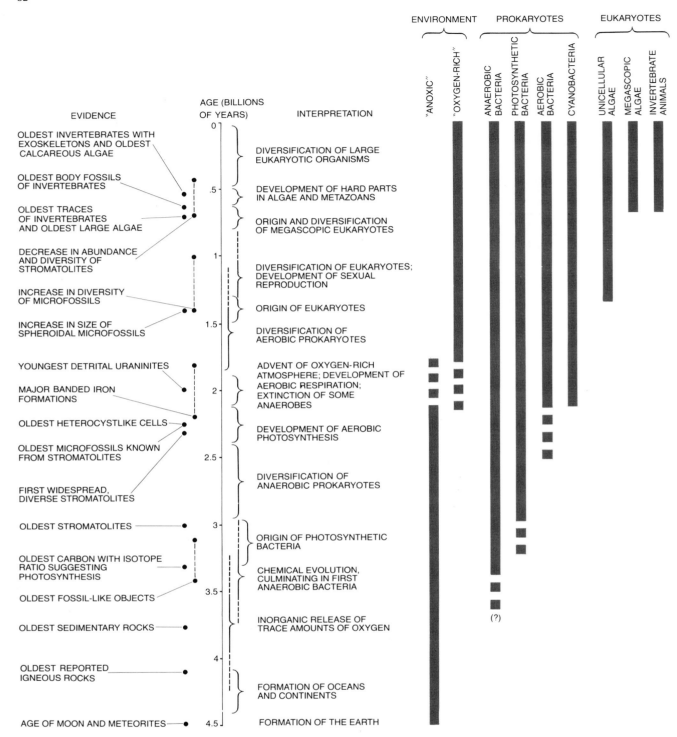

MAJOR EVENTS in Precambrian evolution are presented in chronological sequence based on evidence from the fossil record, from inorganic geology and from comparative studies of the metabolism and biochemistry of modern organisms. Although the conclusions are tentative, it appears that life began more than three billion years ago (when the earth was little more than a billion years old), that the transition to an oxygen-rich atmosphere took place roughly two billion years ago and that eukaryotes appeared by 1.5 billion years ago.

life, some of them recognizable antecedents of modern plants and animals.

In style and in tempo evolution in the Precambrian was distinctly different from that in the later, Phanerozoic era. The Precambrian was an age in which the dominant organisms were microscopic and prokaryotic, and until near the end of the era the rate of evolutionary change was limited by the absence of advanced sexual reproduction. It was an age in which the major benchmarks in the history of life were the result of biochemical and metabolic innovations rather than of morphological changes. Above all, in the Precambrian the influence of life on the environment was at least as important as the influence of the environment on life. Indeed, the metabolism of all the plants and animals that subsequently evolved was made possible by the photosynthetic activities of primitive cyanobacteria some two billion years ago.

Pre-Cambrian Animals

5

by Martin F. Glaessner
March 1961

Until recently the fossils of organisms that lived earlier than the Cambrian period of 500 to 600 million years ago were rare. Now a wealth of such fossils has been found in South Australia

The successive strata of sedimentary rock laid down on the earth's crust in the course of geologic time preserve a rich record of the succession of living organisms. Fossils embedded in these rocks set apart the last 60 million years as the Cenozoic era—the age of mammals. The next lower strata contain the 150-million-year history of the Mesozoic—the age of reptiles. Before that comes the still longer record of the Paleozoic, which leads backward through the age of amphibians and the age of fishes to the age of the invertebrates. Then, suddenly and inexplicably, in the lowest layers of the Paleozoic the record of life is very nearly blotted out. The strata laid down 500 to 600 million years ago in the Cambrian period of the Paleozoic era show a diversity of primitive marine life: snails, worms, sponges and the first animals with segmented legs, the trilobites and their relatives. But the record fades at the bottom of the Cambrian. The greater part of the journey to the beginning of sedimentation, at least another 2,000 million years, still lies ahead. Yet apart from algae and a few faint traces of other forms, the Pre-Cambrian strata have yielded almost no fossils and have offered no clues to the origin of the Cambrian invertebrates.

The geological record necessarily becomes more obscure the further back it goes. The older rocks have been more deeply buried and more strongly deformed than the younger rocks. They have undergone longer exposure to the heat and pressure and the mineralizing solutions by which fossils are commonly destroyed. One can find fossils, however, in greatly deformed younger sediments, including metamorphic rocks, which have been even more thoroughly reworked by geologic processes than some older rocks. What is more, no rock-de-forming process affects the entire surface of the globe. Some Pre-Cambrian formations have escaped extreme alteration, just as the lower Cambrian rocks are altered in some places and not in others.

The abrupt termination of the fossil record at the boundary between the Cambrian and the Pre-Cambrian has appeared to many observers as a fact or paradox of decisive importance. They have advanced many different explanations for the mystery, from cosmic catastrophes to the postulate of an interval of time without sedimentation; from the assumption of a lifeless ocean to the thought that all Pre-Cambrian organisms may have lived at the surface of the sea and none on its bottom, or all in the deep sea and none on its shores.

The need for such speculation has at last been obviated by the discovery in the Ediacara Hills in South Australia of a rich deposit of Pre-Cambrian fossils. The first finds at this site were made in 1947 by the Australian geologist R. C. Sprigg. In sandstones that were

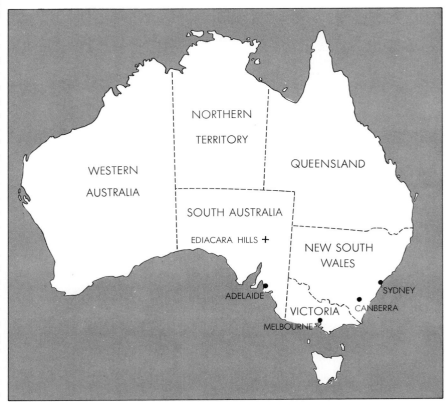

LOCATION OF PRE-CAMBRIAN FOSSIL BED is in the Ediacara Hills (*cross*), some 300 miles north of Adelaide. The geologist R. C. Sprigg made the first discoveries there in 1947.

thought to belong to the lowest strata of the Cambrian he came upon varieties of fossil jellyfish. Sprigg's find was followed up by other geologists and by students under the leadership of Sir Douglas Mawson, who found some plantlike impressions that appeared to be algae. Some time later two private collectors, Ben Flounders and Hans Mincham, brought to light not only large numbers of presumed fossil jellyfish but also segmented worms, worm tracks and the impressions of two different animals that bear no resemblance to any known organism, living or fossil. These discoveries prompted the South Australian Museum and the University of Adelaide to undertake a joint investigation of the region. Re-examination of the geology now showed that the fossil-bearing rocks lie well below the oldest Cambrian strata. This finding, taken together with the nature of the fauna represented in the fossils and their evident relationship to certain fossils discovered in South Africa before World War I and more recently in England, established that all these fossils date from the Pre-Cambrian era.

To date some 600 specimens have been collected in the Ediacara Hills. The fauna include not only jellyfish representing at least six and probably more extinct genera but also soft corals related to the living sea pens; segmented worms with strong head shields; odd bilaterally symmetrical animals resembling certain other types of living worm; and the two animals that look like no other living thing.

All the Ediacara animals were soft-bodied; none had hard shells, and their soft tissues were strengthened by nothing more than spicules: needles of calcium carbonate that served as a primitive support. All, of course, lived in the sea, some fixed to the bottom, some crawling and others free-floating or swimming. Their preservation is due to rather unusual, though not unique, conditions. The animals lived or were stranded in mud flats in shallow waters. Their impressions or their bodies were molded in the shifting sands that washed over the flats and were preserved as molds or casts in sandstone, mostly on the lower surfaces of sandstone beds. The resulting rich and varied assemblage of fossil animals gives the first glimpse of the marine life of the Pre-Cambrian era. It is a glimpse not merely of several types of animal but also of an association of creatures living together in the sea.

The soft-bodied nature of these fossils

justifies the characterization of the Pre-Cambrian as the "age of the jellyfish." The term jellyfish, however, applies to a number of highly diverse and only remotely related forms, of which the most common belong to the coelenterate phylum. These are animals that alternately take the free-swimming medusoid, or jellyfish, form and the sedentary polyp form. Sprigg concentrated on the medusoid jellyfish among his finds. He arranged some of them in two classes and four orders that have living representatives and placed the more commonly occurring specimens, which he called *Dickinsonia*, in a more problematic position with respect to living forms. But further study has indicated that none of the Pre-Cambrian medusae can be tied with any confidence to living orders, suborders or families.

Greater interest perhaps attaches to the leaf- or frondlike stalked fossils that Sprigg apparently took to be algae. The stalk is some 12 inches long and three-quarters of an inch wide. The body measures some nine inches long and four and a half inches wide; it is characterized by transverse ridges branching off from either a tapering median field or a median zigzag groove and divided in turn by longitudinal grooves [*see bottom illustration on page 67*]. No living algae display such structures. The true nature of these fossils appears in specimens

that show the impressions of spicules in the stalk and along the lower edges of the side branches. These suggest the spicules of otherwise soft alcyonarian corals living today and identify the fossil fronds as animals of the coelenterate phylum rather than as plants.

One group of modern corals—the sea pens (*Pennatulacea*)—has a similar arrangement of spicules, along with the stalk and side branches. Thus the fossils appear to be sea pens, which are normally rare in the geological record. The differences between the Pre-Cambrian sea pens and the modern animals are remarkably small, considering the 600 million years of evolution that separate them. In the living sea pens the frond is either deeply dissected into movable side branches or it forms an entire plate-like body. In the fossil the lateral ridges are separated by furrows and not by open slits. Coral polyps that occupy the surfaces of the fronds and stalks in modern sea pens are so small that they would not be apparent in the rather coarse sandstone casts of the fossils.

The Australian frond fossils are similar to those discovered before World War I by German geologists in Southwest Africa. Those fossils were named *Rangea* and *Pteridinium*. The Pre-Cambrian fossil discovered recently in England and named *Charnia masoni* also re-

PRE-CAMBRIAN SEASHORE AREA, reconstructed from fossils found in South Australia, supported several types of animal. Some are shown stranded in dried-up mudholes (*A*) and on sand of beach, where they were fossilized. Others appear (*lower left*) in sand and

sembles certain of the fossil Australian sea pens. The English fossil seems to possess a circular disk with concentric ribs at the end of the stalk opposite the frond. Although the connection between these two structures is uncertain, it may be that this fossil represents the two alternating coelenterate forms, that is, the free-swimming medusa and the branching colony of small polyps that remains fixed to the ocean bottom. In this case one might speculate that the Pre-Cambrian sea pens grew from free-swimming, solitary medusae. But this is as yet pure guesswork about the reproductive processes of long-dead organisms. Further discoveries may prove or disprove the connection between fronds, stalks and disks.

The most spectacular finds in the Pre-Cambrian strata of South Australia were small annelid worms named *Spriggina floundersi* after their discoverers, Sprigg and Flounders. They had a narrow, perfectly flexible body up to one and three-quarter inches long, a stout horseshoe-shaped head shield and as many as 40 pairs of lateral projections (parapodia) ending in needle-like spines. A pair of fine threads projected backward along the sides of the body from the lateral horns of the head shield, and another thread probably grew from the segment behind it [*see illustration at top right p. 66*]. Although such worms no longer

exist, they resemble the living marine *Tomopteridae*, which have similar but wider heads, transparent narrow bodies and parapodia ending in flat paddles [*see illustration, page 68*]. These modern worms, because of their special paddle adaptation to the free-swimming life, had not been considered primitive or of ancient origin. It now appears, however, that they are directly descended from extremely ancient forms. The shape of the head of the Pre-Cambrian worms suggests the possibility of a relationship between them and the arthropods, such as the now extinct trilobites, which first appear in large numbers in the Cambrian. All of these later animals represent a considerable advance over the primitive anatomical organization of the coelenterates.

The most common fossil at the Ediacara site, the *Dickinsonia*, represented by more than 100 specimens, may also be related to living worms. The fossilized bodies are quite remarkable. They are more or less elliptical in outline, bilaterally symmetrical and are covered with transverse ridges and grooves in a distinctive pattern. The size of the bodies and the number of ridges vary so much that Sprigg attempted to distinguish species by counting the ridges. One recently discovered specimen has some 20 ridges; a larger one may have had as many as 550. The animals range in

length from a quarter of an inch up to two feet. The numerous impressions of wrinkled and folded-over specimens indicate that all were soft-bodied, for there are none of the fractures that would be apparent if the creatures had possessed shells. These animals vaguely resemble certain flatworms living today. There is also one genus of annelid worm with a strikingly similar pattern of ridges formed by extensions of its parapodia. This similarity proves little or nothing, especially since no traces of eyes, legs or intestines are preserved in the fossils, but it provides some hope of finding out what these strange creatures were.

There is possibly less hope of placing in the family tree of the animal kingdom the two completely novel forms discovered in the Ediacara Hills. One had a shield- or kite-shaped body with a ridge that looked like an anchor. It was named *Parvancorina minchami* [*see illustration at top right p. 67*]. The first specimen was tiny, but others found later measure up to one inch in length. Some show faint oblique markings within the shield on both sides of the mid-ridge, as if the animal had had legs or gills underneath. Here again folded and distorted specimens occur, proving that their bodies were soft.

The other entirely new creature is even stranger. Named *Tribrachidium*, it has three equal, radiating, hooked and

water as though seen in an aquarium. They are jellyfish-like creatures (1); the wormlike *Dickinsonia* (2); the segmented worm *Spriggina floundersi* (3) and worm trails (4); *Parvancorina* (5), which resembles no other known animal; *Tribrachidium* (6), another unknown type; the sea pens *Rangea* and *Charnia* (7); hypothetical algae and sponges (8), and a worm in a sand burrow (9).

tentacle-fringed arms. Nothing like it has ever been seen among the known millions of species of animals. It recalls nothing but the three bent legs forming the coat of arms of the Isle of Man.

Considered together, the South Australian fossils suggest a rough and incomplete picture of conditions in the late Pre-Cambrian. Of course, such a group of fossils constitutes no more than a small, biased sample of the life of the time. Animals buried together in slabs of sandstone did not necessarily live together. Some, if they really are medusae, were floating in the sea. Others, like the annelid worm *Spriggina*, with its numerous legs and sinuously curving body, were free-swimming. *Dickinsonia* was probably also a free-swimming form, apparently along with *Parvancorina*. Scattered miniature treelike stands of sea pens, waving their flexible fronds, must have covered parts of the shallow sea floor. Elsewhere earthworm-like annelids, which have left only their tracks, crawled over and through the sediment, feeding on the decaying organic matter in it. Other worms inhabited the U-shaped burrows that have been found, consuming tiny creatures in the sediment and possibly also marine plankton, which left no traces in the rock. The fixed, three-rayed spread of tentacles of the strange *Tribrachidium* may be similar to the plankton-fishing structures around the mouth of the living brachiopods (lamp shells), bryozoa (lace corals) and some worms. If that is correct, *Tribrachidium* may have been a bottom dweller, possibly occupying low, conical, ridged cups, of which a few impressions have been found.

Bundles of impressions of needle-shaped spicules also occur in the Ediacara strata. Since spicules are characteristic of sponges, these sessile, bottom-dwelling animals may have been present. Snails and small crustaceans, as well as various protozoa (radiolarians and foraminifera), may also have existed at that time, but they would have been too small or too fragile to be preserved. Plant life likewise left no traces here.

The worm tracks are the only fossils indicating without doubt that the animals lived where their remains are found. Thus the *Spriggina* worms, the *Dickinsonia* and *Parvancorina* may have lived near or on the sedimentary beds. They are represented by individuals varying in size and growth stage, which indicates accidental death rather than transport from afar and later burial. On the other hand, the jellyfishes were probably stranded and the soft corals torn from their anchorage before they came

PRE-CAMBRIAN FOSSILS preserved in sandstone are seen in these eight photographs. This is *Dickinsonia costata*, shown actual size.

SEGMENTED WORM *Spriggina floundersi*, shown about twice actual size, resembles certain segmented worms living today.

JELLYFISH *Spriggia annulata* is one of the many types of this organism that have been found. The fossil is very slightly enlarged here.

ANOTHER JELLYFISH, *Medusina mawsoni*, is shown nearly three times actual size. Jellyfish were the first fossils found.

to rest on the bottom.

The sandstone in which the fossils are found shows ripple marks and other evidences of currents, which would have had to be rather strong to transport the coarse grains of sand. Thus it is difficult at first to see how imprints of delicate, soft-bodied creatures could have been preserved. Careful study of the fossils has yielded an explanation. Only a very few of the animals came to rest on the shifting sand. Most of them came down on mud flats or on patches of fine clay that settled out of the water during calmer periods. Some of the mud patches dried out, possibly between tides, and developed deep cracks. The next high tide or shifting current covered them with a layer of sand. The lower surfaces of such sandy layers preserved the clay surfaces in the form of perfect casts,

showing the wrinkles in the clay and the cracks formed by drying as well as the shapes of the animals stuck in the clay. The sand grains were cemented by silica solutions and turned to quartzite in the transformation from soft sediment to hard rock. The clay changed to thin slatelike streaks of the mineral sericite and was compacted almost beyond recognition. Since the sericite inclusions are small and irregular, the rock does not split along their surfaces as slate would. Only the slow, natural weathering in the arid climate of South Australia can open up the rock along the vital sericitic partings where the fossils occur. Slabs of quartzite of all sizes remain in place, projecting from the hillsides until they break off. They often turn over when moving downhill and their lower surfaces become exposed to the infre-

quent rain. Then the weathering causes them to reveal their wonderful riches of Pre-Cambrian animals. But if the rocks are not collected, the fossils are ultimately worn away by the weather and by the sand drifting in on the wind from the adjoining desert plains.

The age of the fossil-containing rocks cannot be determined directly in years because it does not contain radioactive minerals suitable for dating. Fortunately in the Ediacara Hills one can follow the stratification in unbroken sequence upward until the first undoubtedly Cambrian fossils are reached in dolomitic limestone 500 feet above the Pre-Cambrian level. These fossils in the limestone are typical of the lowest Cambrian strata elsewhere and are quite unlike the strange fossil organisms in the quartzite

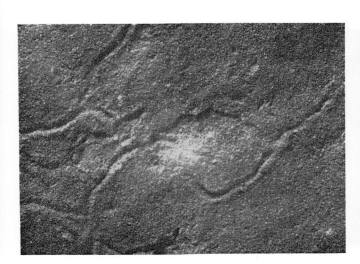

WORM TRAILS, approximately actual size, provide proof that fossilized worms lived in the area where they were preserved.

UNKNOWN TYPE OF ANIMAL, *Parvancorina minchami*, here enlarged nearly three diameters, resembles no other known organism.

SEA PEN *Rangea arborea* left this imprint, shown here twice actual size. The fossil resembles some of the living sea pens.

ANOTHER SEA PEN, *Charnia*, is shown actual size. Viewing photographs upside down may give fuller idea of animals' appearance.

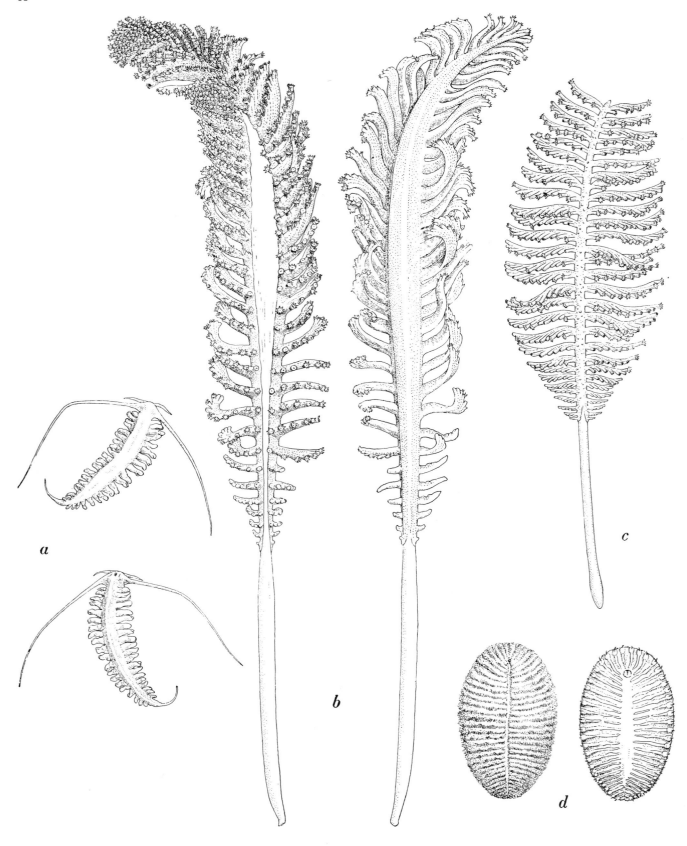

FOUR LIVING ANIMALS that resemble some of the Pre-Cambrian fossils from South Australia are a segmented worm, *Tomopteris longisetis* (*a*), seen in dorsal and ventral views; sea pens *Pennatula rubra* (*b*), shown front and back, and *Pennatula aculeata* (*c*); and the worm *Spinther citrinus* (*d*), which looks like the many specimens of *Dickinsonia* in the Pre-Cambrian rocks.

below. The quartzites higher up in the Cambrian strata do not contain any fossils of the type now known from the Pre-Cambrian, and the dolomites and limestones lower down contain no Cambrian fossils. From this distribution of fossils in the rocks it can be judged that the lack of shells and hard skeletons (other than the spicules) in the Pre-Cambrian animals was not due to any factors in the physical environment. The development of shells in the Cambrian was not a result of a sudden change in the habits or habitats of the animals. Rather, shells appeared as a step forward in biochemical evolution. Calcium metabolism underwent a change that produced hard shells and other skeletal material, providing the protection and mechanical support so important to the more advanced animals.

This is as far as the paleontologist and geologist can take the story today. The biochemist and physiologist may see in it a lead to experimentation that could well open a new chapter in the story of fundamental research in evolution.

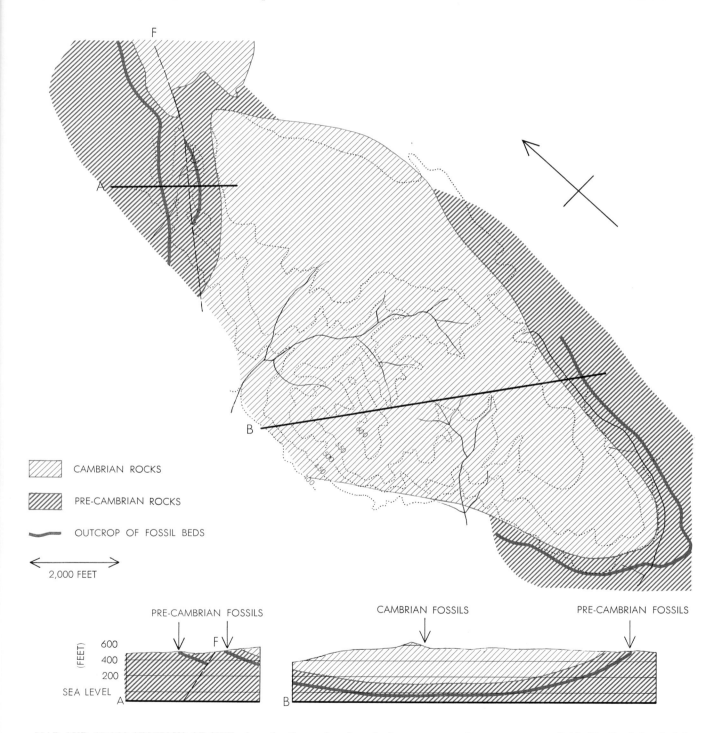

CAMBRIAN ROCKS

PRE-CAMBRIAN ROCKS

OUTCROP OF FOSSIL BEDS

2,000 FEET

PRE-CAMBRIAN FOSSILS

CAMBRIAN FOSSILS

PRE-CAMBRIAN FOSSILS

(FEET) 600 400 200 SEA LEVEL

MAP AND CROSS SECTIONS OF SITE where fossils are found show relative positions of Cambrian and Pre-Cambrian rocks. Lines *A* and *B* indicate locations of cross sections *A* and *B* (*below map*). Broken line *F* is a fault that has caused part of Pre-Cambrian fossil bed to move, creating two outcrops (*left*). Pre-Cambrian bed is under Cambrian rocks, except where its edges come to surface at periphery of Cambrian area. Dotted lines indicate contours. Part of the region shown here has been included in a fossil reserve.

6

The Animals of
the Burgess Shale

by Simon Conway Morris and H. B. Whittington

*The fossils of a rock formation in western Canada are
a rich sample of an animal community in the Mid-
Cambrian. Some of the animals are ancestors of those
living today; others are unique and bizarre*

By far the most numerous fossils representing the first abundant life on the earth are the hard parts of various marine animals without backbones: shells and similar fragments of external skeleton. This makes for a lopsided fossil record. For example, of the 30 or so phyla of animals living today more than half are made up of species with few hard parts or none. As a result the descent of these phyla remains largely undocumented by fossil evidence.

Fortunately the situation is not completely lopsided. A few geological deposits have been discovered that as a result of exceptional circumstances contain exquisitely preserved fossils of animals that are partly or entirely soft-bodied. Here we shall describe one such deposit: the Burgess Shale of western Canada. The great age and the rich variety of the marine invertebrates in the Burgess Shale make it perhaps the best-known of all such deposits. In addition to describing the Burgess Shale fauna we shall attempt to reconstruct the kind of underwater environment these organisms inhabited early in Paleozoic times: some 530 million years ago.

In the fall of 1909 the Secretary of the Smithsonian Institution, Charles Doolittle Walcott, was searching for fossil-bearing rock formations in British Columbia. Following a footpath that ran across the western slope between Wapta Mountain and Mount Field in the southern part of the province, Walcott literally stumbled over a block of shale that had fallen onto the path from the slope above. Examining the easily split rock, he was astonished to find the fossil impressions of a number of soft-bodied organisms preserved in its layers. In a letter to a colleague in Toronto dated November 27, 1909, he dryly reported that he had spent "a few days collecting ... in the vicinity of Field and found some very interesting things."

Walcott returned to the spot the following year to search upslope for the shale stratum that had been the source of his fallen rock. His search was successful: he found two fossil-bearing shale exposures separated by a vertical distance of some 70 feet. He did shallow quarrying in both; the lower exposure proved to be the richer of the two. He shipped back to the District of Columbia thousands of fossil specimens that he removed from what he called his Phyllopod Bed. (The term, little used by paleontologists today, refers to certain fossil arthropods, or joint-legged animals, that are probably ancestral to living crustaceans.)

As Walcott's own work and decades of study by others have shown, the fossils of the Burgess Shale include a great abundance of marine invertebrates: more than 120 species. Some of them belong to the Phylum Porifera: the sponges. This phylum of primitive animals is the only one in the subkingdom of parazoans, a category higher than the subkingdom of the one-celled protozoans but lower than the subkingdom of the many-celled metazoans. Perhaps 10 other species represent metazoan phyla that were unknown before they were found in the Burgess Shale; they are not present elsewhere in the fossil record. The scores of other species that lack hard parts can be assigned to one or another phylum of metazoans with living relatives as follows:

Coelenterates: the phylum that includes such living marine animals as jellyfishes, sea pens and corals. The Burgess Shale coelenterate species number perhaps four.

Echinoderms: the phylum that includes, among others, starfishes, sea cucumbers and crinoids, or sea lilies. At least four species of Burgess Shale echinoderms are recognized.

Mollusks: the phylum that includes, among others, oysters and clams, squids and octopuses and the primitive chitons (of the Class Amphineura). Three Burgess Shale molluscan species are recognized.

Arthropods: the phylum that includes, among a great many others, lobsters, shrimps, crabs and barnacles (all of the Class Crustacea) and the less familiar terrestrial animal *Peripatus* (a member of the Class Onychophora). The Burgess Shale arthropods include several representatives of the long extinct trilobite class, a peripatus-like animal that was aquatic rather than terrestrial, and about 30 other species of arthropods.

Priapulids: a minor phylum of unsegmented marine worms. The living genus, *Priapulus,* gives the phylum its name. Seven species of these now obscure bottom burrowers flourished in the Burgess Shale muds.

Annelids: the phylum that includes earthworms, leeches and a less familiar but very large class of marine worms, the polychaetes. The annelid phylum is represented in the Burgess Shale by six species.

Finally, we find among the Burgess Shale fauna one of the earliest-known invertebrate representatives of our own conspicuous corner of the animal kingdom: the chordate phylum. Among its living representatives (other than vertebrates) are the sea squirts and the peculiar marine animal *Amphioxus.* The chordates are represented in the Burgess Shale by the genus *Pikaia* and the single species *P. gracilens.*

Such a remarkably preserved soft-bodied fauna, representing eight known and 10 or more previously unknown phyla that flourished in the Middle Cambrian, is by itself of great interest to students of the fossil record. In addition to this intrinsic interest, however, the Burgess Shale invertebrates, with their specialized adaptations, have an even wider importance in clarifying the early evolution of the animal kingdom. The only earlier-known soft-bodied animals are representative of late Precambrian time, 700 to 600 million years ago and therefore at least 70 million years earlier than Middle Cambrian times. These are the Ediacara animals, first discovered some 30 years ago in the Ediacara Hills of southern Australia and recognized since then in a number of other places throughout the world. The Ediacara

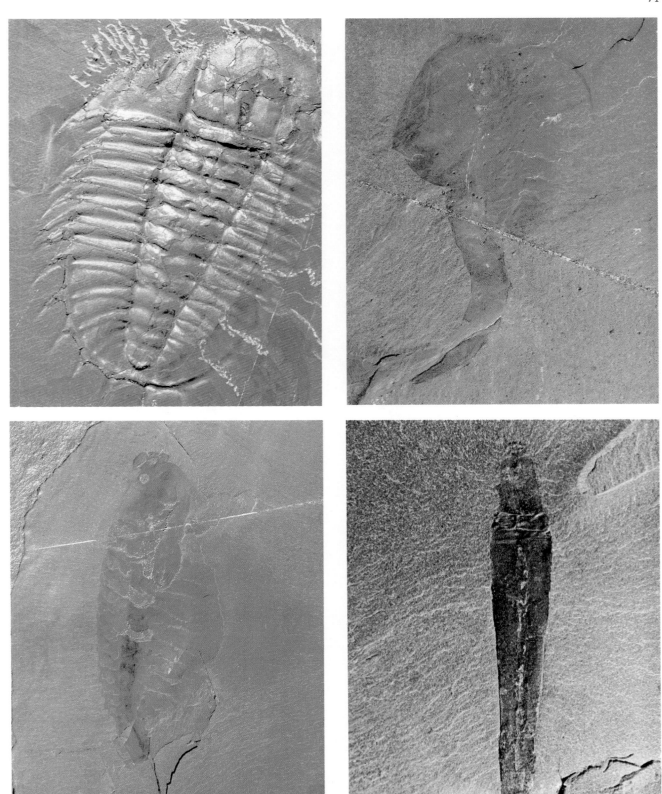

FOUR ANIMALS that lived in the ocean in Middle Cambrian times, some 530 million years ago, are seen in these fossils. At the top left is a trilobite, *Olenoides,* one of the many animals whose anatomy has been preserved in remarkable detail in the silts that solidified to form the Burgess Shale of British Columbia. The specimen is 5.5 centimeters long. Unlike most arthropods, or joint-legged animals, *Olenoides* had unspecialized limbs. At the top right is another Burgess Shale arthropod, *Waptia.* When this bottom-feeding animal was extended, it was four centimeters long. At the bottom left is *Opabinia,* one of about 10 animal species found in the shale that belong to previously unknown phyla. It had five eyes and steered its seven-centimeter body with a vertical tail fin as it swam close to the sea floor in search of food. At the bottom right is one of the many unsegmented marine worms that inhabited the sea floor. It is *Selkirkia,* one of the priapulid phylum. With its projecting proboscis it measured five centimeters. A successful group in the Cambrian, priapulid worms are now rare.

LOWER QUARRY, named the Phyllopod Bed by Charles Doolittle Walcott of the Smithsonian Institution, who first sampled the Bur- gess Shale, shows patches of winter snow. The view looks south. This and a higher shale exposure were requarried for fossils in 1966–67.

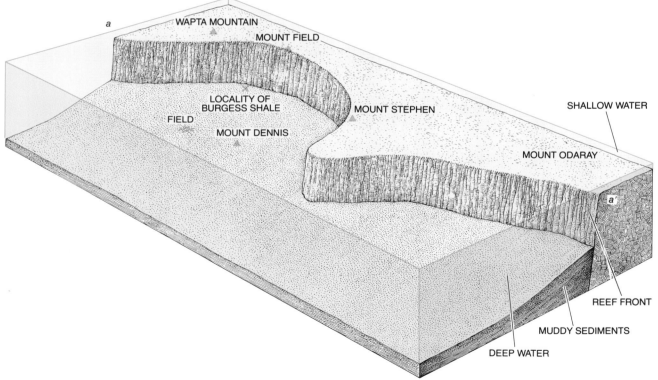

BURGESS SHALE OUTCROP, marked by the colored X in this block diagram of the Middle Cambrian seascape, is a small portion of an extended sea-bottom deposit of silts accumulated at the foot of a deeply embayed algal reef that rose vertically some 530 feet above the shallowest silts. The reef did not rise above sea level but was covered by shallow water. The vertical scale in the block diagram is ex- aggerated by a factor of five, and the distance (a–a') from north to south along the reef is some nine miles. Later uplift and dissection gave rise to the peaks of the Rockies along the border of Alberta and British Columbia, indicated by colored triangles. The reconstruction of the sea floor and the reef is based on the work of I. A. McIlreath of Petro-Canada, one recent investigator of the unique formation.

fauna stands in marked contrast to the Burgess Shale fauna both in the kinds of animals represented (chiefly coelenterates) and in these earlier animals' limited range of specialization.

The event that separates the impoverished Ediacara fauna from the Burgess Shale fauna is an explosive evolutionary diversification of multicelled animals that took place near the beginning of Cambrian time. The fossils of the Burgess Shale thus give us a unique glimpse into the results of this sudden metazoan adaptation relatively soon after it occurred.

In spite of the work done by Walcott and others significant gaps remain in what is known about the Burgess Shale paleoenvironment and how its fauna was preserved. A fuller appreciation of these gaps stimulated a reinvestigation of the site by the Geological Survey of Canada, beginning more than a decade ago. The authorities of the Yoho National Park in British Columbia and the Parks Canada administration in Ottawa granted special permission to collect material from the shale outcroppings. Walcott's quarries were reopened in 1966 and 1967 under the direction of J. D. Aitken of the Geological Survey of Canada. Both the new material collected during these two seasons and a part of the great Burgess Shale collection amassed by Walcott some 60 years earlier then came to us at the University of Cambridge for analysis.

W hat kind of environment did the Burgess Shale fauna inhabit? Studies done by I. A. McIlreath of Petro-Canada and W. H. Fritz of the Geological Survey of Canada show that the animals lived on or in a muddy bottom where sediments had accumulated at the base of a gigantic reef. This structure, made up of material secreted by algae, rose vertically for hundreds of feet from a deep-water basin that was gradually being filled with sediments. Scattered outcrops of the reef front can still be traced for miles across British Columbia. The bottom waters of the basin were apparently limited in circulation, rich in hydrogen sulfide and poor in oxygen. The various invertebrates flourished where the muddy sediments were banked high enough against the reef to be clear of the stagnant bottom waters, about 530 feet below sea level.

The reef-front sediments were not stable. Studies of the shale by D. J. W. Piper of Dalhousie University show that periodic slumping resulted in the flow of mud into the deeper anaerobic waters of the basin. These flows wiped away all the surface tracks and subsurface burrows made by the Burgess Shale fauna. Because the animals trapped in the torrents of mud died during or shortly after their burial they could not leave new traces. This means that the way of life of

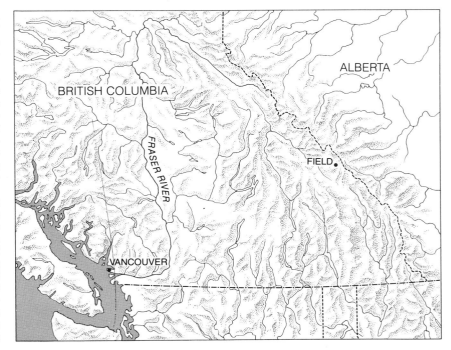

BURGESS SHALE FORMATION is situated some 350 miles northeast of Vancouver near the town of Field, B.C. The fossil-rich formation was found accidentally by Walcott in 1909.

each species must be deduced from a study of their organs of locomotion and from comparisons with living invertebrates of the same kind.

At the same time the catastrophic burials, in anaerobic deposits of fine silt where scavengers could not survive, greatly favored the preservation of the animals' soft parts. As the mud gradually compacted and became hard rock the buried carcasses were flattened and the soft parts were transformed into thin films of calcium aluminosilicate. In general the films are rather dark, but certain parts of most specimens are preserved as highly reflective areas.

Paradoxically, although the animals' soft parts are wonderfully preserved, signs of rotting after burial can often be detected. Many specimens are associated with a black-stained area, a result of the body contents of the carcass seeping out into the surrounding mud. In extreme cases the fossil of a worm consists only of a hollow bag of cuticle because practically all the animal's internal organs have been destroyed by decay. In some worms a subtler indication of decay is the pulling away of body-wall muscles from the cuticle.

Burial in a mud flow has other important effects. For one thing, many of the animals came to be buried at all angles; the shale bedding has therefore preserved them in a variety of orientations that reveal much more of the animals' anatomy than simple horizontal burial does. For another, the fluid sediments that penetrated between the appendages of animals such as arthropods and poly-

chaetes during the turbulent flow of silt were eventually reduced to thin layers of shale. Judicious work with a microchisel enables one to remove these fine layers, thereby revealing further details of a specimen's anatomy that would otherwise remain hidden.

T he composition of the Burgess Shale fauna upsets the conventional notion of what makes up a typical assemblage of Cambrian animals. The fossils found at most Cambrian localities are the exoskeletons of such arthropods as trilobites, the shells of various members of the brachiopod phylum and of such echinoderms as the extinct plate-shelled Eucrinoid class. Animals such as these account for barely 20 percent of the invertebrate genera in the Burgess Shale. Is it justified, then, to regard the Burgess Shale assemblage as the Cambrian norm, at least with respect to the fauna of deeper waters, and to view the other Cambrian faunas as being skewed by the selective fossilization of animals with hard parts?

Since the Burgess Shale represents a single environment that has been frozen for a split second of geologic time, no firm answer can be given to the question. Several factors nonetheless suggest that the Burgess Shale fauna was not untypical of Middle Cambrian times. The scattered occurrence of species similar to those from the Burgess Shale in other Cambrian rocks hints at the existence in this period of a widespread soft-bodied fauna. Furthermore, in some Cambrian fossil assemblages certain rather peculiar species were able to flourish because access by sea to the area of deposition was limited. The Burgess Shale, on the other hand, lay at the edge of the open sea and would have been exposed to colonization by marine larvae floating in from other areas. This circumstance adds weight to the hypothesis that the Burgess Shale fauna approaches the Cambrian norm.

In this connection it should be noted that representatives of certain modern invertebrates that have almost certainly had a long geological history are absent from the Burgess Shale. No species of the platyhelminth phylum, the flatworms whose living members include flukes, tapeworms and planarians, are present. There are also no species of another worm phylum, the Nemertea, which includes the modern proboscis worm, and none of still another, the sipunculid phylum. It may be that such worms are not represented because the reef-front environment was not suitable for them.

Most of the species found in the Burgess Shale can be placed in the ecological framework of a bottom-dwelling marine community that thrived on the muddy sea floor between intervals of slumping. The mud supported an active group of burrowing invertebrates, with priapulid worms predominant. Attached to the sea floor and growing to various heights were a variety of sponges representing at least 15 genera; they fed on food particles suspended in the water. Actively patrolling the sea-floor surface or plowing through the mud in search of food were many species of arthropods. Certain brachiopods occupied a peculiar niche: they attached themselves to the elongated spicules of one of the sponges, *Pirania*. For the brachiopods the advantages are obvious: they lived somewhat above the turbid waters of the sea floor and could capture food particles such as the sponges fed on at these higher levels.

In addition to this community of fixed and mobile surface dwellers and burrowers a number of free-swimming species inhabited the waters along the reef front. Of these animals there are only tantalizing glimpses, in the form of rare specimens buried by chance in the slumped sediments. The different members of this pelagic fauna probably lived at different depths; some among them may have been species swept into the reef-front area from the open sea.

At most Cambrian fossil localities the mineralized exoskeletons of trilobites, the most familiar of all Paleozoic arthropods, are in the majority. In the Burgess Shale, however, trilobites—with one exception—are comparatively unimportant. The exception is *Olenoides,* which is of great significance because in several specimens the appendages have

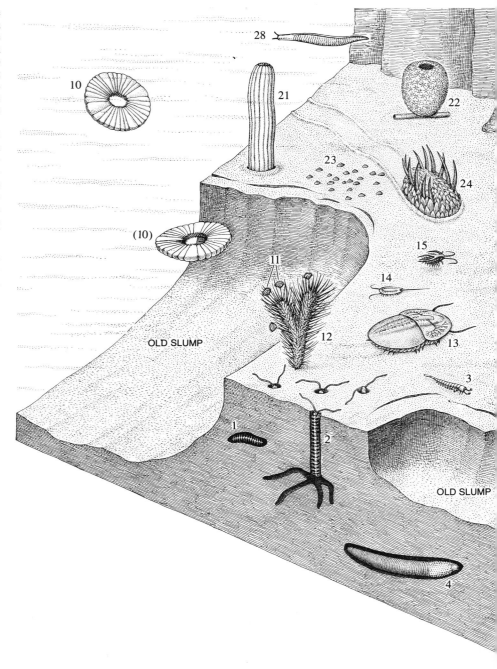

UNDERWATER SCENE where silts slope down from the face of the great reef and the Burgess Shale fauna lived is shown in an idealized reconstruction. No attempt has been made to show the animals in numbers proportional to their fossil abundance. The fauna are identified by number, starting at the bottom left; only about a fifth of the species fossilized in the shale are shown. Most of the immobile animals of the sea floor are sponges: *Pirania* (12), seen with brachiopods attached to its spicules; *Eiffelia* (22); the gregarious *Choia* (25); a gracile species of *Vauxia* (5), with a more robust species at the top right, and *Chancelloria* (27). Three other immobile animals are *Mackenzia* (21), a coelenterate; *Echmatocrinus* (16), a primitive crinoid, seen attached to an empty worm tube, and *Dinomischus* (17), one of the Burgess Shale species that represent hitherto unknown invertebrate phyla. The burrow-dwelling animals are *Peronochaeta* (1), a polychaete worm that fed on food particles in the silt; *Burgessochaeta* (2), a second polychaete that captured food with its long tentacles; *Ancalagon* (4), a priapulid worm pos-

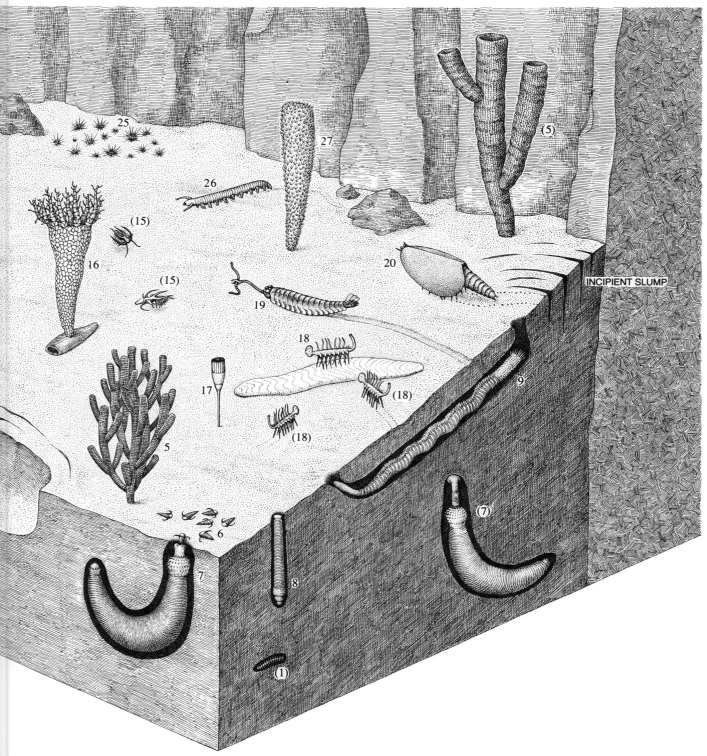

INCIPIENT SLUMP

sibly ancestral to some modern parasites; *Ottoia* (*7*), another priapulid, seen at the center feeding on the mollusk *Hyolithes* (*6*) and at the right burrowing; *Selkirkia* (*8*), a third priapulid, seen here in a burrow front end down, and *Louisella* (*9*), a fourth priapulid that inhabited a double-ended burrow and undulated its body to drive oxygenated water over its gills. *Peytoia* (*10*) is a free-swimming coelenterate shaped like a pineapple ring. The sea-floor-dwelling mollusks in addition to *Hyolithes* are *Scenella* (*23*), its soft parts hidden under "Chinese hat" shells, and *Wiwaxia* (*24*), with its covering scales and defensive spines, seen here plowing a trail through the silt. Among the many arthropod genera of the sea floor are *Yohoia* (*3*), with its distinctive grasping appendages; *Naraoia* (*13*), an atypical trilobite that retained

some larval characteristics; *Burgessia* (*14*), with its long tail spine; *Marrella* (*15*), which may have swum just above the sea floor; *Canadaspis* (*20*), an early crustacean, and *Aysheaia* (*26*), a stubby-legged animal suggestive of the living land dweller *Peripatus*. Other representatives of new phyla seen in addition to *Dinomischus* are *Hallucigenia* (*18*), one preparing to feed on a dead worm and two others approaching it, and *Opabinia* (*19*), seen here grasping a small worm with its single bifurcated appendage. Finally, seen swimming alone at the top left, is *Pikaia* (*28*), the sole representative of the chordate phylum in this Middle Cambrian fauna. *Pikaia* probably used its zigzag array of muscles to propel itself above the sea floor. The phylum of chordates includes the subphylum of vertebrates, which evolved later.

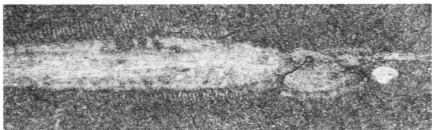

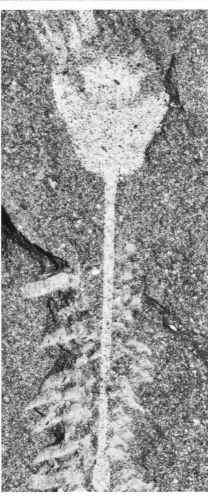

been preserved in detail. *Olenoides* had a pair of slender antennae in front and a pair of cerci, or antennalike structures, in back. The limbs along the length of the animal, up to 16 of them, were all similar in construction. The coxa, a large unit closest to the body, carried a battery of ferocious-looking spines. Attached to the coxa were two appendages; one was a filamentous gill and the other was a walking leg. *Olenoides* could seize and shred soft food, such as small worms, and pass the fragments along to its mouth. The forward antennae and the rear cerci no doubt supplied the animal with information about both food and potential predators. The fact that the primitive limbs of this trilobite are all similar is in marked contrast to the arrangement in many fossil and living arthropods whose limbs are variously modified and specialized.

About 40 percent of the Burgess Shale fauna consists of arthropods. Both in the number of species and the number of individual specimens the soft-bodied representatives of the phylum outrank the hard-shelled trilobites. Many of these "nontrilobites" have had their appendages preserved in remarkable detail; some of them must have been effective predators and scavengers. The most abundant is *Marrella,* an arthropod with a wedge-shaped head that bore two pairs of long hornlike spines curving to the rear. *Marrella* sensed the sea-floor environment with a pair of antennae and swept food particles toward its mouth with adjacent feathery appendages. Its score or more of side limbs were jointed, and a filamentous gill branched from each limb.

The next most abundant of the nontrilobites is *Canadaspis.* All but the hind part of its body was concealed by a double shell. Careful removal of this covering reveals the underlying appendages, which are remarkably like those of certain living crustaceans. Still another arthropod was one of the larger predators of the sea floor. This is *Sidneyia,* an animal whose distinctive limbs recall those of the living horseshoe crab, *Limulus.* It has been possible to identify some of the gut contents of *Sidneyia* as being fragments of brachiopod shells, evidence

UNIQUE ANIMALS of the Burgess fauna include the four seen in the photographs at the left. At the top is *Hallucigenia,* **shown in the illustration on the preceding two pages. Second from the top is** *Nectocaris,* **a streamlined animal with conspicuous fin rays. At the bottom left is** *Amiskwia,* **a gelatinous worm with prominent fins. At the bottom right is** *Dinomischus,* **the stalked animal also shown reconstructed on the preceding two pages. Each of these species and six or more others represent hitherto unknown phyla of invertebrates.**

that this arthropod was able to crush hard-bodied prey.

Sidneyia was not the largest of the Burgess Shale arthropods. The shale contains the impressions of isolated large limbs; if they are in proportion, they may have belonged to an animal as much as a meter long. The name *Anomalocaris* has been assigned to these fossils, which possibly represent one of the largest of all Cambrian invertebrates.

The most interesting of all the Burgess arthropods is *Aysheaia,* an animal with a pudgy body and stubby limbs. When Walcott first published a photograph of this fossil half a century ago, a number of zoologists wrote to him to point out how much this Middle Cambrian invertebrate resembled *Peripatus,* an animal with eight genera of relatives (comprising two families) in the small class of onychophores within the arthropod phylum. *Peripatus* was a land animal and *Aysheaia* was a marine form; nevertheless, *Aysheaia* surely represents the kind of ancestor that could have given rise to such living arthropods as myriapods and insects.

Although some of the nontrilobite arthropods in the Burgess shale, such as *Aysheaia,* are reminiscent of later forms, most of them cannot be placed in any recognized group. They have no obvious relatives either among the other Burgess Shale species or among the arthropods of later times. Because they exhibit a surprisingly wide array of anatomical features, indicating a high degree of specialization, they are evidence of a hitherto unsuspected adaptive radiation of arthropods in Cambrian times. It appears that the numerous stocks that arose during this period of rapid evolution were mostly unsuccessful. It is interesting to note that the animals that were to become dominant in later geological history generally have only a minor position in the various Cambrian faunas; a hypothetical observer would have been hard-pressed to predict just which groups had the flexibility necessary for long-term biological success.

Of the Burgess Shale animals other than arthropods the representatives of six phyla are particularly noteworthy. Among the echinoderms the class of holothurians, the group that includes the sea cucumbers, was once thought to be widely represented. Now only one animal, *Eldonia,* is so classified. Unlike the great majority of the species in its class, *Eldonia* had a jellyfishlike body and a pair of oral tentacles. These animals probably swam through the water in shoals, using their tentacles to capture food. Another Burgess Shale echinoderm, the sea lily *Echmatocrinus,* is the earliest crinoid in the fossil record; as might be expected, it shows a number of primitive features.

The species of the coelenterate phy-

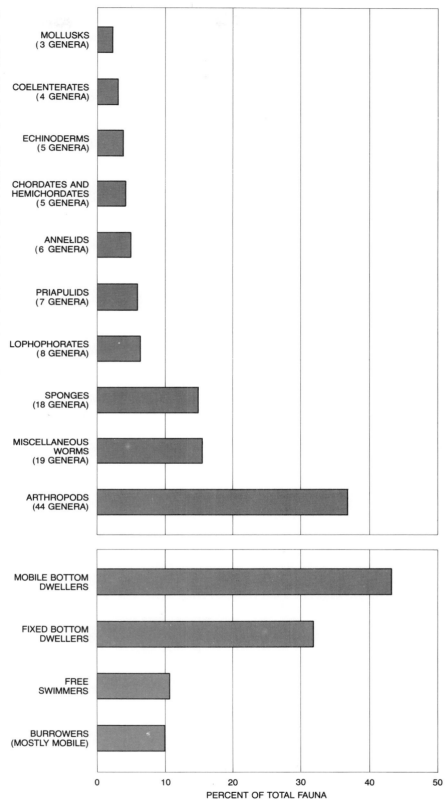

BURGESS SHALE GENERA currently number 119. The percent of the total assigned to various phyla is indicated in the upper part of this bar chart. Nearly 40 percent of the total are arthropods; only 14 of the 44 arthropod genera are trilobites. Worms other than priapulids and annelids (19 genera) and sponges (18 genera) make up another 30 percent of the total; mollusks are the most poorly represented. In terms of habitat, as the lower set of bars indicates, more than 40 percent of the Burgess Shale animals wandered the sea floor and more than 30 percent were rooted in the silt. Most of the burrowing animals also moved freely, although some remained fixed. Burrowers were slightly outnumbered by animals that swam above the sea floor.

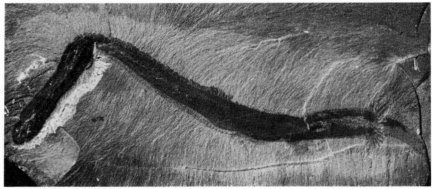

TWO PHYLA OF WORMS in the Burgess Shale fauna are the familiar annelids and the less common priapulids. A typical Burgess Shale annelid is *Canadia,* a polychaete worm *(top);* its setae, bundles of fine bristles that were organs of locomotion, are preserved in detail. A typical priapulid worm *(bottom)* is *Louisella,* also reconstructed in illustration on pages 74 and 75.

lum are among the most primitive of the metazoans. For example, in the late Precambrian fossil assemblage from the Ediacara Hills the coelenterates predominate. In contrast, the several Burgess Shale coelenterates, some resembling jellyfishes and others resembling sea pens, seem to have played a rather limited role in the community. On the other hand, the Burgess Shale sponges, the most primitive of all the animals present, were prominent members of the community. They were abundant and varied in form; some species grew on the sea floor in thickets.

The various Burgess Shale "worms" were mainly assigned by Walcott to the annelid phylum in general and to the class of polychaetes in particular. It is now realized that many of them belong to other phyla. Nevertheless, it is among the polychaete worms that some of the

ANCESTRAL ARTHROPOD with a striking resemblance to the living onychophore *Peripatus* is this remarkably preserved invertebrate, *Aysheaia.* Cambrian arthropods such as *Aysheaia* could have been ancestral to such living members of that phylum as the myriapods and insects.

most spectacular examples of soft-body preservation are to be found: the setae, or bundles of fine bristles, that were these animals' organs of locomotion have been particularly well preserved as bright, reflective films in the shale. One of the polychaetes, *Canadia,* apparently did not live in a burrow but spent much of its time swimming close to the sea floor. Another, *Burgessochaeta,* was probably a more typical burrower, taking refuge in the muddy bottom and searching for food around the burrow entrance with its long tentacles.

Today the priapulid phylum is of interest only to a handful of specialists. These worms, however, were an important group in Cambrian times, and two priapulids present in the Burgess Shale are particularly noteworthy. One of them, *Ottoia,* is the most abundant of the group. It has been preserved in such detail that muscles are clearly visible and the gut content of some specimens can be analyzed. *Ottoia* fed on two kinds of shellfishes: brachiopods and hyolithids. The hyolithids, possibly members of the mollusk phylum, had a conical shell that was capped by a protective operculum, or lid, when the animal was fully withdrawn. The teeth of *Ottoia* were not strong enough to break open the shell, and so the hyolithids were swallowed whole and their soft parts were digested as the shells passed through the priapulid's gut unscathed. These shellfishes were not *Ottoia*'s only food. A unique specimen contains within its gut the remnants of another worm of the same species, showing that (as with some living priapulid worms) *Ottoia* could be cannibalistic.

Parasitologists take considerable interest in another Burgess Shale priapulid. It is *Ancalagon,* which may be ancestral to the living group of spiny-headed worms, the Acanthocephala, that seem to have been parasites for millions of years. These parasitic worms have no gut and absorb nourishment through their body wall while they are lodged in the intestine of their host. If evolution is hypothetically reversed and the worms are reendowed with the organs necessary for a free-living existence, the reconstructed animal is remarkably like *Ancalagon.*

Two other supposed worms, once considered to be polychaetes, are *Wiwaxia* and *Pikaia.* The body of *Wiwaxia* was covered with large scales. Long spines that curved upward and outward along the animal's back evidently were protection against predators. That the spines actually were protective is indicated by the fact that in some specimens of *Wiwaxia* they have been snapped off. An inhabitant of the sea floor, *Wiwaxia* nourished itself by scraping off fragments of food with a rasping organ. The rasp resembles the radula, or horny toothed tongue, of certain living mollusks. Is *Wiwaxia* a primitive mollusk? If it is, the details of its remarkably preserved anatomy will throw new light on the early evolution of this highly successful phylum of invertebrates.

What about *Pikaia,* formerly considered a polychaete worm? Some 30 well-preserved specimens show a prominent rod along the animal's back that appears to be a notochord, the cartilagelike stiffening organ that gives the chordate phylum its name. In addition to this key anatomical feature the blocks of muscle in *Pikaia* form a zigzag pattern that is comparable to the musculature of the primitive living chordate *Amphioxus* and of fishes. Although *Pikaia* differs from *Amphioxus* in several important respects, the conclusion that it is not a worm but a chordate appears inescapable. The superb preservation of this Middle Cambrian organism makes it a landmark in the history of the phylum to which all vertebrates, including man, belong. There are possible instances of even earlier chordates from Lower Cambrian formations in California and Vermont but none is as rich in detail.

Perhaps the most intriguing problem presented by the Burgess Shale fauna is the 10 or more invertebrate genera that so far have defied all efforts to link them with known phyla. They appear to be the only known representatives of phyla whose existence had not even been suspected. Their origins must lie in Precambrian obscurity, where the initial metazoan diversification began. The peculiarity of these novel animals is exemplified by the aptly named *Hallucigenia.*

This animal propelled itself across the sea floor by means of seven pairs of sharply pointed stiltlike spines. Seven tentacles arose from the upper surface of the animal's body; at the end of each tentacle was a pair of strengthened tips. Did the tentacles gather food? If they did, did each tentacle act as an individual mouth with a direct connection to the animal's alimentary canal? There are more questions than answers, but a valuable clue to the animal's behavior is preserved in a specimen from a Harvard University collection. There one can see more than 15 individual *Hallucigenia* associated with a large worm. There seems little doubt that, having detected the carcass of the worm, these odd animals had congregated to scavenge it.

Compared with *Hallucigenia* a second unique animal, *Opabinia,* seems almost orthodox. Its five eyes were arranged across its head, so that it was probably able to avoid predators with ease as it swam close to the sea floor, steering itself with a vertical tail fin. *Opabinia* fed by capturing prey with a grasping organ that projected forward.

Alternative approaches to problems

MOLLUSK REPRESENTATIVE, *Hyolithes,* **had a cone-shaped shell that was capped by a protective lid. One of the burrowing worms,** *Ottoia,* **preyed on these mollusks but was not able to break the shell open. The worm digested** *Hyolithes'* **soft parts and excreted its shell.**

PROBABLE MOLLUSK, *Wiwaxia,* **with its cover of large scales and array of long protective spines, was first placed among the polychaete worms of the Burgess Shale. Its rasplike feeding organ, similar to a mollusk's radula, suggests that it belongs to the molluscan phylum instead.**

of functional design are evident among these unusual invertebrates. For example, for a worm with a fluid-filled body cavity one problem is that muscular contraction in one part of the body will distort the shape of the rest of the body. In annelid worms the problem has been solved by dividing the body cavity into a series of watertight compartments. *Banffia,* a unique Burgess Shale worm, developed an alternative solution. The stiffened front half of its body was separated from the more saclike back half by a prominent constriction at the midpoint. The constriction appears to have damped the hydrostatic fluctuations set up by the locomotor muscles of the animal's front half, thereby minimizing the distortion of its unstiffened back half.

Some representatives of new phyla have been preserved by the dozen. Others, particularly the free-swimming inhabitants of the higher water levels that would seldom be trapped by slumping mud, are quite rare. One such animal is the worm *Amiskwia;* judging by its prominent fins, it was probably quite an active swimmer. Another animal, *Nectocaris,* a fast-swimming predator, had enormous eyes and evidently propelled its streamlined body by rapid lateral flicks of its body. Prominent dorsal and ventral fins, stiffened by numerous fin rays, helped to keep the animal stable as it was swimming.

Conodonts, or "cone teeth," are enigmatic fossils that resemble tiny teeth; they are found in formations ranging in age from the latest Precambrian to the Triassic, a span of almost 400 million years. Although they look like teeth, they cannot have acted as such because they show no signs of wear. What soft-bodied animal had conodonts and for what purpose has long been an unanswered question. Another rare pelagic invertebrate preserved in the Burgess Shale, *Odontogriphus,* may be that animal. The tentacular feeding apparatus of the animal, another unique representative of a hitherto unknown phylum, incorporates a set of minute conical objects that appear to be conodonts. Since conodonts cannot have acted as teeth, the hypothesis has been advanced that they were some kind of support for the feeding tentacles. Was the feeding apparatus of *Odontogriphus* and of animals like it the source of the conodonts so copiously distributed throughout the Paleozoic and the earliest Mesozoic fossil record? Possibly so.

As more is learned about the Burgess Shale fauna the picture of Cambrian life will gain a new perspective, particularly with respect to the explosive evolution of the metazoans. For example, the wide range of arthropods, with their distinctive and different groupings of anatomical characteristics, is already

such that a single phylum seems too small to hold them all. The adaptive radiation of the Cambrian invertebrates can be seen as the initial response to the availability of a very wide variety of marine ecological niches. Hence many Cambrian animals seem to be pioneering experiments by various metazoan groups, destined to be supplanted in due course by organisms that are better adapted. The trend after the Cambrian radiation appears to be the success and the enrichment in the numbers of species of a relatively few groups at the expense of the extinction of many other groups.

An additional possibility is suggested by the Burgess Shale fauna itself. Some groups of major stature in Cambrian times, such as the priapulid worms, may have fared badly against later competitors and only escaped extinction by migrating into marginal niches that were either unattractive or unavailable to other metazoans. One such manifestation of movement into a marginal niche is the scaling down of body size. This miniaturization may well be how some priapulids managed to survive. An alternative escape route is to become parasitic; the priapulids that appear to have given rise to the parasitic spiny-headed worms could be an example of the alternative. In any event the Burgess Shale fauna affords both a marvelous glimpse of evolution in action during this brief interval of Middle Cambrian times and a stern reminder of how impoverished and distorted the fossil record is. The study of these soft-bodied animals illuminates many hitherto unsuspected aspects of the history of life.

STIFFENING ROD, or notochord, runs partway along the back of the early chordate *Pikaia.* The animal's head, seen in more detail in the illustration below, is at the right. The pattern of its musculature resembles that of fishes and of the living primitive chordate *Amphioxus.* A reconstruction of the free-swimming chordate appears in the illustration on pages 74 and 75.

FRONT END of *Pikaia* is seen enlarged in this photograph, making visible the animal's pair of sensory tentacles and behind them a short row of small appendages. Earlier Cambrian formations preserve the remains of possible chordates, but none compare with *Pikaia* in detail.

III

INTERPRETING FOSSILS

INTERPRETING FOSSILS III

INTRODUCTION

The fossil record in post-Precambrian rocks that accumulated during the Phanerozoic interval of geologic time is abundant and diverse in aggregate, but at any one time or place it may be quite patchy and badly biased toward one kind of organism or another. The reasons for this are rather simple. On the one hand, during the great span of time represented by the Phanerozoic eon, which includes the Paleozoic, Mesozoic, and Cenozoic eras, there have been innumerable opportunities for animals and plants to become buried and preserved in accumulating sediments. On the other hand, organisms that have mineralized skeletons and live in areas where active sedimentation occurs have a much greater potential for being fossilized than soft-bodied creatures living in places undergoing erosion. So, while it is true that the fossil record contains a marvelous variety of past life, the sampling of that life over the ages has been sporadic and incomplete. Despite this, paleontologists nevertheless can infer much about ancient organisms and their environments from their fossil remains and from the rocks in which they are imbedded.

Some fossils, such as calcareous algae, shelly invertebrates, and vertebrates, occur as more or less unaltered hard parts. That is, after death, the organism may be buried; decomposition of the soft tissue leaves behind part or all of the mineralized skeleton. The relative resistance of the hard parts to chemical decomposition and physical disintegration permits them to survive subsequent compaction and lithification of the surrounding sediment with little or no change. Of course, percolating water in the enclosing sediment or rock can introduce chemical solutions that may dissolve and replace the original mineral matter. Yet the detailed structure of the fossil is often preserved. Unaltered or chemically replaced hard parts include such things as fish scales, crab legs, horse teeth, alligator scutes, dinosaur vertebrae, clam shells, stony corals, echinoid spines, algal crusts, and human jawbones.

Sometimes percolating waters will deposit mineral matter—usually silica or calcium carbonate—around and within the tissues (hard or soft), thereby "embalming" them with resistant material. Bones of vertebrates, for example, are commonly found with material precipitated within the natural pore spaces. Plants, too, may be thoroughly encased in silica—not only the outer parts of the plants but even the fine interstices within.

But what about animals and plants that do not have resistant hard parts or that are not protected by later precipitated mineral matter? Organisms without any hard, mineralized skeletal parts more often just rot or wear away before being buried. Or, if such organisms are buried, microorganisms and circulating water decay and oxidize the remaining organic matter. In rare instances, however, the soft tissues may be preserved as compressed carbon-ized films, because the surface on which the dead organism came to rest lacked sufficient oxygen to support bacterial decomposition or oxidation. Thus, anaerobic muds may contain the soft-bodied remains of organisms that drifted down from aerated waters above, yielding rich fossiliferous layers in black shales many millions of years later.

At other times, soft-bodied organisms may be buried quickly in finer-grained sediments, which then lithify enough to mold the shape of the organism. Although later removal of the soft tissues occur, the form of the organism may be retained within the surrounding rock. Hard parts, too, can be dissolved away from the enclosing rock but leave behind a detailed mold of the original material.

Besides these direct kinds of evidence of life—unaltered and altered hard parts, compressions, and molds—we can also find indirect evidence of animals and plants: tracks, trails, burrows, and root markings. Although these so-called trace fossils are sometimes difficult to identify in terms of the parent organism that made them, they do have the advantage of being found in place. That is, whereas a shell or tooth or bone can be transported some distance from where its owner lived and died before being buried, trace fossils, by their very nature, cannot be transported. Hence, when we find them, we can be sure that that is where the organism lived. Moreover, such traces can often tell us something about the way the organism functioned or behaved. For example, some invertebrate burrows clearly indicate that the animal was moving through and feeding on the sediment; others tell us that the animal was sitting in a burrow but feeding outside it in the overlying water.

In this section we consider a variety of fossil occurrences—from the large massive stony deposits called reefs to the microscopic skeletons of floating protozoans; from the tracks and trails of deep sea creatures to lumbering dinosaurs—and see how they are interpreted to reconstruct the habits and habitats of ancient organisms. These articles demonstrate the great range of phenomena being dealt with when we study fossils; they also suggest the kinds of evolutionary problems paleontologists investigate. Only by such specific consideration of the fossil record can we determine what *in fact* happened in the history of life, rather than what *might* have happened.

"The Evolution of Reefs" by Norman D. Newell explores the theme of the reef ecosystem and shows how this ecosystem became established during life's great expansion in the late Precambrian and early Cambrian and has stayed with us ever since. Although there have been many comings and goings of different kinds of organisms throughout the Phanerozoic, the basic organization of a "reef" has remained intact. Thus, while there has been considerable turnover in the actors, the scenario has not changed much during the long run of this evolutionary play. However, there have been times in the past when reef communities have been greatly restricted, even collapsed. Newell attributes these times of stress to the dramatic changes in climate and size of shallow seas that accompanied the fragmentation and reassembly of the continents through plate tectonics. Although reefs are only a part of the total marine environment, their history does provide a good summary of the evolutionary highlights of calcareous algae and shelly invertebrates throughout the Phanerozoic.

"Fossil Behavior," by Adolf Seilacher, discusses how we can learn about the way marine invertebrates behaved in their environment from their preserved traces in sedimentary rocks. Such tracks and trails record foraging for food on and within the sea floor; simple movements across the sediment, pursuing who knows what errands; or burrowing down into sands and muds either for shelter or for nutrient-rich layers. Sequences of similar traces through geologic time exhibit trends in more efficient use of sediments for their food content or in burrow construction. Seilacher notes that certain types of traces tend to be associated with particular kinds of habitats.

Today within paleontology, the field of ichnology studies trace fossils. Much of the impetus for the development of this research in North America came from the efforts of Seilacher himself. Trace fossils not only add information about faunas, environments, and evolution in rocks that contain hard-part fossils but also extend our observations into many rocks that lack such hard parts, and so for years were thought to be "unfossiliferous." As with the Precambrian, the fossil record of the Phanerozoic has been greatly expanded as we learn to look for these very different kinds of fossils.

The next paper considers the rich fossil record of microscopic shells found in deep-sea sediments. "Micropaleontology," by David B. Ericson and Goesta Wollin, reviews the kinds of organisms leaving such shells, including plants, such as siliceous diatoms and calcareous coccolithophorids, and animals, such as siliceous radiolarians and calcareous foraminiferans. These fossils are abundant, very small, and sufficiently diverse over the last hundred million years that they prove to be very useful in determining the geologic age of deep-sea deposits and the environmental conditions of the associated oceanic water masses.

Two very important results have come from the study of such microfossils. The first is that oceanic sediments range in age from the Jurassic to the recent and that, in general, as one proceeds toward the continents and away from midoceanic ridges where new crust is formed, ocean sediments become older and older. This observation supports the concept of sea-floor spreading away from the oceanic ridges and the subduction of sea floor beneath the continents. Subduction has apparently consumed any and all pre-Jurassic oceanic sediment. The second important result is the documentation of changing global climates over the last several hundred thousands of years, change that periodically culminated in glaciation in the high latitudes. The patterns of climate change further suggest that we may experience another significant global cooling some 20,000 years from now. Rather than being at the end of the glacial cycles, we are very likely simply between cycles.

Ericson and Wollin's article illustrates well the useful but not infallible methodology of studying present-day organisms to infer past environmental conditions, another example of the "present as a key to the past." Thus, by knowing how extant species of plankton, or floating marine microorganisms, are controlled by the temperature and salinity of the surrounding waters, we can determine past oceanic conditions by examining what species are present at successively older levels of deep sediment cores. We can thereby derive climatic curves going back well into the dawn of the Pleistocene. And, if we care to, we can project these curves forward in time to estimate how the oceans might change in the future. By doing so, we use the past as a key to the future. Other earth scientists find the Pleistocene very helpful in seeing how present-day processes and phenomena become recorded in the geologic record, going from the recent to the near past and then extrapolating forward in time to predict what the near future might be like. In this way, geology, a historical science, is taking on some of the attributes of a predictive science.

"Dinosaur Renaissance," by Robert T. Bakker, nicely illustrates the devious paths paleontologists must follow to reconstruct the life of extinct organisms. Using such diverse evidence as bone microstructure, predator–prey ratios, and geographic distribution of living and fossil reptiles, Bakker comes to the conclusion that dinosaurs were warm-blooded; that is, unlike modern reptiles, they maintained a high internal body temperature. Dinosaurs survive today through their direct descendants, the birds. Still another line of reptiles, which led to the mammals, is also considered by Bakker to have been warm-blooded. Not only are these ideas important on their own, but Bakker's paper also shows paleontological reasoning at its best: weaving together different lines of independent evidence from the recent and the past into a coherent, internally consistent hypothesis. There is no *direct* way of telling from their scattered remains, hundreds of millions of years after the fact, if dinosaurs were indeed warm-blooded. As with most fossils, paleontologists must rely on indirect inferences to build their case for one interpretation or another. Just as art is often an original recombination of commonplace materials and images, so too is good historical science that brings together in an imaginative way seemingly unrelated observations and concepts. We should also note that, as in the Newell article, reconstruction of ancient continental configurations plays an important part in the Bakker hypothesis. The development of global plate tectonics in the 1960s provides the theory that makes possible such continental reshuffling and that thereby contributes to fundamental paleontologic theory.

SUGGESTED FURTHER READING

Bakker, R. T. 1975. "Experimental and Fossil Evidence of the Evolution of Tetrapod Bioenergetics," in *Perspectives of Biophysical Ecology*, D. Gates and R. Schmerl, ed. New York: Springer Verlag. A somewhat difficult, but important, article that examines various levels of metabolism and thermoregulation in spiders, amphibians, reptiles, birds, and mammals. Bakker's approach is ingenious and thoughtful, nicely tying together how the different organisms exploit their energy resources.

Cline, R. M., and J. D. Hays, ed. 1976. *Investigation of Late Quaternary Paleoceanography and Paleoclimatology*. Boulder, Colo.: Geological Society of America Memoir 145. A technical—but largely understandable to the beginner—series of scientific reports about how different groups of microfossils in deep-sea cores reveal the nature and distribution of ancient climates during the last half million years or so.

Frey, R. W., ed. 1975. *The Study of Trace Fossils*. New York: Springer-Verlag. A compendium of articles describing the nature and interpretive value of a wide variety of animal and plant traces found in marine, freshwater, and terrestrial environments.

Heckel, P. 1974. "Carbonate Buildups in the Geologic Record," *Society of Economic Paleontologists and Mineralogists Special Publication*, vol. 18, pp. 90–154, Tulsa, Okla. An in-depth treatment of the geological history of reefs in terms of their stratigraphy and sedimentology, contributing organisms, and environmental controls; abundant comprehensive references.

Raup, D. M., and S. M. Stanley. 1978. *Principles of Paleontology*, 2nd ed. San Francisco: W. H. Freeman and Company. In a dozen chapters, these paleontologists cover a broad spectrum of topics from the identification and description of fossil specimens, interpretation of hard-part morphology, and stratigraphic correlation to paleoecology, evolutionary rates, and patterns of evolution.

Thomas, R. D. K., and E. C. Olson, ed. 1980. *A Cold Look at the Warm-Blooded Dinosaurs*. Boulder, Colo.: Westview Press, American Association for the Advancement of Science Symposium 28. As the title indicates, a critical appraisal of presumed endothermy in dinosaurs.

7

The Evolution of Reefs

by Norman D. Newell
June 1972

The community of plants and animals that builds tropical reefs is descended from an ecosystem of two billion years ago. The changes in this community reflect major events in the history of the earth

To a mariner a reef is a hazard to navigation. To a skin diver it is a richly populated underwater maze. To a naturalist a reef is a living thing, a complex association of plants and animals that build and maintain their own special environment and are themselves responsible for the massive accumulation of limestone that gives the reef its body. The principal plants of the reef community are lime-secreting algae of many kinds, including some whose stony growths can easily be mistaken for corals. The chief animal reef-builders today are the corals, but many other marine invertebrates are important members of the reef community.

This association of plants and animals in the tropical waters of the world is the most complex of all ocean ecological systems; as we shall see, it is also the oldest ecosystem in earth history. Its closest terrestrial counterpart, in terms of organization and diversity, is the tropical rain forest. Both settings evoke an image of exceptional fertility and exuberant biomass. Both are dependent on light in much the same way; the sunlight filters down through a stratified canopy, and the associations at each successive level consist of organisms whose needs match the available illumination and prevailing conditions of shelter. There is even a parallel between the birds of the rain forest and the crabs and fishes of the reef. Both play the part of lords and tenants, yet their true role in the history and destiny of the community is essentially passive.

It is a common belief that a reef consists principally of a rigid framework composed of the cemented skeletons of corals and algae. In reality more than nine-tenths of a typical reef consists of fine, sandy detritus, stabilized by the plants and animals cemented or otherwise anchored to its surface. Physical and biochemical processes that are little understood quickly convert this stabilized detritus into limestone. The remains of dead reef organisms make a substantial contribution to the detritus. This major component of the reef, however, has a fabric quite different from the upward-growing lattice of stony algal deposits and intertwined coral skeletons that forms the reef core.

The Reef Community

The interaction of growth and erosion gives the reef an open and cavernous fabric that in an ecological sense is almost infinitely stratified and subdivided. In the dimly lit bottom waters at the reef margin, rarely more than 200 feet below the surface, caves and overhanging ledges provide shelter for the plants and animals that thrive at low levels of illumination. From the bottom to the surface is a succession of reef-borers, cavern dwellers, predators and detritus-feeders, each living at its preferred or obligatory depth, that includes representatives of nearly every animal phylum. Near and at the surface the sunlit, oxygen-rich and turbulent waters provide an environment that contributes to a high rate of calcium metabolism among the myriads of reef-builders active there.

The most familiar of the reef animals, the corals, are minute polyps that belong to the phylum Coelenterata. The polyps live in symbiosis with zooxanthellae, microscopic one-celled plants embedded in the animals' tissue, where they are nourished by nitrogenous animal wastes and, through photosynthesis, add oxygen to the surrounding water. Experiments show that the zooxanthellae promote the calcium metabolism of the corals. The corals themselves are carnivores; they feed mainly on small crustaceans and the larvae of other reef animals.

The limestone-secreting algae—blue-green, green and red—are the principal food base of the reef community, just as the plant life ashore nourishes terrestrial herbivores. The algae are distributed across the reef in both horizontal and vertical zones. The blue-green algae are most common in the shallows of the tidal flat, an area where red algae are absent. The green algae are predominantly back-reef organisms and the reds are mostly reef and fore-reef inhabitants [*see illustration on page 97*].

The other important members of the reef community are all animals. Next in significance to the corals as reef-builders are several limestone-secreting families of sponges, members of the phylum Porifera. The phylum Protozoa is represented by a host of foraminifera species whose small limy skeletons add to the deposits in and around reefs. Several species of microscopic colonial animals of the phylum Bryozoa also contribute their limestone secretions, as do the spiny sea urchins and elegant sea lilies of the phylum Echinodermata, the bivalves of the phylum Brachiopoda and such representatives of the phylum Mollusca as clams and oysters, all of whose accumulated skeletons and shells contribute to the reef limestones.

Many organisms in the community do not add significantly to the reef structure; some burrowers and borers are even destructive. The marine worms that inhabit the reef are soft-bodied and thus incapable of contributing to the reef mass. The hard parts of such reef dwellers as crabs and fishes are systematically consumed by scavengers. A few fragments may escape, but except for such passive and minor contributions to the reef detritus these organisms are not reef-builders.

The reef community is adapted to a low-stress environment characterized by the absence of significant seasonal change. The mean winter temperature of the water where reefs grow is between 27 and 29 degrees Celsius and the difference between summer and winter monthly mean temperatures is three degrees C. or less. The water is clear (so that the penetration of light is at a maximum), agitated (so that it is rich in oxygen) and of normal salinity. Even under these ideal circumstances many reef organisms (for example corals) do not grow at depths greater than 65 feet. This adaptation to freedom from stress makes the community remarkably sensitive to environmental change.

The fossil record documents hundreds of episodes of sweeping mass extinction, some continent-wide and others worldwide. These times of ecological disruption have simultaneously affected such disparate organisms as ammonites at sea and dinosaurs ashore, the plants on land and the protozoans afloat among the

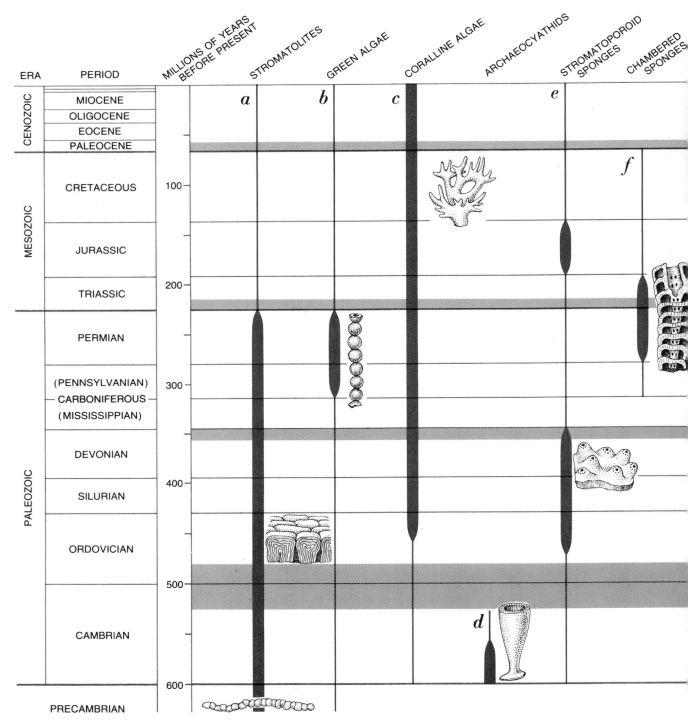

FOUR COLLAPSES (*bands of color*) have altered the composition of the reef community since the initial association between plant and animal reef-builders was established nearly 600 million years ago. This was when a group of spongelike animals, the archaeocyathids (*d*), appeared among the very much older reef-forming algal stromatolites (*a*) at the start of the Paleozoic era. In less than 70 million years the archaeocyathids became extinct; their demise marks the first community collapse. A successor community arose in mid-Ordovician times. Its members included coralline algae (*c*); the first corals, tabulate (*h*) and rugose (*i*); stromatoporoid sponges (*e*), and communal bryozoans (*m*). The group flourished almost to the end of the Devonian period, some 350 million years ago, the

oceanic plankton. The causative phenomena underlying the disruptions must therefore be unlike the ordinary, Darwinian causes of extinction—natural selection and unequal competition—that tend to affect species individually and not en masse.

For generations, in a reaction to the biblical doctrine of catastrophism that dominated 18th-century geology, scholars have viewed the apparent lack of continuity in fossil successions with skepticism. The breaks in the record, they proposed, were attributable to inadequate collections or to accidents of fossil preservation. At the same time, certain pioneers—T. C. Chamberlin and A. W. Grabau in the U.S. and Hans Stille in Germany—saw the breaks in fossil continuity as reflections of real events and sought a logical explanation for them. These men were eloquent proponents of a theory of rhythmic pulsations within the earth: diastrophic move-

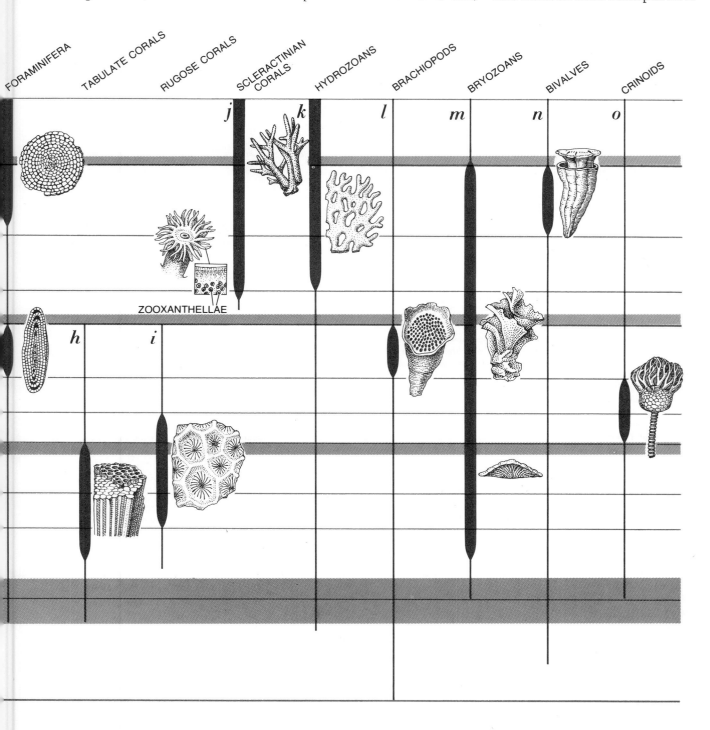

time of the second collapse. Its successor, some 13 million years later, contained a new sponge group (f) and increased numbers of green algae (b), foraminifera (g), brachiopods (l) and crinoids (o). These reef-builders flourished until the end of the Paleozoic era and the third collapse. The next resurgence occupied most of the Mesozoic era. It was marked by the appearance of modern corals (j) and a dramatic upsurge of a mollusk group, the rudists (n), that became extinct at the time of the fourth community collapse some 65 million years ago. Draining of shallow seas during the Cenozoic era produced cooler climates and led to formation of the Antarctic ice cap. Both developments have been factors in restricting the successor community in diversity and distribution.

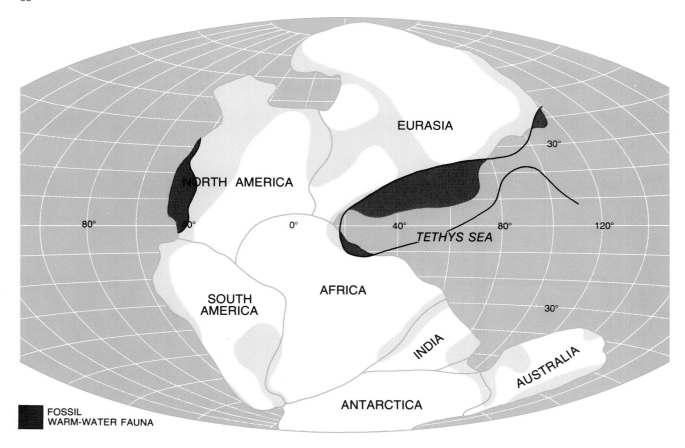

FOSSIL
WARM-WATER FAUNA

TWO FACTORS that have affected world geography and climate are the movement of the continental plates and their greater or lesser invasion by shallow seas. The status of both factors during three key intervals in earth history is shown schematically here and on the opposite page. Near the end of the Paleozoic era (a) the continental plates had gathered into a single land area. Many of the reef-building species were then pantropical in distribution. By the end of the Mesozoic era (b) sea-floor spreading had separated the continental plates. The Atlantic Ocean had become enough of a barrier to the migration of reef organisms between the Old and New World to allow the evolution of species unique to each region. At the end of the Mesozoic era the shallow seas encroaching on the continents were drained completely. Early in the Cenozoic era (c) the shallow seas had been reestablished in certain zones, but the total continental area above water was larger than in the Mesozoic. The change resulted in a trend toward greater seasonal extremes of climate; the distribution of tropical and subtropical organisms, however, remained quite broad, as palm-tree fossils show.

ments that had been accompanied by significant fluctuations in sea level and consequent disruptive changes in climate and environment.

The lack of any demonstrable physical mechanism that might have produced such simultaneous worldwide geological revolutions kept a majority of geologists and paleontologists from accepting the proposed theory of the origins of environmental cycles. Today, however, we have in plate tectonics a demonstrated mechanism for changes in sea level and shifts in the relative extent of land and sea such as Chamberlin and his colleagues were unable to provide [see the article "Plate Tectonics," by John F. Dewey; Offprint 900]. Significant changes in the volume of water contained in the major ocean basins, produced by such plate-tectonic phenomena as alterations in the rate of

lava welling up along the deep-ocean ridges or in the rate of set-floor spreading, have resulted sometimes in the emergence and sometimes in the flooding of vast continental areas. The changing proportions of land and sea meant that global weather patterns alternated between mild maritime climates and harsh continental ones.

Now, reef-building first began in the earth's tropical seas at least two billion years ago. As we have seen, the modern reef community is so narrowly adapted to its environment as to be very sensitive to change. It seems only logical to expect that the same was true in the past, and that changes in reef communities of earlier times would faithfully reflect the various rearrangements of the earth's land masses and ocean basins that students of plate tectonics are now documenting. The expectation is justified;

whereas many details of earth history will be clarified only by future geological and paleontological research, the record of fossil reef communities accurately delineates a number of the main catastrophic episodes.

As one might expect, the oldest of all types of reef is the simplest. Algae alone, without associated animals, are responsible for limestone reef deposits, billions of years old, that were formed in the seas of middle and late Precambrian times.

The Precambrian algae produced extensive accumulations of a distinctively laminated limestone that are found, flanked by aprons of reef debris, in rock formations around the world. Geologists call these characteristic limestone masses stromatolites, from the Greek for "flat" and "stone." The microscopic organisms that built the stromatolites are rarely

b

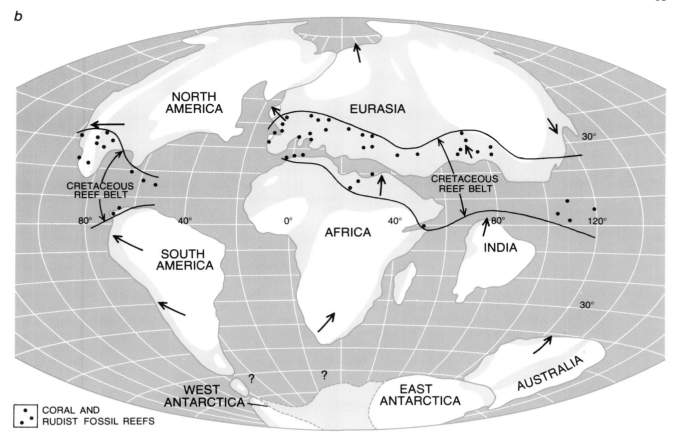

CORAL AND
RUDIST FOSSIL REEFS

c

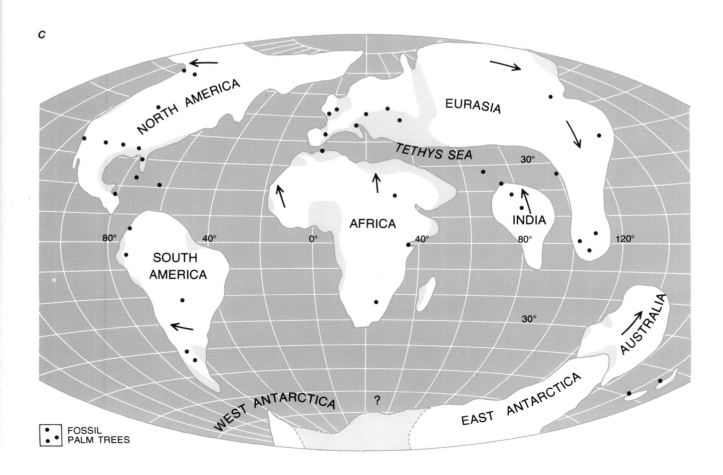

FOSSIL
PALM TREES

preserved as fossils, but they must surely have been similar to the filamentous blue-green algae that form similar masses of limestone today.

The accomplishments of these Precambrian algal reef-builders were not inconsiderable: individual colonies grew upward for tens of feet. They did so by trapping detrital grains of calcium carbonate and perhaps by precipitating some of the lime themselves. The resulting fossil bodies take the form of trunklike columns or hemispherical mounds.

As I outline the evolution of the reef community the reader will note that I often speak of first and last appearances. This does not mean, of course, that the organism being discussed was either instantly created or instantly destroyed. Each had an extensive evolutionary heritage behind it when some chance circumstance provided it with an appropriate ecological niche. Similarly, the decline and extinction of many major groups of organisms can be traced over

periods of millions of years, although in numerous instances the time involved was too short to be measured by the methods now available.

Enter the Animals

The long Precambrian interval ended some 600 million years ago. The opening period of the Paleozoic era, the Cambrian, saw the first establishment of a reef community. The stromatolites' first partners were a diverse group of stony, spongelike animals named archaeocyathids (from the Greek for "ancient" and "cup"). In early Cambrian times these stony animals rooted themselves along the stromatolite reefs, grouped together in low thickets or scattered like shrubs in a meadow. It is not hard to imagine that the vacant spaces within and between these colonies provided shelter for the numerous bottom-feeding trilobites that inhabited the Cambrian seas. Not every reef harbored archaeocyathids, however; some reefs of early and

middle Cambrian times are composed only of stromatolites.

By the end of the middle Cambrian, some 540 million years ago, the archaeocyathids had vanished. No single cause for this extinction, the first of the four major disasters to overtake the reef community, can be identified. One imaginable cause—competition from another reef-building animal—can be ruled out completely. The seas remained empty of any kind of reef-building animal throughout the balance of the Cambrian and until the middle of the Ordovician period, some 60 million years later. All reefs built during this long interval were the work of blue-green algae alone.

Fossil formations in the Lake Champlain area of New York, a region that lay under tropical seas in middle Ordovician times some 480 million years ago, contain the first evidence of a renewed association between reef-building plants and animals. The community that now arose was a rather complex one. Stromatolites continued to flourish and a second kind

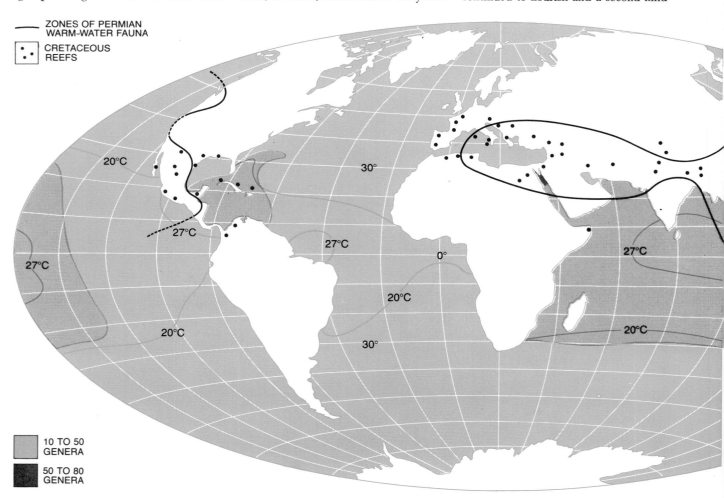

ZONES OF PERMIAN WARM-WATER FAUNA

CRETACEOUS REEFS

10 TO 50 GENERA

50 TO 80 GENERA

REEF-BUILDERS TODAY are confined to a narrow belt, mainly between 30 degrees north and 30 degrees south latitude (*light gray*), and even within this belt the greatest number of species are found where the minimum average water temperature is 27 degrees Celsius. Paleozoic and Mesozoic reef fossils, however, are found in areas far outside today's limits (*black lines and dots*). This suggests that the true Equator in those times lay well to the north of the equators shown on the maps on the preceding pages. The asym-

of plant life appeared: the coralline red alga *Solenopora*, a direct progenitor of the modern coralline algae. Colonial bryozoans, previously insignificant in the fossil record, now assumed an important role in the expanding reef community. Animal newcomers included a group of stony sponges, the stromatoporoids, some shaped like encrusting plates and others hemispherical or shrublike in form. These calcareous sponges were to play a major role in the community for millions of years. The most significant new animals, however, in light of subsequent developments, were certain stony coelenterates: the first of the corals. The intimate collaboration between algae and corals, apparently unknown before middle Ordovician times, has continued (albeit with notable fluctuations) to the present day.

These new arrivals and the other corals that appeared during the Paleozoic era were mainly of two types. In one type the successive stages of each polyp's upward growth were recorded as a

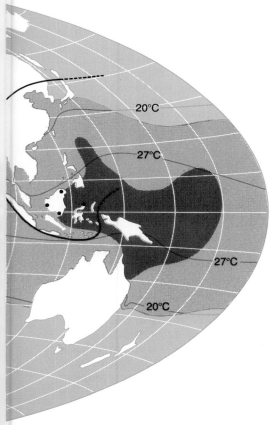

metry of the fossil-reef belt with respect to the present belt suggests that much of a once wider array has been engulfed by subduction along continental-plate boundaries.

series of parallel floors that subdivide the stony tube sheltering the animal; these organisms are called tabulate corals. The conical or cylindrical stone tube that sheltered the second type has conspicuous external growth wrinkles on its surface; these corals are called rugose.

A little more than 350 million years ago, near the end of the Devonian period, worldwide environmental changes caused a number of mass extinctions. Among the victims were many previously prominent marine organisms, including several groups in the reef community. That community now underwent a major retrenchment. Up to this point the tripartite association between algae, sponges and corals that first appeared in Ordovician times had proliferated for 130 million years without significant disturbance. The environmental alterations that nearly wiped out the reef community at the end of that long and successful period of radiation remain unidentified, although one can conjecture that a change from a mild maritime climate to a harsh continental one probably played a part. In any event the episode was severe enough so that only scarce and greatly impoverished reef communities, consisting in the main of algal stromatolites, survived during the next 13 million years. Not until well after the beginning of the Carboniferous period was there a community resurgence.

Some 115 million years passed between the revival of the reef community in Carboniferous times and the end of the Paleozoic era; the interval includes most of the Mississippian and all of the Pennsylvanian (the two subdivisions of the Carboniferous) and the closing period of the Paleozoic, the Permian. The revitalized assemblage that radiated in the tropical seas during this time continued to include stromatolites, numerous bryozoans and brachiopods and a dwindling number of rugose corals. Except for these organisms, however, the community bore no great resemblance to its predecessor of the middle Paleozoic. Both the stromatoporoid sponges and the tabulate corals are either absent from Carboniferous and Permian reef deposits or are present only in insignificant numbers.

Two new groups of calcareous green algae, the dasycladaceans and codiaceans, now attained quantitative importance in the reef assemblage. As if to match the decline of the stromatoporoid sponges, a second poriferan group—the calcareous, chambered sphinctozoan sponges—entered the fossil record. At the same time a group of echinoderms—the crinoids, or sea lilies—assumed a

larger role in the reef community. As the Paleozoic era drew to a close, the crinoids and the brachiopods achieved their greatest diversity; their skeletons preserved in Permian reef formations number in the thousands of species.

The Third Collapse

Half of the known taxonomic families of animals, both terrestrial and marine, and a large number of terrestrial plants suffered extinction at the end of the Paleozoic era. The alteration in environment that occurred then, some 225 million years ago, had consequences far severer than those of Devonian times. In the reef community the second successful radiation—based principally on a new tripartite association involving algae, bryozoans and sphinctozoan sponges—came to an end; reefs are unknown anywhere in the world for the first 10 million years of the Mesozoic era.

What was the cause of this vast debacle? There is little enough concrete information, but analogy with later and better-understood events encourages the conjecture that once again unfavorable changes in climate and habitat were major factors. In late Paleozoic times all the continents had come together to form a single vast land mass: Pangaea. Continental ice sheets appeared in the southern part of this supercontinent, a region known as Gondwana, in Carboniferous and early Permian times. These glaciers are concrete evidence of cooler climates; paleomagnetic studies show that the glaciated areas were then all located near the South Pole. Any relation between these early Permian ice sheets and the widespread late-Permian extinctions, however, is not yet evident. What is probably more significant is evidence that, at least during a brief interval, all the shallow seas that had invaded continental areas were completely drained at the end of the Paleozoic era. Serious climatic consequences must have resulted from the disappearance of a mild, primarily maritime environment.

In late Paleozoic times a wide tropical seaway, the Tethys, almost circled the globe. The only barrier to the Tethys Sea was formed by the combined land masses of North America and western Europe, which were then connected. The tongue of the Tethys that eventually became the western Mediterranean constituted one end of the seaway. The opposite end invaded the west coast of North America so deeply that great Permian reefs arose in what is now Texas.

The Mediterranean extremity of the Tethys Sea was the setting for a significant development when, after 10 million years of eclipse, the reef community was once again revitalized. There, in mid-Triassic times, a new group of corals, the scleractinians, made its appearance. The scleractinians were the progenitors of the more than 20 families of corals living in the reef community today. At first the new coral families, six in all, were represented in only a few scattered reef patches found today in Germany, in the southern Alps and in Corsica and Sicily. Even by late Triassic times, some 200 million years ago, the new corals were still subordinate as reef-builders to the calcareous algae.

The Mesozoic Community

During the 130 million years or so of Jurassic and Cretaceous times the reef community once again thrived in many parts of the world. The stromatoporoid sponges, all but extinct since the community collapse of Devonian times, returned to a position of some importance during the Jurassic. The new coral families steadily increased in diversity and reached an all-time peak in the waters bordering Mediterranean Europe during the Cretaceous period. In that one region there flourished approximately 100 genera of scleractinians; this is a greater number than can be found worldwide today. The reef community in this extremity of the Tethys Sea was also rich in other reef organisms. It included the two groups of sponges and such reef-builders as sea urchins, foraminifera and various mollusks. In addition a hitherto minor group of coralline red algae, the lithothamnions, now began to play an increasingly important role. By this time the stromatolites, important reef-builders throughout the Paleozoic, were no longer conspicuous members of the reef community.

Early in the Cretaceous period, some 135 million years ago, there was an unfavorable interval: reefs are unknown in the fossil record for some 20 million years thereafter. This pause merely set the stage, however, for a further efflorescence. Both in the Mediterranean area and in the waters of the tropical New World, hitherto unknown or insignificant coral families appeared. The present Atlantic Ocean was just beginning to form. Regional differences between reef communities in the Old World and the New that appeared at this time testify to the growing effectiveness of the Atlantic deeps as a barrier to the ready migration of reef organisms.

An extraordinary evolutionary performance was now played. Certain previously obscure molluscan members of the reef community, bivalves known as rudists, abruptly came into prominence as primary reef-builders. The next 60 million years saw a phenomenal rudist radiation that brought these bivalves to the point of challenging the corals as the dominant reef animals. Along the sheltered landward margins of many fringing barrier reefs rudists largely supplanted the corals. Their cylindrical and conical shells were cemented into tightly packed aggregates that physically resembled corals, and many of the aggregates grew upward in imitation of the coral growth pattern. Before the end of the Cretaceous period some 65 million years ago the rudists had attained major status in the reef community. Then at the close of the Cretaceous they quite abruptly died out everywhere.

Ever since the pioneer days of geology in the 18th century the end of the Cretaceous period has been known as a great period of extinctions. Nearly a third of all the families of animals known in late Cretaceous times were no longer alive at the beginning of the Cenozoic era. The reef community was not exempt; in addition to the rudists, two-thirds of the known genera of corals died out at this time. Forms of marine life other than reef organisms were also hard hit. The ammonites, long a major molluscan group, had suffered a decline during the final 10 million years of the Cretaceous; by the end of the period they were all extinguished. The belemnites, another major group of mollusks, declined sharply, as did the inoceramids, a diverse and abundant group of clams that had previously flourished worldwide. Among the foraminifera, the free-floating, planktonic groups suffered in particular.

The environmental changes that devastated life at sea also took their toll of land animals. Perhaps the most spectacular instance of extinction ashore involved the group that had been the dominant higher animals during most of the Mesozoic era: the dinosaurs. Of the 115-odd genera of dinosaurs found in late Cretaceous fossil deposits, none survived the end of the period. The concurrent breakdown of so many varied communities of organisms clearly suggests a single common cause.

What was the source of this biological crisis? We have now come sufficiently close to the present to have a better grasp of the evidence. Throughout almost all of Mesozoic times life both on land and in the sea seems to have been remarkably cosmopolitan. A broad belt of equable climate extended widely in both directions from the Equator and there was no very evident segregation of organisms into climatic zones. The earth has always been predominantly a water world; the total land area may never have exceeded the present 30 percent of the planet's surface and has often been as little as 18 percent. In the late Cretaceous almost two-thirds of today's land area was submerged under shallow, continent-invading seas. It appears that during times of extensive inundation such as this one there was nothing like the blustery global circulation of air and strong ocean currents of today.

Paleoclimatology reveals the contrast between contemporary and Cretaceous conditions. By measuring the proportions of different isotopes of oxygen present in the carbonate of fossil foraminifera and mollusks it is possible to calculate the temperature of the water when the animals were alive. Today the temperature of deep-ocean water is about three degrees C. Cesare Emiliani of the University of Miami has shown that early in the Miocene epoch, some 20 million years ago, the bottom temperature of the deep ocean was about seven degrees C. He finds that in the Oligocene, 10 million years earlier, the temperature was about 11 degrees and that in late Cretaceous times, 75 million years ago, it was 14 degrees. He suggests that the onset of cooling may have been lethal to dinosaurs. In any event it is clear that cold bottom water has been accumulating in the deep-ocean basins since early in the Cenozoic era.

The accentuated seasonal oscillations in temperature and rainfall at the end of the Cretaceous period, a trend that evidently started when the shallow continental seas began to drain away into the deepening ocean basins, have been credited by Emiliani and also by Daniel I. Axelrod of the University of California at Davis and Harry P. Bailey of the University of California at Riverside with the simultaneous reduction in numbers or outright extinction of many other animal and plant species at that time. Genetically adapted to a remarkably equable world climate over millions of years, many of these Mesozoic organisms were ill-prepared for the extensive emergence of land from under the warm, shallow continental seas and the climate of accentuated seasonality that followed. Perhaps we should be surprised not that so many Cretaceous organisms died but that so many managed to adapt to the new conditions and thus survived.

Near the end of the Paleocene epoch, some 10 million years after the great

FOSSIL BARRIER REEF was built in upper Devonian times by a community of marine plants and animals living in the warm seas that covered part of Australia more than 350 million years ago. Ex- posed by later uplift and erosion, the reef now forms a belt of jagged highlands, known as the Napier Range, in Western Austra- lia. A stream has cut a canyon (*foreground*) through the reef rock.

YOUNGER FOSSIL REEF, built some 250 million years ago in the Permian period, forms a rim of rock 400 miles long surrounding the Delaware Basin (*right*) on the border between Texas and New Mexico. Most of the reef is buried under later deposits, but one exposure forms this 40-mile stretch of the Guadalupe Mountains. El Capitan (*foreground*), a part of the reef front, is 4,000 feet high.

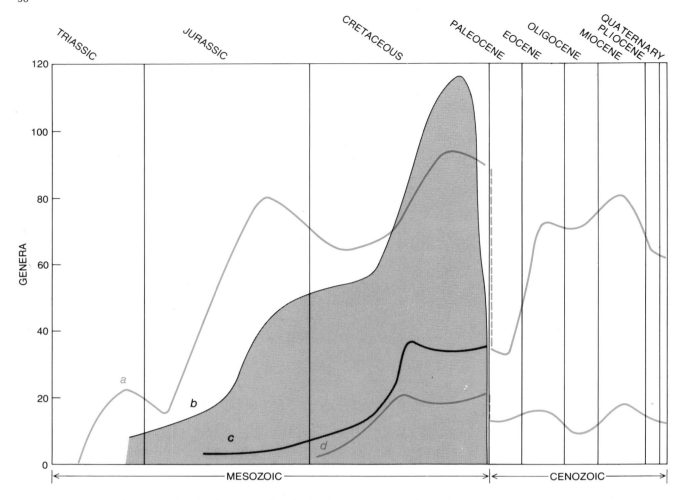

MESOZOIC INCREASE in the diversity of ocean and land animal genera reached a peak in the Cretaceous period. It was followed both by extinctions and by severe reductions in the number of genera that chanced to survive. At sea the explosively successful rudists (c) and ashore the long-dominant dinosaurs (b) became extinct. Corals (a) and globigerine foraminifera (d) fell in numbers.

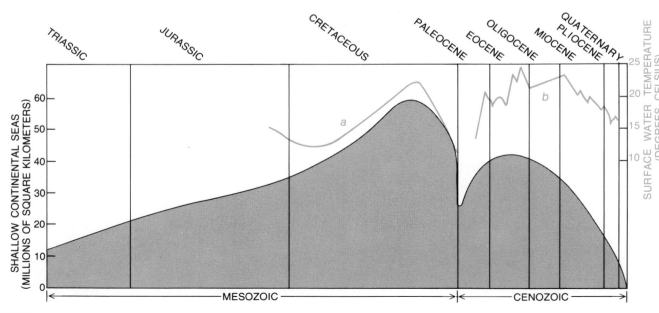

SHRINKING OF SEAS that had invaded large continental areas (gray) began near the end of the Mesozoic era and accelerated in the Miocene epoch. Oxygen isotopes in belemnite (a) and forami- nifera (b) skeletons show that the ocean has become progressively cooler since early in Miocene times (color). A relation seems to exist between a larger land area and greater seasonal extremes in climate.

Cretaceous collapse, a reef community sans rudists reappeared in the tropical seas. The following epoch, the Eocene, saw a new radiation of the scleractinian corals. Several genera that now appeared worldwide were unknown earlier in the fossil record; many of them are still living today.

The Cenozoic Decline

A sharp reduction in coral diversity that began in late Eocene times and lasted throughout the Oligocene epoch seems to reflect a continued increase in seasonality of climate and a substantial lowering of mean temperatures over large areas of the globe. Nonetheless, communities built around a bipartite association of corals and coralline algae continued to build extensive reefs in the Gulf of Mexico and the Caribbean area, in southern Europe and in Southeast Asia. Continued sea-floor spreading and deepening of the Atlantic basin, which enhanced the effectiveness of the Atlantic as a barrier to migrating corals, are evidenced in increasing differences between Caribbean and European coral species in late Oligocene fossil deposits.

By now the earth had been free from continental glaciers for almost 200 million years, but a change in the making was to have profound consequences for world climate: the Antarctic ice cap was becoming established. Fossil plant remains and foraminifera from nearby deep-sea deposits both testify that even in early Miocene times the climate in much of the Antarctic continent was not much different from the blander days of

the early Cenozoic, when palm trees grew from Alaska to Patagonia. At this time, moreover, Antarctica was some distance away from its present polar position. Nonetheless, mountain glaciers had begun to appear there millions of years earlier, in Eocene and Oligocene times. Sands of that age, produced by glacial streams and then rafted out to sea on shelf ice, are found in offshore deep-sea cores. The cooling trend was well established before the end of the Miocene epoch. In the Jones Mountains of western Antarctica lava flows of late Miocene age overlie consolidated glacial deposits and extensive areas of glacially scoured bedrock.

With the formation of the Antarctic ice cap some 15 to 20 million years ago a factor came into being that strongly influences world weather patterns to this day. The ice cap energizes the world weather system. In the broad reaches of open ocean surrounding Antarctica the surface water is aerated and cooled until it is too heavy to remain at the surface. The cold water sinks and spreads out along the sea floor, following the topography of the bottom. The result is a gravity circulation of cold water from Antarctica into the world's ocean basins, with a consequent lowering of the mean ocean temperature and cooling of the overlying atmosphere. The energetic interactions of the atmosphere above with the ocean and land below, in turn, strongly influence global wind patterns and worldwide weather. Today's climate is the product of a long cooling trend, marked by ever greater seasonal extremes; the trend became strongly ac-

centuated when the Antarctic ice cap came into being late in the Miocene epoch.

This event and others in the Cenozoic era are faithfully recorded in terms of changes in the reef community. For example, in spite of the development of new barriers to the migration of reef organisms during the Mesozoic era, such as the Atlantic deep, the reef community had remained predominantly cosmopolitan up to the close of the Cretaceous. By Miocene times, however, what had once been essentially a single pantropical community was effectively divided into two distinct biogeographic provinces: the Indo-Pacific province in the Old World and the Atlantic province in the New World.

In the Old World the increasingly unfavorable climate had eliminated the reef community in European waters. It was during Miocene times, when Australia reached its present tropical position, that the Old World reef-builders first began to colonize the shallows of the Australian shelf; in terms of maximum diversity, the headquarters of the Indo-Pacific province today lies in the Australasian region where seasonal contrasts in water temperature are minimal.

Like the Atlantic deep, the deep waters of the eastern Pacific formed a generally effective barrier to the migration of reef organisms from the Indo-Pacific province into the hospitable tropical waters along the west coast of the Americas, principally around Panama. In Miocene times this Pacific coastal pocket was still connected to the reef-rich Caribbean, the headquarters of the At-

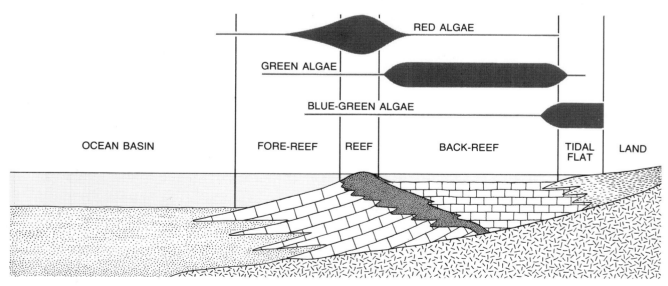

REEF ANATOMY centers on a rigid core (*color*) composed of the cemented skeletons of algae and corals. In this diagram growth of the core began at right and continued upward and outward. Most of the bulk of the reef consists of wide areas of stabilized detritus that are continuously being converted into the limestone that comprises the fore-reef and back-reef. The sandlike detritus is stabilized by the growth of the animals and plants in the reef community. The zones occupied by different algae are indicated here.

lantic province. Contact between the Atlantic province and the small Pacific enclave continued until the two areas were separated during the Pliocene epoch by the uplifting of the Isthmus of Panama.

The Pliocene saw a further contraction of the world tropics. The community of reef-builders gradually retreated to its present limits, generally south of 35 degrees north latitude and north of 32 degrees south latitude. Rather than serving as a center for new radiations, this tropical belt essentially became a haven. The epoch that followed, the Pleistocene, was marked by wide fluctuations in sea level and sharp alternations in climate that accompanied a protracted series of glacial advances and retreats. Oddly enough, the reef community was scarcely affected by such ups and downs; the main reasons for this

apparent paradox seem to be that neither the total area of the deep tropical seas nor their surface water temperature underwent much change during the Pleistocene ice ages.

Is a fifth collapse in store for the most complex of ocean ecosystems? It would be foolhardy to respond with a flat yes or no. If, however, the past is prologue, the answer to a collateral question seems clear. The question is: Will a reef community in any event survive? The most significant lesson that geological history teaches about this complex association of organisms is that, in spite of the narrowness of its adaptation, it is remarkably hardy. At the end of the long interlude that followed each of four successive collapses, the reef community entered a new period of vigorous expansion. Moreover, without exception each revitalized aggregate of reef-builders in-

cluded newcomers in its ranks.

The conclusion is inescapable that even during the times most unfavorable to the reef ecosystem the world's tropical oceans must have had substantial refuge areas. In these safe havens many of the threatened organisms managed to adapt and survive while others seem to have crossed some evolutionary threshold that had prevented their earlier appearance in the community. Today the Atlantic and Indo-Pacific provinces are two such refuge areas.

As long as the cooling trend that began in Cretaceous times does not entirely destroy these tropical refuge areas, one long-range conclusion seems firm. Any collapse of the present reef community will surely be followed by an eventual recovery. The oldest and most durable of the earth's ecosystems cannot easily be extirpated.

Fossil Behavior

by Adolf Seilacher

August 1967

Some fossils represent the tracks or burrows of ancient animals. Such fossils can seldom be identified with a particular animal, but they do show how the animal behaved and something of how behavior evolved

Most of what is known about the evolution of plants and animals has been learned from fossils. One might think that this information is limited to the anatomical changes in living organisms, but such is not the case. There is a class of fossils that provides evidence on animal behavior. These fossils consist not of animal remains but of fossilized animal tracks and burrows.

The great majority of such fossils are markings made in the soft sediments of the ocean floor by ancient invertebrates: ancestral marine worms, starfishes and sea snails, extinct arthropods such as the trilobites, and the like. The tubes and tunnels, trails and feeding marks left by these animals are preserved as raised or depressed forms in layers of sediment that gradually became rock. Paleontologists call such forms "trace fossils"; they classify them in taxonomic arrays and assign them names. Geologists find them useful as indicators of age in formations of sedimentary rock that do not contain the usual kind of fossil. Trace fossils are also clues to the interrelations of organisms and environments—the ecology— in the ancient oceans. This article, however, is concerned with what trace fossils reveal about the behavior of the animals that produced them and how such behavior evolved over periods of millions of years.

The most obvious question to ask about trace fossils is: What animal is responsible for a given track or burrow? This question is one of the most difficult to answer. Except in the case of some trilobites and a few other arthropods, clear-cut "fingerprints" preserved in tracks and burrows are rare or difficult to recognize. As far as the identity of the animal that made them is concerned, many trace fossils may remain a mystery forever. Such fossils can nevertheless be sorted out and classified according to the behavior that gave rise to them. Many of their differences are functional; for example, a hole that was dug as a shelter would differ from a hole made by a sediment-feeding animal in the course of a meal. Similarly, an animal that feeds on the surface of the ocean bottom produces one kind of trail when it is foraging and a distinctly different kind when it is trying to elude a predator.

I have found it useful to put trace fossils in groups representing five activities [*see illustrations on pages 101 and 102*]. The first group consists of crawling tracks, indicative of nothing more than simple motion. The second is made up of foraging tracks, marks left by animals that moved along the ocean bottom (or just below it) in the course of feeding. The third consists of feeding burrows, as distinct from foraging tracks, made by animals that tunneled well into the bottom sediments. The fourth contains "resting" tracks made by animals that took temporary refuge by burying themselves in sandy bottoms. The last group is composed of dwelling burrows, the permanent shelters of animals such as marine worms that lived and fed without moving from place to place but gathered their food from outside the burrow.

One way of reconstructing fossil behavior is to devise a model of it in the form of a program of commands such as might be written for a computer. The validity of the model can be tested by determining if the sequence of commands will produce actions that are compatible with the fossil evidence. A simple example is provided by "pipe-rocks," seacoast sediments that have now become sandstone and are named for the abundance of vertical dwelling burrows they contain. The pipe-rock burrows, known in trace-fossil taxonomy as *Scolithos*, appear to be the work of animals with a simple pattern of behavior. A model program for the *Scolithos* animals might consist of two commands. The first would be "Dig down vertically for n times your length," the second "Avoid crossing other burrows." Behavior responsive to these two commands would suffice to produce pipe-rocks.

The trace fossils left by animals that feed on sediments, either by grazing on the bottom or by tunneling into it, show particularly prominent behavior patterns. The nutrients in a given area or volume of sediment are best extracted by orderly movements rather than random ones. Efficient sediment-feeders produce regular winding or branching tracks in which repetition of the same kind of turn produces an intricate pattern. Some 40 years ago the German paleontologist Rudolf Richter pointed out the significance of these patterns in a pioneering behavioral analysis of the trace fossil *Helminthoida labyrinthica*.

The animal responsible for the *H. lab-*

$\longrightarrow$

FOSSIL TRACKS on the following page suggest how the behavior of animals that lived on the ocean floor evolved in the Paleozoic era (from 600 million to 230 million years ago). At left are primitive "scribbles." The track at top was made during the Ordovician period by a wormlike animal; the track in the middle, during the Cambrian period, possibly by a snaillike animal; the track at bottom, during the Cambrian by a trilobite. At right are more advanced spirals and meanders. The one at top was made during the Mississippian period by a wormlike animal; the one in the middle, during the Pennsylvanian period, possibly by a snaillike animal; the one at bottom, during the Ordovician by a trilobite.

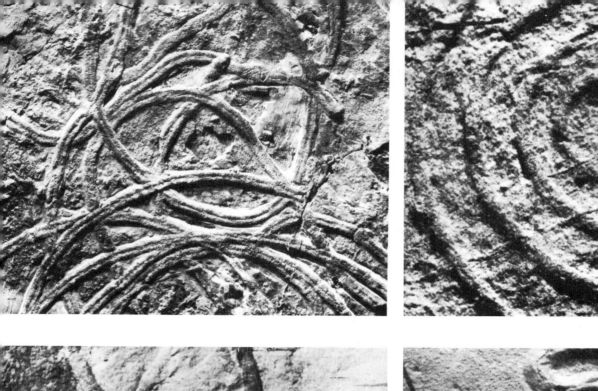

yrinthica markings was a particularly efficient sediment-feeder. Its tunnels are found in fine-grained sedimentary rocks in the Alps and in Alaska, usually silts and marls of Cretaceous and Eocene age (between 135 million and 36 million years ago). To rephrase Richter's analysis as a model program, it appears that the animal obeyed only four commands. The first command was "Move horizontally, keeping within a single stratum of sediment." Obedience to this command is evident in the fact that the animal's tunnels lie within the horizontal laminations of sediment. The second command was "After advancing one unit of length make a U-turn." Here obedience to the command is apparent in the "homostrophy," or uniformity of turning, that is characteristic of the animal's tunnels. The third command was "Always keep in touch with your own or some other tunnel." Biologists call obedience to such a command "thigmotaxis," meaning an involuntary approaching movement in response to the stimulus of contact with some object. That this command was followed is evident in the closeness of the tunnels to one another. The fourth command was "Never come closer to any other tunnel than the given distance *d.*" Obedience to this command, which is termed "phobotaxis," would have prevented the animal from digging across another tunnel; crossed tunnels are in fact absent from the fossils.

The model program does not, of course, specify the sensory responses that would have allowed the animal to follow its orders. It is possible, however, to guess what these responses were. For three of the commands nothing more complex seems to be needed than positive or negative chemotaxis: either approach to a chemical stimulus or avoidance of it. Each of the laminations in the sediment would probably have had, so to speak, a characteristic flavor that

would have served to guide the animal's horizontal movements and thus would have resulted in obedience to the first command. Obedience to the third and fourth commands is somewhat harder to understand until one considers that as the animal moves it churns up sediment along the sides of its tunnel. Such areas of disturbance are visible in some of the fossils. It seems likely that the animal could chemically distinguish between the disturbed sediment of an adjacent tunnel and the undisturbed sediment in the opposite direction.

The second command, on which the animal's turning maneuver depends, seems to require something other than a chemical stimulus. Its successful execution must be related to the fact that the *H. labyrinthica* animal was shaped like a worm. The length of its body could thus serve as a measuring rod. When it had dug a straight section of tunnel far enough forward, its tail would emerge from the last U-turn and straighten out. Tail-straightening was probably the only cue required for the animal's head to start its next U-turn.

One fact of real life that is neglected in model command programs is that a need to disobey commands must frequently arise. If the tunnel being dug by the *H. labyrinthica* animal had penetrated between two other tunnels, for example, obedience to the third and fourth commands would have left it trapped if it had not violated the first command and moved upward or downward. In such circumstances disobedience is required for survival.

Evidence that the commands were not inflexible is found in the way the animal's meandering turns were executed. The length of the individual "lobes"—the straight passages between the U-turns—are not uniform; some are shorter than average and some are longer. One can

assume that a short lobe was formed when contact with some unidentified object triggered obedience to the fourth, or "avoid," command before the entire "forward march" portion of the second command had been executed. A long lobe could have been formed if some accidental bend in the straight part of the tunnel misled the animal into delaying its next U-turn. There is no way to test the hypothesis concerning short lobes. Examination of many fossils shows, however, that in a majority of cases long lobes are associated with secondary bends.

The same behavioral program operating in different animals can be expected to give rise to a wide variety of patterns because of the differences in the animals' modes of locomotion, feeding, responses to commands and even simpler differences such as body length and turning ability. The *H. labyrinthica* animal, for instance, could make very tight turns. The animal that produced the trace fossil *Helminthoida crassa*, however, was apparently less agile: its turns tend to be shaped like a teardrop. Judging by these loose meanders, the *H. crassa* animal was also less sensitive to thigmotactic commands.

A further example of such meandering trace fossils should be mentioned because it demonstrates how a relatively small change in a command program can result in a highly complicated meander. In the Alps and Alaska and other regions of folded sedimentary rocks there are series of a particular kind of sandstones called graywacke. On the bottom of graywacke layers one finds the complicated trails of *Spirorhaphe*. If one draws a loose inward spiral, makes a U-turn at the center and spirals outward again in the space between the inward whorls, one has traced the path taken by the *Spirorhaphe* animal.

I thought at first that it would be very

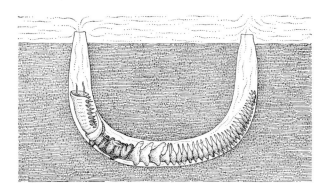

DWELLING BURROW differs from other trace fossils because it represents the shelter of an animal that is stationary and may even

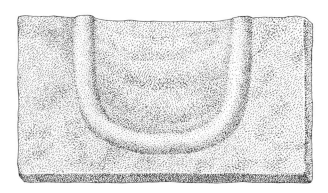

be anchored. The animal illustrated here is a polychaete worm that feeds by filtering particles of food from the water around it.

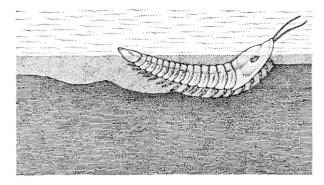

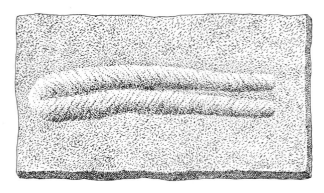

CRAWLING TRACKS comprise one of the five main categories of "trace fossils," which record the various activities but seldom the identity of the animals that made them. The examples illustrated on this page, however, are all attributable to trilobites.

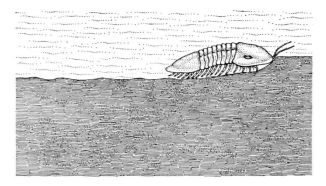

FORAGING TRACKS differ from crawling tracks by recording not simple motion but an animal's search for food over the surface of the ocean floor. The illustration of fossil tracks and burrows on these pages and on pages 103 and 104 were made by Thomas Prentiss.

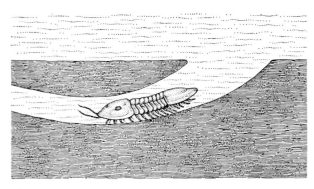

FEEDING BURROW is made by an animal that tunnels into the sediments of the ocean floor and eats the bits of organic matter it extracts from the silt. The burrow also provides shelter. The meander and the spiral illustrated on page 103 are feeding burrows.

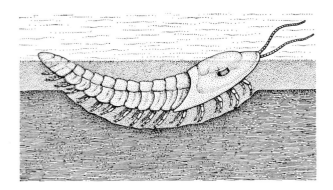

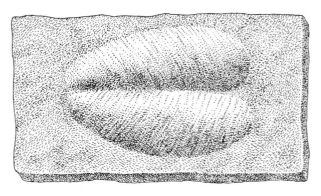

RESTING TRACKS are the marks produced when an animal, often a scavenger, makes a shallow, temporary hiding place just below the surface of the ocean floor, usually where the bottom is sandy. Their shape generally corresponds to the outline of the animal.

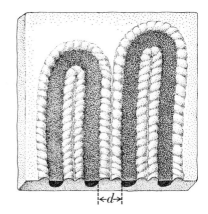

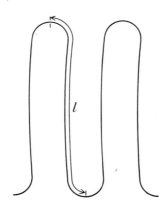

HORIZONTAL TRAIL (*left*), composed of tight meanders, was made by an unknown wormlike animal. The diagram in perspective shows how the animal's passage through the sediments on which it fed churned up the silt on each side of its tunnel. The width of this churned area established the critical distance ("*d*" *in perspective diagram*) that separates adjacent trails. The average distance between one of the animal's U-turns and the next ("*l*" *in plan view*) seems to have been determined by the length of its body.

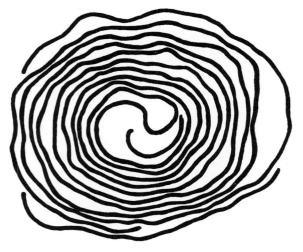

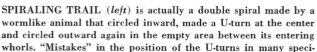

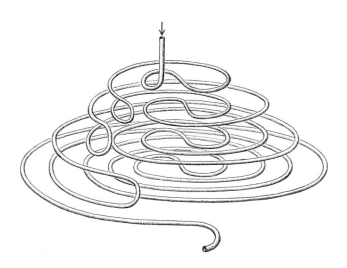

SPIRALING TRAIL (*left*) is actually a double spiral made by a wormlike animal that circled inward, made a U-turn at the center and circled outward again in the empty area between its entering whorls. "Mistakes" in the position of the U-turns in many specimens and the presence of short alien trails nearby (*short segment in center at left*) made author aware that the fossils preserved only a single, horizontal segment of what had been an elaborate, three-dimensional burrow in the silt (*hypothetical outline at right*).

difficult to devise a command program for the motions required to make such a double spiral. In examining a number of *Spirorhaphe* specimens, however, I noted small "mistakes" in several of them. In each case the animal had been forced to reverse its course too soon because the center of its spiral was already occupied by an alien U-turn (for which no inward and outward spirals could be seen). The meaning of these mistakes soon became apparent: each *Spirorhaphe* fossil preserves only one horizontal layer of what had been a multifloored, three-dimensional tunnel [*see lower illustration above*]. The "alien" U-turns are the earlier work of the same *Spirorhaphe* animal tunneling down from above. The spiral portions of the earlier whorls, which lay at a higher level, have not been preserved.

Although the three-dimensional tunnels are complex in appearance, the program required to produce them is simpler than the one required for the one-story pattern. Only two commands need be added to the four that produce horizontal meanders. The first additional command is "After spiraling inward make a U-turn and keep in contact with the adjacent tunnel on the horizontal." The second is "After spiraling outward turn down and keep in contact with the adjacent tunnel vertically." The wormlike *Spirorhaphe* animal could distinguish between inward and outward spirals by means of a cue similar to the one responsible for homostrophy among the animals that meander horizontally. As the animal spiraled inward its head end would be more tightly curved than its tail end. As it spiraled outward after

making its U-turn the tail end would be more tightly curved.

What are the ways in which the evolution of behavior among trace-fossil animals can be detected? One is the comparative study of groups of fossils that appear to be both closely related and closely associated in space and time. One such group is the Graphoglypt family; two of the trace fossils I have mentioned, *Spirorhaphe* and *Helminthoida crassa*, belong to this group. Graphoglypt trace fossils are alike in their time range, being found mainly in formations of Cretaceous to Tertiary age (between 135 million and some two million years ago). They are found in the same kinds of rock and are preserved in the same way. It seems probable, in view of the factors they have in common, that they were made by closely related animals, al-

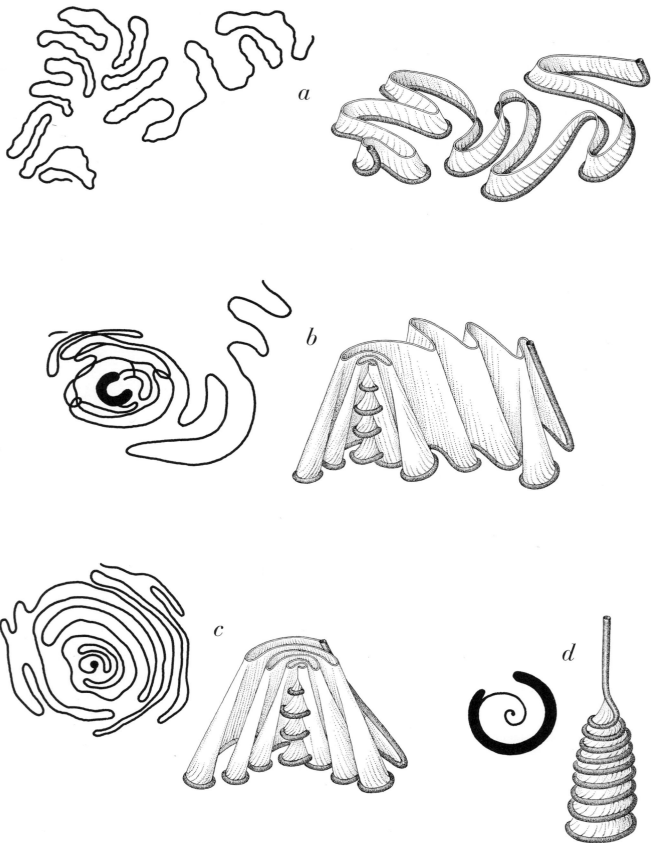

CHANGE IN BEHAVIOR over 150 million years is apparent in the trace fossil *Dictyodora*, the trail left by an animal that seems to have possessed a siphon. In Cambrian times it fed by meandering a few millimeters down in the sediment (*a*). The path produced during its foraging is at left; a three-dimensional reconstruction of the entire fossil track is at right. Later the animal left the upper sediment, eating its way deeper along a corkscrew-like path and then meandering in a restricted manner ("*b*" and "*c*"; *path at left, reconstruction at right*). By middle Mississippian times the animal no longer meandered (*d*) but only corkscrewed deep into the silt.

though what animals they were we do not know. When the group is examined as a whole, the various Graphoglypt genera and species can be arranged in a few lines of descent that have one trend in common. In terms of behavioral programming, the lines of descent show a gradual shift from rigid behavior patterns to patterns that gave the animals freedom to adapt to local circumstances. This trend toward flexibility enabled the animals to forage more efficiently, a factor that is of obvious value in terms of natural selection.

A second example of behavioral evolution is apparent when trace fossils are examined in a more general way to see what changes have taken place over millions of years in behavior that deals with one specific biological task. The obvious task to study is foraging, and we quickly find that several methods of foraging other than meandering exist. The simplest of them resembles the scribbling of children: the tracks form a series of circles with slightly offset centers. The program for scribbling has one simple command: "Keep heading to one side but don't stay in the same track." Scribbling, however, covers an area much less efficiently than meandering does.

A somewhat more complex method of foraging, and one more efficient than scribbling, consists of moving in an outward spiral. The tighter the spiral, the more efficient the coverage. The program for tight spiraling has two commands. One is "Keep circling in one direction"; the other is "Keep in touch with the spiral whorl made earlier." This program is simpler than the four-command program for meandering, but spiraling is less efficient than meandering. The strips between the spirals remain unexploited.

Several unrelated groups of trace fossils provide evidence that the progress from simple to complex forms of foraging is an evolutionary one [*see illustration on page 100*]. In Cambrian times, at the beginning of the Paleozoic era some 600 million years ago, none of the animals of the ocean floor had evolved the meandering method. Scribbling was practiced by some trilobites and by what were probably varieties of marine snails. In a heterogeneous group of trails attributed to various unknown worms good scribbles appear just after the end of the Cambrian, in Ordovician times (between 500 million and 425 million years ago). Eventually, however, scribbling disappears altogether and worms, snails and trilobites all begin to forage by meandering, with spiraling as an occasional alternative method. Complex meanders and dense double spirals are features of

ANTLER-SHAPED BURROWS are the work of an anchored animal of the Cretaceous period, roughly 100 million years ago. The lobes in the sediment that the animal excavated for their food content were narrow and much of the adjacent sediment was left unexploited.

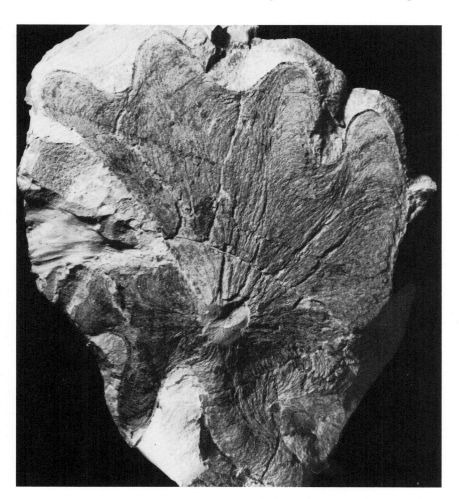

SKIRTLIKE BURROW is the work of the same kind of animal some 40 million years later, in the Tertiary period. One large excavation is more efficient than the series of lobes made in Cretaceous times. Early in the life of the animal, however, its burrow is Cretaceous in design (*bottom left*), showing that behavioral ontogeny can recapitulate phylogeny.

comparatively late geologic history, not appearing until Cretaceous times, at the end of the Mesozoic era (between 135 million and 63 million years ago). They indicate further progress in the foraging efficiency of sediment-feeders.

There is a well-documented instance in which an anatomical change in a trace-fossil animal was accompanied by a change in its behavior. The fossil *Dictyodora* is the work of an unknown sediment-feeder that tunneled into the ocean floor and filled its tunnel behind it. The *Dictyodora* animal, however, seems to have been equipped with a long, thin siphon that allowed it to maintain contact with the water above it. As the animal traveled in loose meanders through the sediments, its siphon, dragging behind it, left its own marks in the ooze.

From Cambrian to Devonian times, between 600 million and 350 million years ago, *Dictyodora* animals foraged only a few millimeters down in the sediment. Their siphons were short. This particular ecological niche, however, was within reach of many smaller competitors. By Mississippian times (beginning about 350 million years ago) the *Dictyodora* animals had become adapted to feeding at a deeper and less crowded level. Their siphons had become longer and their meandering trails were well below the reach of their sediment-feeding contemporaries.

As the *Dictyodora* animals evolved anatomically they also altered their behavior. In earlier millenniums they had apparently not begun to feed until they had worked their way down to a specific foraging level. The Mississippian animals, however, ate their way deep into the sediments, leaving a corkscrew trail behind them, before they began their horizontal meanders. Moreover, the first several meanders no longer followed the older loose pattern. Instead they curled concentrically around the initial corkscrew.

As time passed the *Dictyodora* animals' behavior became simpler. Trace fossils from later Mississippian formations in East Germany show a high percentage of *Dictyodora* burrows in which loose meanders are altogether absent, concentric meanders taking their place. Trace fossils from southern Austria, which are possibly even younger, show a further behavioral change. Most of the burrows have no meanders at all; the central corkscrew trail has simply been drilled deeper. It is doubtful that the *Dictyodora* animals remained in the same burrow all their lives. During their reproductive cycle, for example, the animals probably rose to the surface of the sediment. Thereafter they would have had to dig new burrows, if they did not begin a new mode of life altogether.

Trace fossils include the burrows of a few animals that led a sedentary existence. These burrows, of course, are a record of the animals' entire life. One such fossil is *Zoophycos;* it is the burrow of an unknown wormlike animal that foraged through the sediments by constantly shifting the lobes of its U-shaped tunnel. The two openings of the tunnel remained fixed in the sediment, but thin concentric layers inside the lobes record the tunnel's shifts of position. *Zoophycos* burrows have been found in sedimentary rocks as old as the Ordovician. In Cretaceous and Tertiary times (from 120 million to some two million years ago) the *Zoophycos* animal seems to have taken over the *Dictyodora* animal's ecological niche in the deep sediments that have since become Alpine rocks.

When the burrows of a *Zoophycos* animal of Cretaceous and Tertiary times are compared, an evolution in behavior is readily apparent. The earlier burrow consists of narrow lobes that join to form an antler-like pattern [*see top illustration on preceding page*]. The sediment that lay between the lobes remained exploited. In Tertiary *Zoophycos* burrows, however, the individual lobes have fused into a continuous, skirtlike pattern inside which all the sediment has been explored [*see bottom illustration on preceding page*]. A much more effective coverage of a food-bearing area has resulted from what was probably a relatively small change in the *Zoophycos* animal's behavioral program.

Surprisingly this more efficient behavior is found only in the portions of the burrow occupied by the adult Tertiary animal. The behavior of the young animal remained like that of its Cretaceous ancestor, producing the same narrow lobes. Thus we see that the old adage about ontogeny recapitulating phylogeny can be applied to an animal's behavior as well as to its anatomy.

Scanty though the trace-fossil material now available is, it gives one hope that the early evolution of behavior patterns will become as valid an area of study as the evolution of anatomical structures. Further insights into fossil behavior should not only add a new aspect to paleontological research but also help to counter the belief that paleontologists are concerned only with dead bodies and have no real comprehension of ancient life.

Micropaleontology

by David B. Ericson and Goesta Wollin
July 1962

Some fossils are so small they can be identified only with the aid of a magnifying glass or a microscope. They label sedimentary layers and provide excellent clues to ancient changes in climate

The words "fossil" and "paleontology" usually evoke pictures of dinosaur bones or other good-sized pieces of vertebrate skeleton. This article is concerned with micropaleontology, which deals with fossils of entirely different magnitude. They are the shells, or, more properly, the skeletons, of minute aquatic animals. None of these microfossils can be recognized without the help of a strong magnifying glass; some must actually be examined in the electron microscope. Their minuteness makes them especially useful for geological research. They can be brought up unbroken—and in enormous numbers—by a narrow coring pipe or even by an oil-well drill. Found both in the ocean floor and in dry-land formations that were once covered by water, microfossils have long served oil prospectors as stratigraphic markers. More recently they have begun to furnish important information about processes of change in the structure of the earth's surface and in the earth's climate.

To meet the needs of the micropaleontologist an organism must have characteristics other than small size. First, of course, the organism must build a skeleton, or some hard part durable enough to fossilize under normal conditions of sedimentation. The fossils in a group should be distinguished as to genus and preferably species. This implies some complexity of organization. Species that flourished during the shortest periods of time, and over the greatest geographical area, are the best geological indicators. They make it possible to differentiate sharply among layers of sedimentary rock of different age and to match strata in widely separated parts of the earth.

In order to obtain evidence of past climate and other environmental conditions the investigator must think of fossils as once living animals with special-

ized adaptations to their particular surroundings. He then tries to reconstruct those surroundings by analogy with the ecological requirements of near relatives of the ancient organisms that are alive today. As might be expected, the method becomes more difficult as the evolutionary distance between fossils and living organisms increases. A paleoecologist must be a good detective and find meaning in all sorts of apparently trivial and irrelevant observations. Certain potentially useful microfossils are just beginning to attract serious study. Previously they were ignored because they are so very small. The coccoliths, disk-shaped plates of calcium carbonate, are one example. They are planktonic, which means that the ocean currents in which they float have carried them to many quarters of the globe and that they have been settling to the bottom for some 500 million years. One cubic centimeter of sediment can contain 800 million of them; they must be enlarged 800 diameters to be seen at all. Because they were found in the topmost layer of sediment on the ocean floor, it was realized that the organism depositing them must be extant. For a long time what the organism was remained a mystery; the finest collecting nets could not catch it. Eventually it was located not in a man-made trap but in the filtering apparatus of the common marine animal called *Salpa.* Recently the electron microscope has revealed that coccoliths possess an astonishingly complex structure. When they have been more thoroughly described and classified, they should be helpful in correlating sedimentary strata from continent to continent as well as in the oceans.

Another group of tiny organisms, now extinct, were the star-shaped discoasters. A little larger than coccoliths, their world-wide distribution in deeper sedi-

ments suggests that they too were planktonic. Although known for almost 100 years, they have not been applied in stratigraphy because of the mistaken notion that all known species lived continuously from 60 million years ago to the time of their disappearance. In reality they evolved fairly rapidly, and some species make excellent guide fossils. The earliest discoaster stars had as many as 24 rays, or points. By 20 million years ago the number of rays had declined to six, and the last forms had only five slender rays. The exact time of their extinction is unknown, but if, as is suspected, it was just before the Pleistocene (the geological period of the last ice age), five-rayed discoasters will occupy an important position as reliable indicators for this important boundary in geologic time.

In their day-to-day work micropaleontologists deal chiefly with diatoms, radiolaria, conodonts, ostracodes and foraminifera. All of these except the diatoms are large enough to be studied in a low-power microscope at a magnification of between 30 and 100 diameters—an important consideration when hundreds of samples must be examined each day, as is the practice in some oil-company laboratories.

Among the microfossils the diatoms and the radiolaria compete for the first place in beauty. Both secrete shells of opal (silica combined with some water). In the lacy skeletons of radiolaria the opal looks like clear spun glass; in diatom shells it takes on a jewel-like quality, often displaying a many-colored fire that must be seen to be appreciated. Many species of diatoms live exclusively in fresh water, whereas all radiolaria live in salt water. In consequence diatom fossils are somewhat more informative; from the species found in a sedimentary

FORAMINIFERA are the most important microfossils in the study of earth history. Various planktonic species produced the fossil shells shown here at a magnification of about 20 diameters.

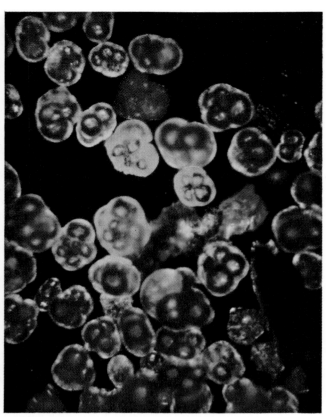

GLOBIGERINA PACHYDERMA, a foraminifer, lives in the Arctic Ocean. In this photomicrograph light from below shows details of chambers in the shells. Magnification here is some 60 diameters.

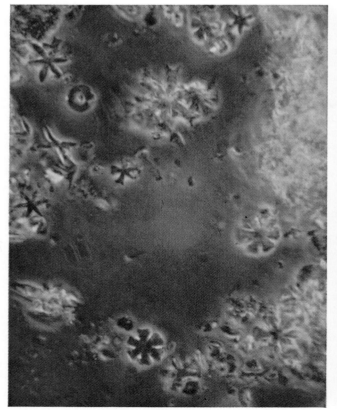

DISCOASTERS, extinct perhaps a million years, are useful as guide fossils. Forms with six rays shown here lived about 40 million years ago. They are magnified approximately 1,000 diameters.

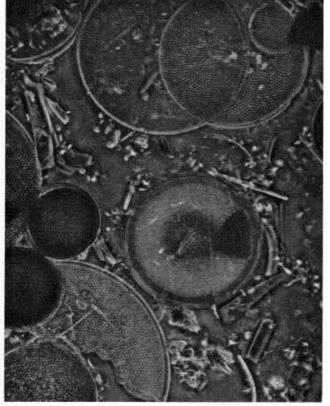

DIATOMS are sometimes abundant in sediments and sometimes missing. These were illuminated by polarized light. Magnification is 200 diameters. Photomicrographs were made by Roman Vishniac.

layer it is usually possible to tell whether the sediment was formed in a lake or in the sea. There are a great many species of diatoms and radiolaria; diatoms can occur locally in such fantastic numbers that they produce fairly thick layers of sediment, known as diatomite, consisting almost entirely of their fossils. The opal skeletons of both diatoms and radiolaria, however, tend to dissolve in water, and they cannot be counted on to show up in a particular region. Often they are missing from just those places where an index fossil is most needed.

Conodonts are small tooth-shaped or platelike objects with one or more points. Like vertebrate teeth, they are made of calcium phosphate. Although many "genera" and "species" have been described since they were discovered more than a century ago, no one knows what kind of animal produced them. Whatever it was, it became extinct some 240 million years ago during the Triassic period. The variation in the form of conodonts from level to level in older sediments makes them particularly use-

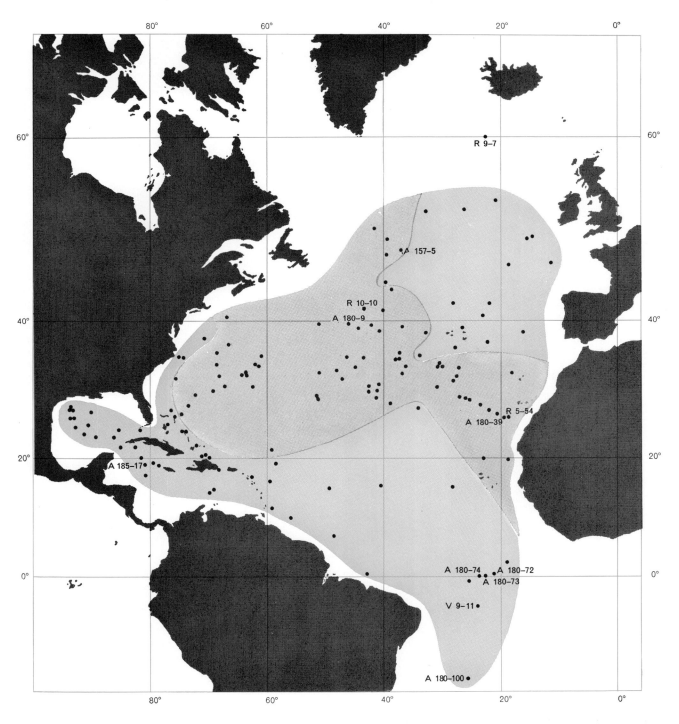

COILING DIRECTIONS of living *Globorotalia truncatulinoides,* a planktonic foraminifer, define three provinces in the Atlantic Ocean. Most of the spiral-shaped shells of this species found on the sea floor coil to the left in the gray region and to the right in the colored areas. The dots mark sites where cores have been taken. The letters and numbers identify the cores diagramed on the following pages. Core V 9-11, brought up just south of the Equator, carries the Pleistocene record back at least 600,000 years.

ful to the petroleum geologist. Because of their small size they often come up undamaged in rock cuttings from oil wells or exploratory holes.

The ostracodes present no mysteries. These odd little relatives of crabs and lobsters are very much alive today and flourish wherever there is enough water, whether it is fresh, brackish or salty.

They are the only crustaceans that have two valves, or shells, which makes them look like tiny clams. The valves vary in length from half a millimeter (a fiftieth of an inch) to three or four millimeters. They first appear in the geological column in sediments deposited at least 450 million years ago in the Cambrian period of the early Paleozoic era.

In the course of evolution the shells have varied greatly in shape and ornamentation, and many species have lived only for short intervals of geologic time. Knowledge of the present distribution of living genera in open salt water, sounds, estuaries, lagoons and lakes helps in analyzing past conditions when similar genera show up in ancient sediments.

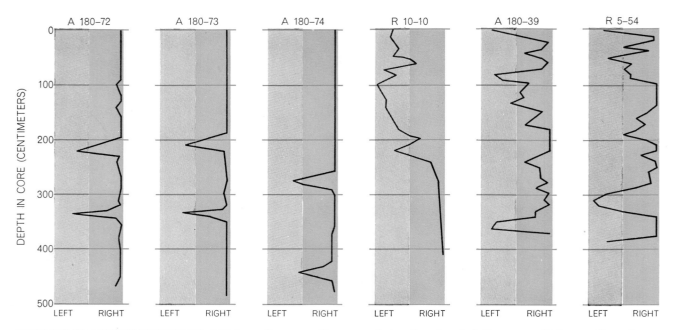

CHANGES IN COILING DIRECTION of *Globorotalia truncatuli-noides* with depth in cores make it possible to correlate cores from different locations. The correlation is readily apparent in the three cores A 180–72, –73 and –74. Cores A 180–39 and R 5–54 also show correlation. Samples of shells were studied every 10 centimeters or so from top of cores to bottom. Variation in each diagram is from 100 per cent left-coiling at left to equal ratio in the middle to 100 per cent right-coiling at right. The older shells are deeper.

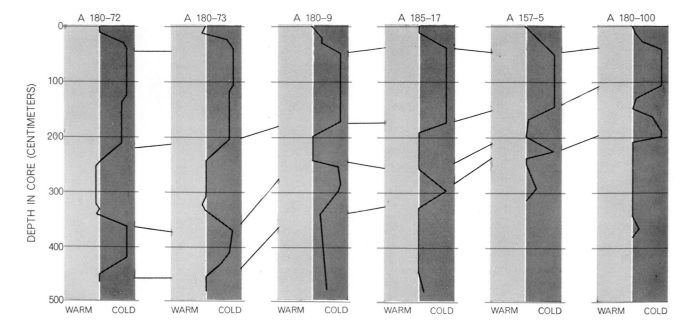

CLIMATE CURVES are based on relative number of warm- and cold-water forms of planktonic foraminifera found in each of six deep-sea sediment cores. "Warm" and "Cold" indicate warm and cold climates compared with present-day climate, which is vertical line in center of each diagram. Thin lines connect faunal changes believed to have occurred at the same time in the various locations. Obviously rates of sedimentation have differed widely. Such curves provide the data for establishing a chronology of the ice ages.

The ostracodes have only one drawback: they are not nearly so numerous as the universal favorites among microfossils, the foraminifera.

The very popularity of foraminifera greatly enhances their usefulness. Over the years a mountain of information of all sorts has accumulated, and most of it is readily available. Descriptions of genera and species have been brought together in an enormous catalogue published by the American Museum of Natural History, which now includes 69 volumes and is growing constantly as descriptions of new species are added.

The foraminifera are protozoa, or single-celled animals, that build tests (shells) of various materials. On the basis of the mode of construction the animals are divided into two large groups. Calcareous species construct tests of calcium carbonate precipitated directly from sea water; arenaceous species build their shells of sand grains, flakes of mica, sponge spicules or even small discarded calcareous tests of other species, any of which they can cement together with secretions of calcium carbonate or iron salts. The size of the shell varies widely with the species. Some long-extinct giant foraminifera exceeded 15 centimeters (six inches) in diameter. The majority of fossil foraminifera range from .2 millimeter to two millimeters.

The architectural unit of the test is the chamber. A few species have only one chamber, but most build anywhere from two to several hundred. On this basic principle of structure the foraminifera improvise endlessly. To duplicate all the strangely shaped chambers and their intricate arrangements would tax the ingenuity of a topologist. Geometric versatility has made possible all the thousands of different species that have come and gone during the past 500 million years.

Almost all the foraminifera species live on the ocean bottom. Although some attach themselves permanently to rocks, most of them potter along at a few millimeters per hour, pushing themselves by means of pseudopodia extruded through minute pores in the shells. Clearly this way of life does not make for wide distribution. Beginning some 100 million years ago in the Upper Cretaceous period, however, a few types became planktonic. Although they constitute only about 1 per cent of the known species, the enormous volume of living space open to them has permitted great proliferation: individuals of the planktonic forms make up almost 99 per cent of the

fossils found in ocean sediments. In some places accumulation of the shells has produced thick deposits of chalk. The white cliffs of Dover and Normandy are such deposits, now uplifted and partly eroded. Today large areas of the bottom of all the oceans are receiving a slow but constant rain of discarded tests of planktonic foraminifera, which make up on the average 30 to 50 per cent of all bottom sediments.

Shells of the important planktonic species have fairly simple forms. The dominant theme is a series of chambers arranged in a spiral. As the animal grows it adds chambers of steadily increasing size. In most species the general form resembles that of a small shell. Like snail shells, some tests coil to the left and others coil to the right, the two kinds being mirror images of each other.

Because of the rapid succession of distinctive species throughout successive geological ages, foraminifera are ideally suited to the needs of the petroleum

geologist, who must deal with many kinds of thick sedimentary rocks, some of them heavily folded and faulted. If one were asked to invent a class of ideal fossils for identifying and matching strata, it would be hard to improve on the foraminifera. It is small wonder that most of the hundreds of paleontologists working for oil companies devote full time to the study of these microfossils.

Foraminifera are a good deal more than tags for sedimentary rocks. To workers in pure geology, and particularly to those in the hybrid branch known as marine geology, they furnish invaluable keys to the remote past. So many samples of ocean sediment have been examined by now and their microfossils classified that it is possible to chart the approximate distribution of the most common species of planktonic foraminifera. The charts show that some species live only in low latitudes, others are most abundant in middle latitudes and

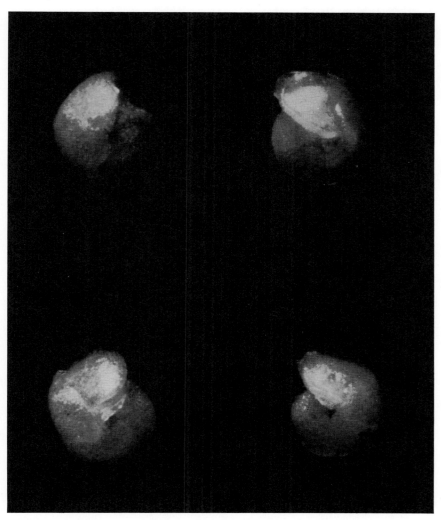

LEFT- AND RIGHT-COILING SHELLS of *Globorotalia truncatulinoides* appear at left and right respectively. Coiling direction is apparently associated with climatic conditions.

still others live in high latitudes. (One species, *Globigerina pachyderma*, ranges up to the North Pole.) Evidently water temperature plays an important part in the distribution of the various species.

If so, the animals must live near the surface; only there does the temperature vary significantly with latitude. The fact that living foraminifera are caught in plankton nets towed through the photic zone (the sunlit top 100 meters of the ocean) seems to bear this out. But the surface samples long presented something of a problem. Their shells are thin-walled and transparent, whereas shells of the same species taken from the bottom are almost all heavily encrusted with calcium carbonate or calcite.

Until recently it was supposed that

the material precipitates on the empty tests after they settle to the bottom. This hypothesis had fatal flaws, however. For one thing, oceanographers generally agree that at depths of several thousand meters in the ocean calcium carbonate dissolves far more rapidly than it precipitates. Our laboratory at the Lamont Geological Observatory of Columbia University finally undertook a close examination of the distribution of calcite on individual shells, which furnished the clue to the correct explanation. We discovered that calcite is thickest on the earliest formed chambers and diminishes steadily in the chambers farther out on the growth spiral. This can only mean that the living foraminifera precipitate the calcite and that the thin-walled spec-

imens in the photic zone are immature.

Recently Allan W. H. Bé of Lamont has caught heavily encrusted tests containing living foraminifera by towing plankton nets at depths of more than 500 meters. They provide the final proof that at least some species mature and reproduce well below the photic zone. Evidently the embryos rise to the photic zone, fatten on diatoms and other photosynthetic organisms there and then sink to a lower level, where they complete their life cycle. The pattern does not conflict with the idea that temperature variations in the upper layers of water determine the geographical distribution of various species; each individual passes a critical period of its life in the photic zone. On the other

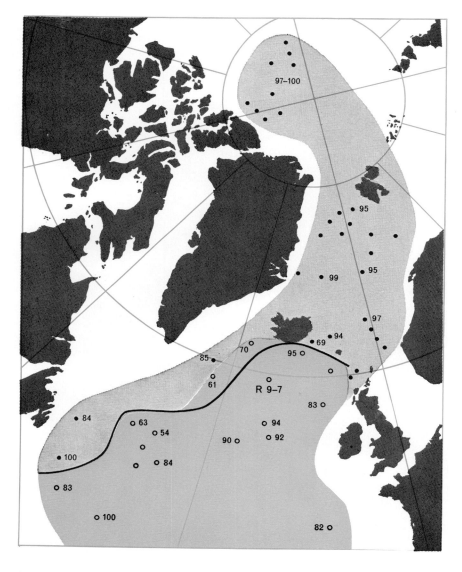

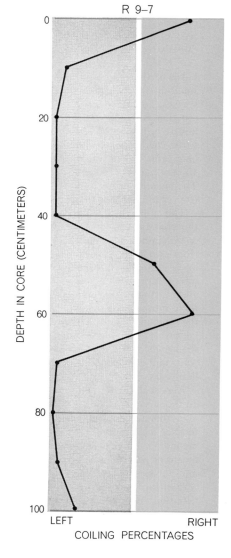

CLIMATE AND COILING DIRECTION of *Globigerina pachyderma* seem to be closely correlated in northernmost Atlantic and adjacent seas. Right-coiling of living pachyderma (*colored area on map*) is associated with warmer water, left-coiling (*gray area on map*) with coldest water. Position of the isotherm marking an average surface temperature of 7.2 degrees centigrade in April (*black line on map*) closely follows border of right-left provinces. Open circles indicate dominance of right-coiling at tops of cores; closed circles mean left-coiling. Percentages of shells coiling in dominant direction are shown for some cores. Curves at right

hand, it suggests a possible explanation for a rather puzzling distribution of the species *Globorotalia menardii*. Fossils of this group are extremely abundant in sediments south and southwest of the Canary Islands, whereas they are entirely absent from the region to the north and northeast. Yet the Canaries current flows southwestward through the area and would sweep all *G. menardii* out of the region south of the islands if they spent their entire lives in the photic zone. We believe that the animals sink below the Canaries current as they approach maturity and enter a deep countercurrent that returns them and the new generation of embryos to the northeast. As yet no one has attempted to detect the countercurrent directly, but mathe-matical analysis of deep circulation in the Atlantic suggests it should be there.

The discovery that foraminifera live part of their lives deeper than 500 meters has an important bearing on recent attempts to estimate ancient water temperatures from measurements of oxygen isotopes in fossil shells. The technique is based on the fact that in warmer waters there is a slight preponderance of heavier isotopes in the shells and in cooler waters a slight preponderance of lighter isotopes. If the relative abundance of the isotopes is to be significant, the depth at which the shells incorporated their oxygen must be known. Deep waters at every latitude are quite cold.

Since warm-water and cold-water species have not changed greatly in the past million years or so, the micropaleontologist can use the fossils to obtain a quite objective picture of changing climatic conditions during the period. Cores of sediment from ocean bottoms furnish a continuous record of geological and climatic events in contrast with the garbled record available from the distorted layers of rock on dry land. From our study of planktonic foraminifera in more than 1,000 cores in our laboratory, we feel that we have been able to arrive at the first accurate set of dates for the most recent glacial and interglacial periods. Carbon-14 analysis of the fossils shows that the last ice age ended 11,000 years ago instead of 20,000 years ago, as had previously been thought. The new date, now generally accepted, may change some ideas about the rate of human evolution. Judging from the average rate of sedimentation, the last part of the last glaciation began approximately 60,000 years ago, after an interstadial period (an interglacial of short duration) of 30,000 years [*see illustration on page 114*]. The preceding glacial period lasted only 20,000 years, whereas the interglacial before it appears to have gone on for 110,000 years. This is as far back as reliable core data go, although a single core from the equatorial Atlantic appears to carry the Pleistocene record back at least 600,000 years.

Since the last glaciation ended 11,000 years ago, and since the minimum interglacial period seems to have been 30,000 years, it would seem that man can look forward to at least 20,000 more years of the present mild climate, if not to even warmer weather. If a warmer climate should melt the glaciers that remain today, the sea level would rise, but probably no more than 10 meters or so. This would be a considerable nuisance—it would put much of New York City under

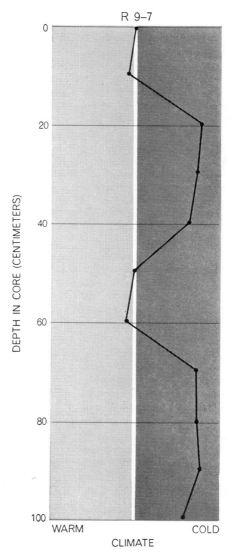

R 9–7

DEPTH IN CORE (CENTIMETERS)

WARM   COLD
CLIMATE

show (*left curve*) relative percentages of left- and right-coiling *G. pachyderma* at 10-centimeter intervals in core R 9–7, and abundance of cold and warm forms of all other planktonic foraminifera in core (*right*).

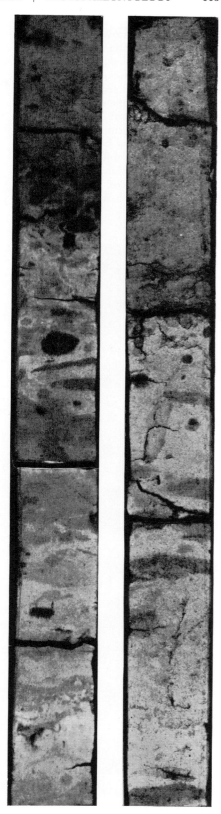

DEEP-SEA CORES represent a vertical cross section of ocean-bottom sediments. These are small parts of cores A 180–72 (*left*) and A 180–73 (*right*). Sometimes layers are quite apparent in cores; at other times no layers can be seen with the naked eye, but the microscope reveals sharp changes in the fossil contents of the undersea sediments.

water, for example—but it would hardly threaten the existence of mankind.

Arriving at accurate dates for the late Pleistocene would have been far easier if it were true, as geologists formerly believed, that marine sediments accumulate everywhere at the rate of one centimeter per 1,000 years. Then each sample core would represent the same time scale. Our studies of hundreds of cores show that the rate of accumulation can be that show, but in many places it is as much as 50, 100 and even 250 centimeters in 1,000 years, depending largely on the underwater topography. "Turbidity" currents, consisting of silt-laden water denser than surrounding water, frequently run down even gentle ocean-bottom slopes, depositing several meters of mud in a few hours in some places; in other places they may scour away sediments accumulated over thousands of years [see "The Origin of Submarine Canyons," by Bruce C. Heezen; SCIENTIFIC AMERICAN, August, 1956]. Another phenomenon responsible for large variations is slumping, in which the sediments simply slide away down a steep slope. The removal of upper layers of sediments by these processes is not all bad, however; it has brought 100-million-year-old fossils near enough to the sea floor to be reached by our coring tubes. So far we have not brought up a core in which the entire upper part of the Pleistocene record has been removed, leaving the older ice-age sediments intact. Several cores of this type, or longer cores from places already sampled, would carry our time scale back to the beginnings of human evolution.

The coiling direction of shells of *Globorotalia truncatulinoides* is proving extremely useful in matching sediments from various locations. In our laboratory we have discovered that the ratio between right-coiling and left-coiling shells varies from place to place in such a way as to define three distinct geographical provinces in the North Atlantic [*see upper illustration on page 110*]. Carbon-14 dating of samples from cores shows that the present pattern of distribution has persisted for about 10,000 years. Evidently some environmental factor has maintained the pattern in spite of the mixing effect of general ocean circulation. Going back in time by determining coiling ratios in fossil samples taken every 10 centimeters in cores, we find that the pattern of distribution changed rather suddenly from time to time during the late Pleistocene, presumably in response to changing currents or shifting water masses. Although we cannot yet say just what the changes were, we can match layers in different cores by means of the coiling-direction ratios. Petroleum geologists in Europe and India have applied this same method to other species to match strata penetrated by oil wells.

In the case of *Globigerina pachyderma,* the dominant direction of coiling follows closely the temperature of surface water in the northernmost Atlantic. Again we find evidence of shifts in the boundary between the left-coiling (cold water) and right-coiling (warm water) populations. Here we believe the shifts were determined directly by temperature changes in the late Pleistocene [*see illustration on pages 112 and 113*]. The coiling changes in a typical core show a period of cold climate preceded by a time of mild climate during which right-coiling was as strongly dominant as it is today. In this lower zone of the core and at the top of the core we find abundant fossils of various other warm-water species; these are absent in the intervening zone of left-coiling. (A direct causal relation between coiling direction and

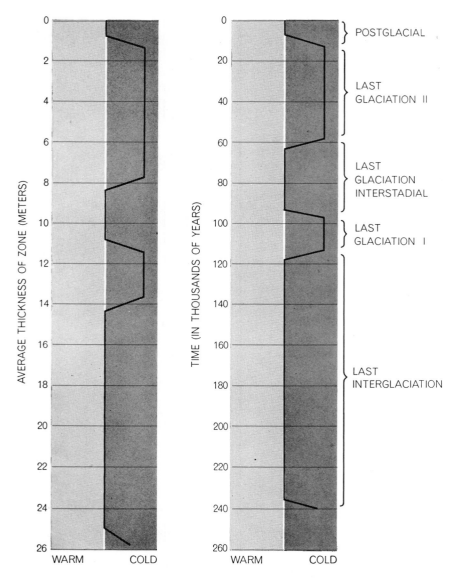

CHRONOLOGY FOR LATE PLEISTOCENE (*right*) is drawn from micropaleontological studies of planktonic foraminifera in 108 deep-sea sediment cores and on extrapolations of the rate of sedimentation from 37 carbon-14 dates in 11 of the cores. The authors consider this the most accurate timetable available for the past 240,000 years. Present-day climate is at center line of diagrams. The average-thickness curve (*left*) shows greater rate of sedimentation during glaciations than during interglacial or interstadial periods. This is caused by lowering of seas, which exposes continental shelf. Rivers thus carry material from continents out to edge of shelf and dump land sediments into deep sea, whereas in warmer times sea is higher and continental sediments are deposited on shelf.

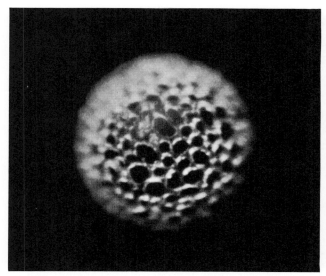

RADIOLARIA produce lacy skeleton that looks like clear spun glass. It is made of opal, which tends to dissolve in water.

OSTRACODES are related to crabs and lobsters but have two valves, or shells, that closely resemble shells of tiny clams.

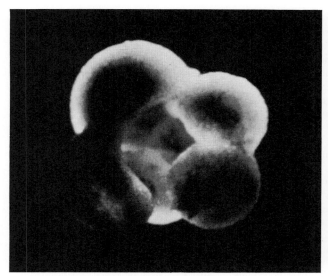

GLOBIGERINA BULLOIDES is found in cool to cold water. This and forms that follow are all species of planktonic foraminifera.

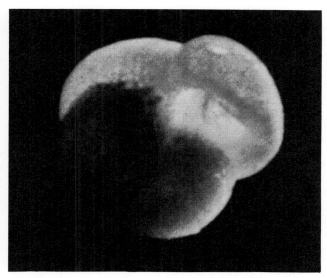

GLOBIGERINA INFLATA is a cool-water form found in middle latitudes. It occurs only in sediments of the Pleistocene epoch.

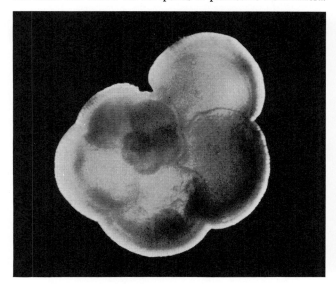

GLOBOROTALIA MENARDII is found in waters of mid-latitudes and tropics. All Pleistocene forms of this species coiled to left.

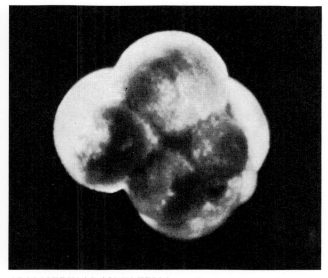

GLOBIGERINA PACHYDERMA is a climate indicator for northern seas. Shells on this page are enlarged 60 to 120 diameters.

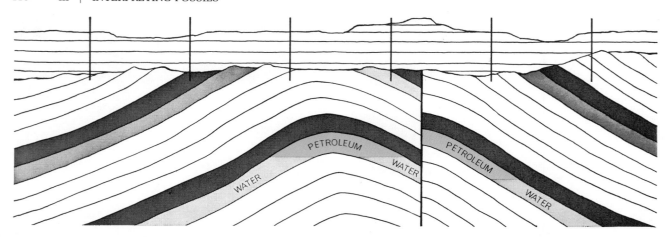

CROSS SECTION OF OIL POOL shows "slim holes," or explora-tory bore holes (*vertical lines*), passing through a series of nearly horizontal sediments and into folded, faulted and partly eroded beds of an earlier era. Two beds of same dark gray tone are both shale, which is likely to lie over oil. Microfossils from bot-toms of slim holes show relative ages of beds and thereby reveal existence of fold and fault. The deeper shale bed is of course much older. Although slim holes have not struck oil, they have shown presence of a dome that in this case contains oil, as well as the fault that has moved part of the oil-bearing stratum upward.

temperature is hardly conceivable. If both the direction of coiling and the temperature tolerance are determined genetically, the two characteristics may be associated because their genes are linked.)

Such changes in coiling direction of *Globigerina pachyderma* at lower levels in sediment cores provide insight into oceanographic conditions in the North Atlantic during the late Pleistocene. For example, left-coiling dominates from top to bottom of all cores taken in the area of present-day left-coiling. This indicates that the inflow of relatively warm Atlan-tic water into the Norwegian Sea was never greater in the late Pleistocene than it is today. Again, the change to left-coiling directly below the tops of cores in the area of present-day right-coiling implies that the inflow of warm

Atlantic water decreased during the last glaciation. As yet we cannot tell whether this resulted from a decrease in energy of the general circulation of the North Atlantic or from the lower sea level that accompanied the glaciations. The latter would make the submarine ridge between Iceland and Scotland a more effective barrier to influx. We hope that further study of the foraminifera in sediment cores will answer this and similar ques-tions. Since ocean circulation, particular-ly that of the North Atlantic, must have had a powerful influence on Pleistocene climates, a better understanding of this circulation may yet provide the basis for a satisfactory theory of the cause of ice ages.

Finally, our microfossil studies have thrown new light on the origin of the Atlantic Ocean. No cores from any ocean

have yielded fossils older than the late Cretaceous period, about 100 million years ago. It is particularly suggestive that the rather thorough sampling of the bottom of the Atlantic during the past 15 years has yielded no older fossils. Can we conclude from this that the At-lantic basin came into existence in its present form at some time during the Cretaceous period? To admit this possi-bility is to question the widely held be-lief in the permanence of continents and ocean basins. For a definitive answer we shall probably have to wait until samples have been raised from one or more deep borings in the Atlantic. In the meantime the accumulating evidence suggests that a drastic reorganization of that part of the earth's surface now occupied by the ocean basins took place about 100 million years ago.

Dinosaur Renaissance

by Robert T. Bakker
April 1975

*The dinosaurs were not obsolescent reptiles but were
a novel group of "warm-blooded" animals. And the
birds are their descendants*

The dinosaur is for most people the epitome of extinctness, the prototype of an animal so maladapted to a changing environment that it dies out, leaving fossils but no descendants. Dinosaurs have a bad public image as symbols of obsolescence and hulking inefficiency; in political cartoons they are know-nothing conservatives that plod through miasmic swamps to inevitable extinction. Most contemporary paleontologists have had little interest in dinosaurs; the creatures were an evolutionary novelty, to be sure, and some were very big, but they did not appear to merit much serious study because they did not seem to go anywhere: no modern vertebrate groups were descended from them.

Recent research is rewriting the dinosaur dossier. It appears that they were more interesting creatures, better adapted to a wide range of environments and immensely more sophisticated in their bioenergetic machinery than had been thought. In this article I shall be presenting some of the evidence that has led to reevaluation of the dinosaurs' role in animal evolution. The evidence suggests, in fact, that the dinosaurs never died out completely. One group still lives. We call them birds.

Ectothermy and Endothermy

Dinosaurs are usually portrayed as "cold-blooded" animals, with a physiology like that of living lizards or crocodiles. Modern land ecosystems clearly show that in large animals "cold-bloodedness" (ectothermy) is competitively inferior to "warm-bloodedness" (endothermy), the bioenergetic system of birds and mammals. Small reptiles and amphibians are common and diverse, particularly in the Tropics, but in nearly all habitats the overwhelming majority of land vertebrates with an adult weight of 10 kilograms or more are endothermic birds and mammals. Why?

The term "cold-bloodedness" is a bit misleading: on a sunny day a lizard's body temperature may be higher than a man's. The key distinction between ectothermy and endothermy is the rate of body-heat production and long-term temperature stability. The resting metabolic heat production of living reptiles is too low to affect body temperature significantly in most situations, and reptiles of today must use external heat sources to raise their body temperature above the air temperature—which is why they bask in the sun or on warm rocks. Once big lizards, big crocodiles or turtles in a warm climate achieve a high body temperature they can maintain it for days because large size retards heat loss, but they are still vulnerable to sudden heat drain during cloudy weather or cool nights or after a rainstorm, and so they cannot match the performance of endothermic birds and mammals.

The key to avian and mammalian endothermy is high basal metabolism: the level of heat-producing chemical activity in each cell is about four times higher in an endotherm than in an ectotherm of the same weight at the same body temperature. Additional heat is produced as it is needed, by shivering and some other special forms of thermogenesis. Except for some large tropical endotherms (elephants and ostriches, for example), birds and mammals also have a layer of hair or feathers that cuts the rate of thermal loss. By adopting high heat production and insulation endotherms have purchased the ability to maintain more nearly constant high body temperatures than their ectothermic competitors can. A guarantee of high, constant body temperature is a powerful adaptation because the rate of work output from muscle tissue, heart and lungs is greater at high temperatures than at low temperatures, and the endothermic animal's biochemistry can be finely tuned to operate within a narrow thermal range.

The adaptation carries a large bioenergetic price, however. The total energy budget per year of a population of endothermic birds or mammals is from 10 to 30 times higher than the energy budget of an ectothermic population of the same size and adult body weight. The price is nonetheless justified. Mammals and birds have been the dominant large and medium-sized land vertebrates for 60 million years in nearly all habitats.

In view of the advantage of endothermy the remarkable success of the dinosaurs seems puzzling. The first land-

118

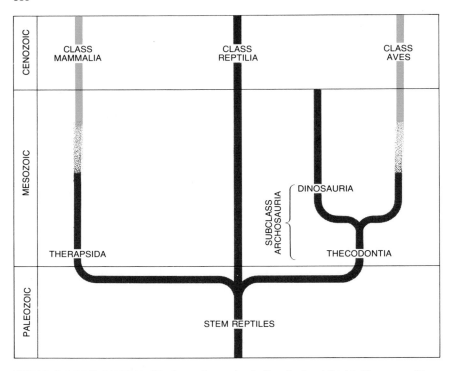

CENOZOIC

CLASS MAMMALIA · CLASS REPTILIA · CLASS AVES

MESOZOIC

DINOSAURIA

SUBCLASS ARCHOSAURIA

THERAPSIDA · THECODONTIA

PALEOZOIC

STEM REPTILES

USUAL CLASSIFICATION of land vertebrates (excluding the Amphibia) is diagrammed here in a highly simplified form. The classes are all descended from the original stem reptiles. Birds (class Aves) were considered descendants of early thecodonts, not of dinosaurs, and endothermy (color) was thought to have appeared gradually, late in the development of mammals and birds. The author proposes a reclassification (see illustration, page 132).

HAIRY THERAPSIDS, mammal-like reptiles of the late Permian period some 250 million years ago, confront one another in the snows of southern Gondwanaland, at a site that is now in South Africa. *Anteosau-*

vertebrate communities, in the Carboniferous and early Permian periods, were composed of reptiles and amphibians generally considered to be primitive and ectothermic. Replacing this first ectothermic dynasty were the mammal-like reptiles (therapsids), which eventually produced the first true mammals near the end of the next period, the Triassic, about when the dinosaurs were originating. One might expect that mammals would have taken over the land-vertebrate communities immediately, but they did not. From their appearance in the Triassic until the end of the Cretaceous, a span of 140 million years, mammals remained small and inconspicuous while all the ecological roles of large terrestrial herbivores and carnivores were monopolized by dinosaurs; mammals did not begin to radiate and produce large species until after the dinosaurs had already become extinct at the end of the Cretaceous. One is forced to conclude that dinosaurs were competitively superior to mammals as large land vertebrates. And that would be baffling if dinosaurs were "cold-blooded." Perhaps they were not.

Measuring Fossil Metabolism

In order to rethink traditional ideas about Permian and Mesozoic vertebrates one needs bioenergetic data for dinosaurs, therapsids and early mammals. How does one measure a fossil animal's metabolism? Surprising as it may seem, recent research provides three independent methods of extracting quantitative metabolic information from the fossil record. The first is bone histology. Bone is an active tissue that contributes to the formation of blood cells and participates in maintaining the calcium-phosphate balance, vital to the proper functioning of muscles and nerves. The low rate of energy flow in ectotherms places little demand on the bone compartment of the blood and calcium-phosphate system, and so the compact bone of living reptiles has a characteristic "low activity" pattern: a low density of blood vessels and few Haversian canals, which are the site of rapid calcium-phosphate exchange. Moreover, in strongly seasonal climates, where drought or winter cold forces ectotherms to become dormant, growth rings appear in the outer layers of compact bone, much like the rings in the wood of trees in similar environments. The endothermic bone of birds and mammals is dramatically different. It almost never shows growth rings, even in severe climates, and it is rich in blood vessels and frequently in Haversian

FEATHERED DINOSAUR, *Syntarsus*, pursues a gliding lizard across the sand dunes of Rhodesia in the early Jurassic period some 180 million years ago. This small dino-

rus (*right*), weighing about 600 kilograms, had bony ridges on the snout and brow for head-to-head contact in sexual or territorial behavior. Pristerognathids (*left*) weighed about 50 kilograms and represent a group that included the direct ancestors of mammals. The reconstructions were made by the author on the basis of fossils and the knowledge, from several kinds of data, that therapsids were endothermic, or "warm-blooded"; those adapted to cold would have had hairy insulation. The advent of endothermy, competitively superior to the ectothermy ("cold-bloodedness") of typical reptiles, is the basis of author's new classification of land vertebrates.

saur (adult weight about 30 kilograms) and others were restored by Michael Raath of the Queen Victoria Museum in Rhodesia and the author on the basis of evidence that some thecodonts, ancestors of the dinosaurs, had insulation and on the basis of close anatomical similarities between dinosaurs and early birds. Dinosaurs, it appears, were endothermic, and the smaller species required insulation. Feathers would have conserved metabolic heat in cold environments and reflected the heat of the sun in hot climates such as this.

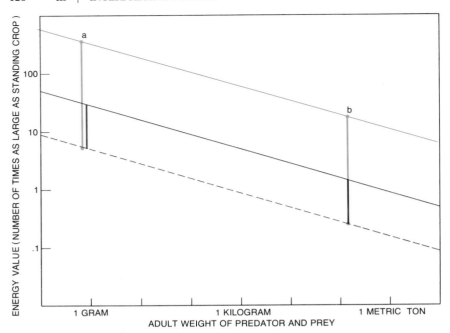

PREDATOR-PREY RATIO remains about constant regardless of the size of the animals involved because of the scaling relations in predator-prey energy flow. The yearly energy budget, or the amount of meat required per year per kilogram of predator, decreases with increasing weight for endotherms (*colored curve*) and for ectotherms (*solid black curve*). The energy value of carcasses provided per kilogram by a prey population decreases with increasing weight at the same rate (*broken black curve*). The vertical lines are proportional to the size of the prey "standing crop" required to support one unit of predator standing crop: about an order of magnitude greater for endothermic predators (*color*) than for ectothermic ones (*gray*), whether for a lizard-size system (*a*) or a lion-size one (*b*).

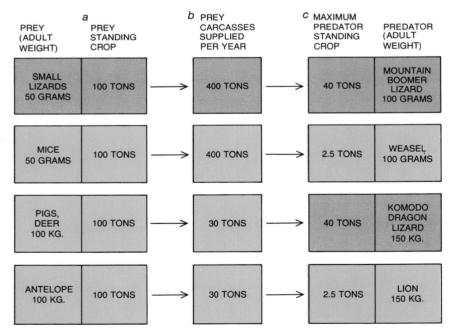

ENERGY FLOW and predator-prey relation are illustrated for predator-prey systems of various sizes. Standing crop is the biomass of a population (or the potential energy value contained in the tissue) averaged over a year. For a given adult size, ectothermic prey (*gray*) produce as much meat (*b*) per unit standing crop (*a*) as endotherms (*color*). Endothermic predators, however, require an order of magnitude more meat (*b*) per unit standing crop (*c*). The maximum predator-prey biomass ratio (*c/a*) is therefore about an order of magnitude greater in an endothermic system than it is in an ectothermic one.

canals. Fossilization often faithfully preserves the structure of bone, even in specimens 300 million years old; thus it provides one window through which to look back at the physiology of ancient animals.

The second analytic tool of paleobioenergetics is latitudinal zonation. The present continental masses have floated across the surface of the globe on lithospheric rafts, sometimes colliding and pushing up mountain ranges, sometimes pulling apart along rift zones such as those of the mid-Atlantic or East Africa. Paleomagnetic data make it possible to reconstruct the ancient positions of the continents to within about five degrees of latitude, and sedimentary indicators such as glacial beds and salt deposits show the severity of the latitudinal temperature gradient from the Equator to the poles in past epochs. Given the latitude and the gradient, one can plot temperature zones, and such zones should separate endotherms from ectotherms. Large reptiles with a lizardlike physiology cannot survive cool winters because they cannot warm up to an optimal body temperature during the short winter day and they are too large to find safe hiding places for winter hibernation. That is why small lizards are found today as far north as Alberta, where they hibernate underground during the winter, but crocodiles and big lizards do not get much farther north than the northern coast of the Gulf of Mexico.

The third meter of heat production in extinct vertebrates is the predator-prey ratio: the relation of the "standing crop" of a predatory animal to that of its prey. The ratio is a constant that is a characteristic of the metabolism of the predator, regardless of the body size of the animals of the predator-prey system. The reasoning is as follows: The energy budget of an endothermic population is an order of magnitude larger than that of an ectothermic population of the same size and adult weight, but the productivity—the yield of prey tissue available to predators—is about the same for both an endothermic and an ectothermic population. In a steady-state population the yearly gain in weight and energy value from growth and reproduction equals the weight and energy value of the carcasses of the animals that die during the year; the loss of biomass and energy through death is balanced by additions. The maximum energy value of all the carcasses a steady-state population of lizards can provide its predators is about the same as that provided by a prey population of birds or mammals of about the same numbers and adult body size.

Therefore a given prey population, either ectotherms or endotherms, can support an order of magnitude greater biomass of ectothermic predators than of endothermic predators, because of the endotherms' higher energy needs. The term standing crop refers to the biomass, or the energy value contained in the biomass, of a population. In both ectotherms and endotherms the energy value of carcasses produced per unit of standing crop decreases with increasing adult weight of prey animals: a herd of zebra yields from about a fourth to a third of its weight in prey carcasses a year, but a "herd" of mice can produce up to six times its weight because of its rapid turnover, reflected in a short life span and high metabolism per unit weight.

Now, the energy budget per unit of predator standing crop also decreases

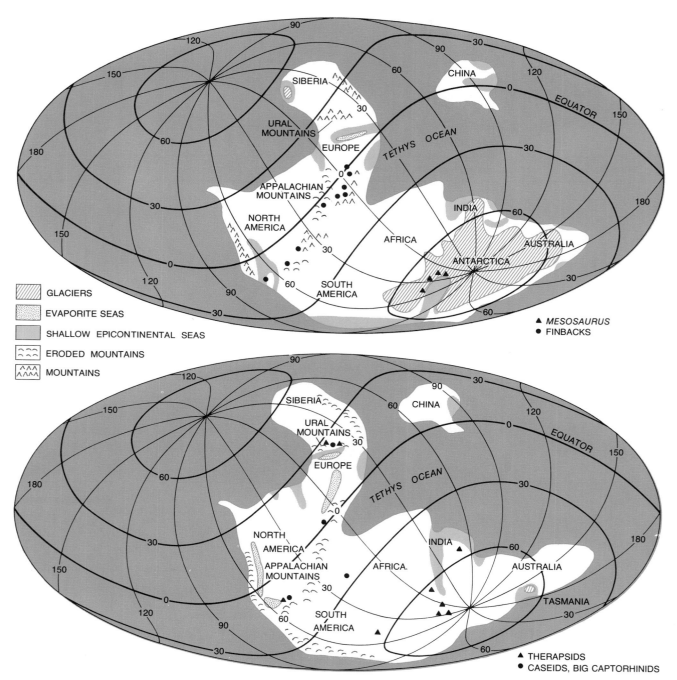

PERMIAN WORLD is reconstructed here (on an oblique Mollweide projection, which minimizes distortion of area) on the basis of paleomagnetic and other geophysical data. All major land masses except China were welded into a supercontinent, Pangaea. In the early Permian (*top*) the Gondwana glaciation was at its maximum, covering much of the southern part of the continent. Big finback pelycosaurs and contemporary large reptiles and amphibians were confined to the Tropics; they had ectothermic bone and high predator-prey ratios. The only reptile in cold southern Gondwanaland was the little *Mesosaurus*, which apparently hibernated in the mud during the winters. The late Permian world (*bottom*) was less glaciated, but the south was still cold and the latitudinal temperature gradient was still steep. Big reptiles with ectothermic bone, caseids and captorhinids, were restricted to the hot Tropics, as large reptiles had been in the early Permian. By now, however, many early therapsids, all with endothermic bone and low predator-prey ratios, had invaded southern Gondwanaland. They must have acquired high heat production and some insulation.

with increasing weight: lions require more than 10 times their own weight in meat per year, whereas shrews need 100 times their weight. These two bioenergetic scaling factors cancel each other, so that if the adult size of the predator is roughly the same as that of the prey (and in land-vertebrate ecosystems it usually is), the maximum ratio of predator standing crop to prey standing crop in a steady-state community is a constant independent of the adult body size in the predator-prey system [*see top illustration page 120*]. For example, spiders are ectotherms, and the ratio of a spider population's standing crop to its prey standing crop reaches a maximum of about 40 percent. Mountain boomer lizards, about 100 grams in adult weight, feeding on other lizards would reach a similar maximum ratio. So would the giant Komodo dragon lizards (up to 150 kilograms in body weight) preying on deer, pigs and monkeys. Endothermic

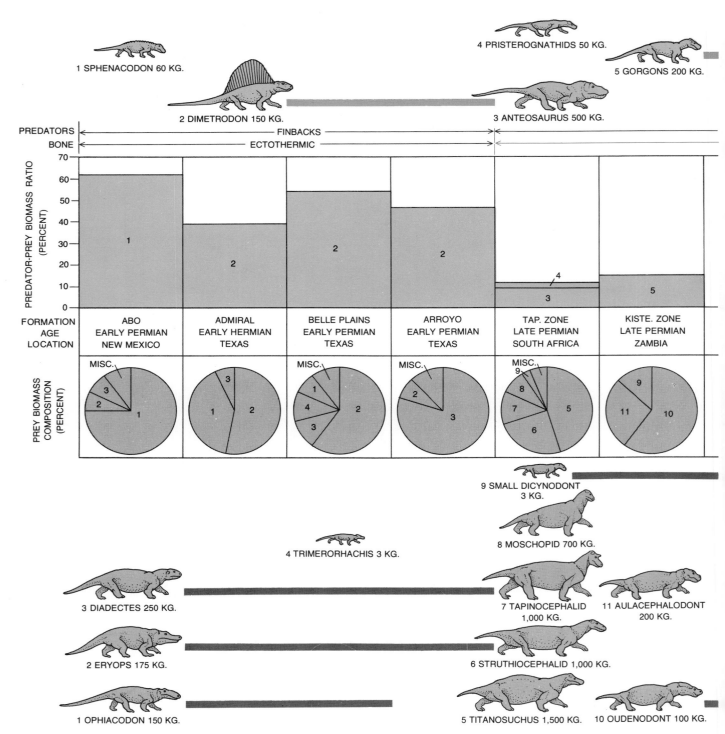

PREDATOR-PREY RATIO AND PREY COMPOSITION are shown on these two pages and the next two pages for a number of fossil communities, each representing a particular time zone and depositional environment. The predator (*top*) and prey (*bottom*) animals involved at each site are illustrated. For each deposit the histogram (*color*) gives the predator biomass as a percent of the

mammals and birds, on the other hand, reach a maximum predator-prey biomass ratio of only from 1 to 3 percent—whether they are weasel and mouse or lion and zebra [*see bottom illustration on page 120*].

Some fossil deposits yield hundreds or thousands of individuals representing a single community; their live body weight can be calculated from the reconstruction of complete skeletons, and the total predator-prey biomass ratios are then easily worked out. Predator-prey ratios are powerful tools for paleophysiology because they are the direct result of predator metabolism.

The Age of Ectothermy

The paleobioenergetic methodology I have outlined can be tested by analyzing

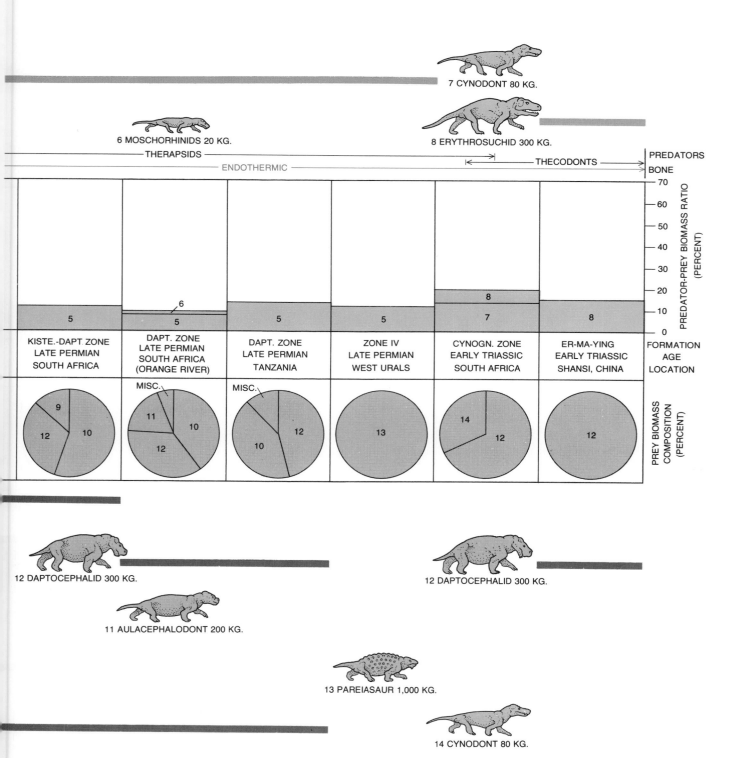

total prey biomass, in other words, the predator-prey ratio. The pie charts give the composition of the available prey. Note the sudden drop in the predator-prey ratios during the transition from the finback pelycosaurs to the early therapsids, which coincided with the first appearance of endothermic bone and also with the invasion of cold southern Gondwanaland by early therapsids of all sizes.

the first land-vertebrate predator-prey system, the early Permian communities of primitive reptiles and amphibians. The first predators capable of killing relatively large prey were the finback pelycosaurs of the family Sphenacodontidae, typified by *Dimetrodon,* whose tall-spined fin makes it popular with car-

toonists. Although this family included the direct ancestors of mammal-like reptiles and hence of mammals, the sphenacodonts themselves had a very primitive level of organization, with a limb anatomy less advanced than that of living lizards. Finback bone histology was emphatically ectothermic, with a low den-

sity of blood vessels, few Haversian canals and the distinct growth rings that are common in specimens from seasonally arid climates.

One might suspect that finbacks and their prey would be confined to warm, equable climates, and early Permian paleogeography offers an excellent op-

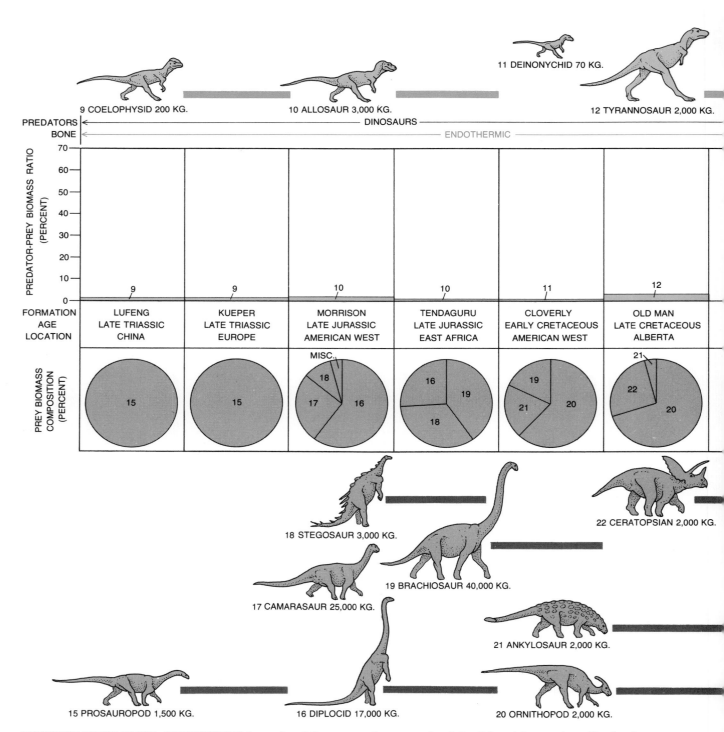

EVIDENCE FROM FOSSIL COMMUNITIES is continued from the preceding two pages. The animals are of course not all drawn to the same scale; their adult weights are given. The drawings are presented with the same limb-stride positions in each to emphasize

portunity to test this prediction. During the early part of the period ice caps covered the southern tips of the continental land masses, all of which were part of the single southern supercontinent Gondwanaland, and glacial sediment is reported at the extreme northerly tip of the Permian land mass in Siberia by Rus-

sian geologists [*see illustration on page 121*]. The Permian Equator crossed what are now the American Southwest, the Maritime Provinces of Canada and western Europe. Here are found sediments produced in very hot climates: thickbedded evaporite salts and fully oxidized, red-stained mudstones. The lati-

tudinal temperature gradient in the Permian must have been at least as steep as it is at present. Three Permian floral zones reflect the strong poleward temperature gradient. The Angaran flora of Siberia displays wood with growth rings from a wet environment, implying a moist climate with cold winters. The

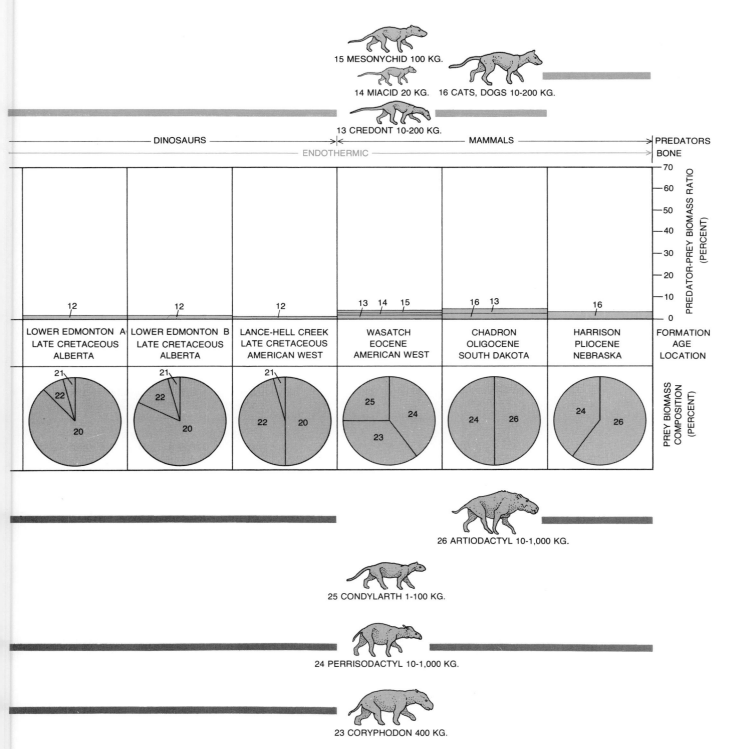

relative limb length. Long-limbed, fast-sprinting vertebrates of large size appeared only with the dinosaurs, in the middle Triassic.

Note the remarkably low predator-prey ratios of the dinosaurs, as low as or lower than those of the Cenozoic (and modern) mammals.

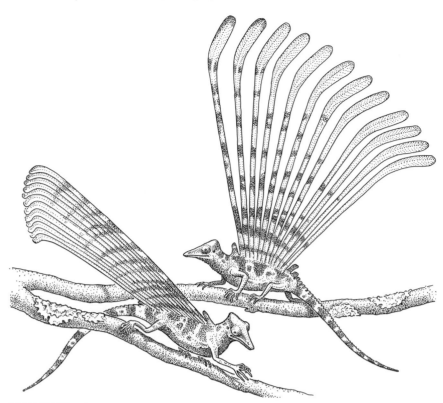

LONGISQUAMA, a small animal whose fossil was discovered in middle Triassic lake beds in Turkestan by the Russian paleontologist A. Sharov, was a thecodont. Its body was covered by long overlapping scales that were keeled, suggesting that they constituted a structural stage in the evolution of feathers. The long devices along the back were V-shaped in cross section; they may have served as parachutes and also as threat devices, as shown here.

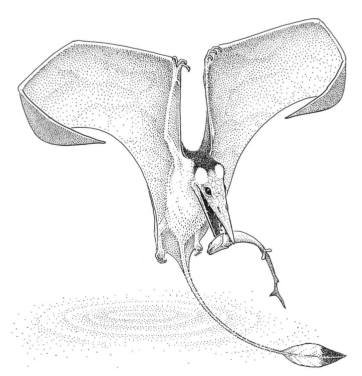

SORDUS PILOSUS, also found by Sharov, was a pterosaur: a flying reptile of the Jurassic period that was a descendant of thecodonts or of very early dinosaurs. Superbly preserved fossils show that the animal was insulated with a dense growth of hair or hairlike feathers; hence the name, which means "hairy devil." Insulation strongly suggests endothermy.

Euramerian flora of the equatorial region had two plant associations: wet swamp communities with no growth rings in the wood, implying a continuous warm-moist growing season, and semiarid, red-bed-evaporite communities with some growth rings, reflecting a tropical dry season. In glaciated Gondwanaland the peculiar *Glossopteris* flora dominated, with wood from wet environments showing sharp growth rings.

The ectothermy of the finbacks is confirmed by their geographic zonation. Finback communities are known only from near the Permian Equator; no large early Permian land vertebrates of any kind are found in glaciated Gondwanaland. (One peculiar little fish-eating reptile, *Mesosaurus*, is known from southern Gondwanaland, and its bone has sharp growth rings. The animal must have fed and reproduced during the Gondwanaland summer and then burrowed into the mud of lagoon bottoms to hibernate, much as large snapping turtles do today in New England.)

Excellent samples of finback communities are available for predator-prey studies, thanks largely to the lifework of the late Alfred Sherwood Romer of Harvard University. In order to derive a predator-prey ratio from a fossil community one simply calculates the number of individuals, and thus the total live weight, represented by all the predator and prey specimens that are found together in a sediment representing one particular environment. In working with scattered and disarticulated skeletons it is best to count only bones that have about the same robustness, and hence the same preservability, in both predator and prey. The humerus and the femur are good choices for finback communities: they are about the same size with respect to the body in the prey and the predator and should give a ratio that faithfully represents the ratio of the animals in life.

In the earlier early Permian zones the most important finback prey were semi-aquatic fish-eating amphibians and reptiles, particularly the big-headed amphibian *Eryops* and the long-snouted pelycosaur *Ophiacodon*. As the climate became more arid in Europe and America these water-linked forms decreased in numbers, and the fully terrestrial herbivore *Diadectes* became the chief prey genus. In all zones from all environments the calculated biomass ratio of predator to prey in finback communities is very high: from 35 to 60 percent, the same range seen in living ectothermic spiders and lizards.

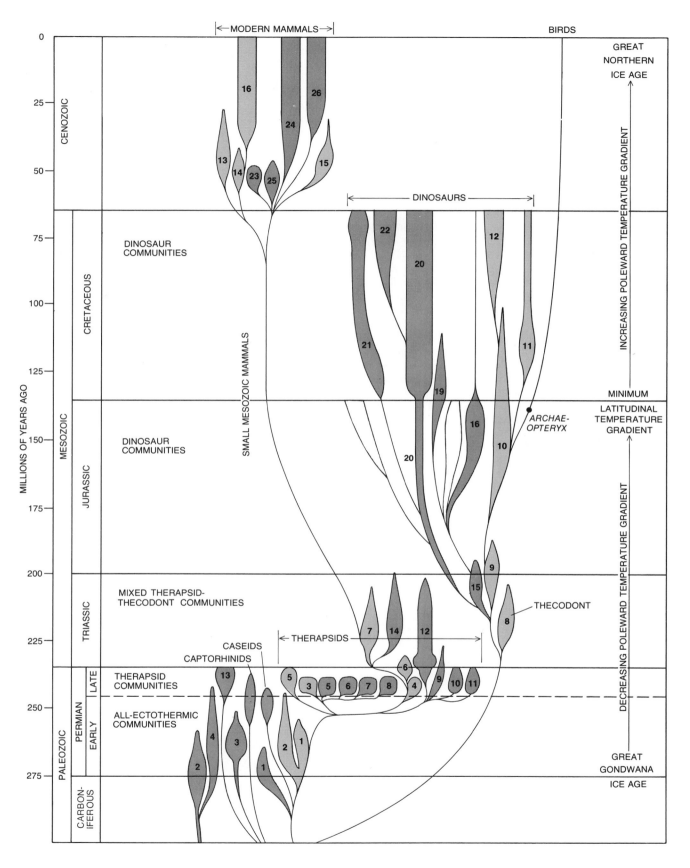

PREDATOR-PREY SYSTEMS of land vertebrates and the paths of descent of successive groups are diagrammed here with the predators (*color*) and the prey animals (*gray*) numbered to refer to the groups named and pictured in the illustrations on pages 122 through 125. The relative importance of the live biomass represented by fossils is indicated by the width of the gray and colored pathways.

All three of the paleobioenergetic indicators agree: the finback pelycosaurs and their contemporaries were ectotherms with low heat production and a lizardlike physiology that confined their distribution to the Tropics.

Therapsid Communities

The mammal-like reptiles (order Therapsida), descendants of the finbacks, made their debut at the transition from the early to the late Permian and immediately became the dominant large land vertebrates all over the world. The three metabolism-measuring techniques show that they were endotherms.

The earliest therapsids retained many finback characteristics but had acquired limb adaptations that made possible a trotting gait and much higher running speeds. From early late Permian to the middle Triassic one line of therapsids became increasingly like primitive mammals in all details of the skull, the teeth and the limbs, so that some of the very advanced mammal-like therapsids (cynodonts) are difficult to separate from the first true mammals. The change in physiology, however, was not so gradual. Detailed studies of bone histology conducted by Armand Riqles of the University of Paris indicate that the bioenergetic transition was sudden and early: all the finbacks had fully ectothermic bone; all the early therapsids—and there is an extraordinary variety of them—had fully endothermic bone, with no growth rings and with closely packed blood vessels and Haversian canals.

The late Permian world still had a severe latitudinal temperature gradient; some glaciation continued in Tasmania, and the southern end of Gondwanaland retained its cold-adapted *Glossopteris* flora. If the earliest therapsids were equipped with endothermy, they would presumably have been able to invade southern Africa, South America and the other parts of the southern cold-temperature realm. They did exactly that. A rich diversity of early therapsid families has been found in the southern Cape District of South Africa, in Rhodesia, in Brazil and in India—regions reaching to 65 degrees south Permian latitude [*see illustration on page 121*]. Early therapsids as large as rhinoceroses were common there, and many species grew to an adult weight greater than 10 kilograms, too large for true hibernation. These early therapsids must have had physiological adaptations that enabled them to feed in and move through the snows of the cold Gondwanaland winters. There were also some ectothermic holdovers from the early Permian that survived into the late Permian, notably the immense herbivorous caseid pelycosaurs and the big-headed, seed-eating captorhinids. As one might predict, large species of these two ectothermic families were confined to areas near the late Permian Equator; big caseids and captorhinids are not found with the therapsids in cold Gondwanaland. In the late Permian, then, there was a "modern" faunal zonation of large vertebrates, with endothermic therapsids and some big ectotherms in the Tropics giving way to an all-endothermic therapsid fauna in the cold south.

In the earliest therapsid communities of southern Africa, superbly represented in collections built up by Lieuwe Boonstra of the South African Museum and by James Kitching of the University of the Witwatersrand, the predator-prey ratios are between 9 and 16 percent. That is much lower than in early Permian finback communities. Equally low ratios are found for tropical therapsids from the U.S.S.R. even though the prey species there were totally different from those of Africa. The sudden decrease in predator-prey ratios from finbacks to early therapsids coincides exactly with the sudden change in bone histology from ectothermic to endothermic reported by Riqles and also with the sudden invasion of the southern cold-temperate zone by a rich therapsid fauna. The conclusion is unavoidable that even early therapsids were endotherms with high heat production.

It seems certain, moreover, that in the

ARCHAEOPTERYX, generally considered the first bird, is known from late Jurassic fossils that show its feather covering clearly. In spite of its very birdlike appearance, *Archaeopteryx* was closely related to certain small dinosaurs (*see illustration on next page*) and could not fly. The presence of insulation in the thecodont *Longisquama* and in *Sordus* and *Archaeopteryx*, which were descendants of thecodonts, indicates that insulation and endothermy were acquired very early, probably in early Triassic thecodonts.

cold Gondwanaland winters the therapsids would have required surface insulation. Hair is usually thought of as a late development that first appeared in the advanced therapsids, but it must have been present in the southern African endotherms of the early late Permian. How did hair originate? Possibly the ancestors of therapsids had touch-sensitive hairs scattered over the body as adaptations for night foraging; natural selection could then have favored increased density of hair as the animals' heat production increased and they moved into colder climates.

The therapsid predator-prey ratios, although much lower than those of ectotherms, are still about three times higher than those of advanced mammals today. Such ratios indicate that the therapsids achieved endothermy with a moderately high heat production, far higher than in typical reptiles but still lower than in most modern mammals. Predator-prey ratios of early Cenozoic communities seem to be lower than those of therapsids, and so one might conclude that a further increase in metabolism occurred somewhere between the advanced therapsids of the Triassic and the mammals of the post-Cretaceous era. Therapsids may have operated at a lower body temperature than most living mammals do, and thus they may have saved energy with a lower thermostat setting. This suggestion is reinforced by the low body temperature of the most primitive living mammals: monotremes (such as the spiny anteater) and the insectivorous tenrecs of Madagascar; they maintain a temperature of about 30 degrees Celsius instead of the 36 to 39 degrees of most modern mammals.

Thecodont Transition

The vigorous and successful therapsid dynasty ruled until the middle of the Triassic. Then their fortunes waned and a new group, which was later to include the dinosaurs, began to take over the roles of large predators and herbivores. These were the Archosauria, and the first wave of archosaurs were the thecodonts. The earliest thecodonts, small and medium-sized animals found in therapsid communities during the Permian-Triassic transition, had an ectothermic bone histology. In modern ecosystems the role played by large freshwater predators seems to be one in which ectothermy is competitively superior to endothermy; the low metabolic rate of ectotherms may be a key advantage because it allows much longer dives. Two groups of thecodonts became large freshwater fish-eaters: the phytosaurs, which were confined to the Triassic, and the crocodilians, which remain successful today. Both groups have ectothermic bone.

(The crocodilian endothermy was either inherited directly from the first thecodonts or derived secondarily from endothermic intermediate ancestors.) In most of the later, fully terrestrial advanced thecodonts, on the other hand, Riqles discovered a typical endothermic bone histology; the later thecodonts were apparently endothermic.

The predator-prey evidence for thecodonts is scanty. The ratios are hard to compute because big carnivorous cynodonts and even early dinosaurs usually shared the predatory role with thecodonts. One sample from China that has only one large predator genus, a big-headed erythrosuchid thecodont, does give a ratio of about 10 percent, which is in the endothermic range. The zonal evidence is clearer. World climate was moderating in the Triassic (the glaciers were gone), but a distinctive flora and some wood growth rings suggest that southern Gondwanaland was not yet warm all year. What is significant in this regard is the distribution of phytosaurs, the big ectothermic fish-eating thecodonts. Their fossils are common in North America and Europe (in the Triassic Tropics) and in India, which was warmed by the equatorial Tethys Ocean, but they have not been found in southern Gondwanaland, in southern Africa or in Argentina, even though a rich endothermic thecodont fauna did exist

DINOSAURIAN ANCESTRY of *Archaeopteryx* (*left*), and thus of birds, is indicated by its close anatomical relation to such small dinosaurs as *Microvenator* (*right*) and *Deinonychus*; John H. Ostrom of Yale University demonstrated that they were virtually identical in all details of joint anatomy. The long forelimbs of *Archaeopteryx* were probably used for capturing prey, not for flight.

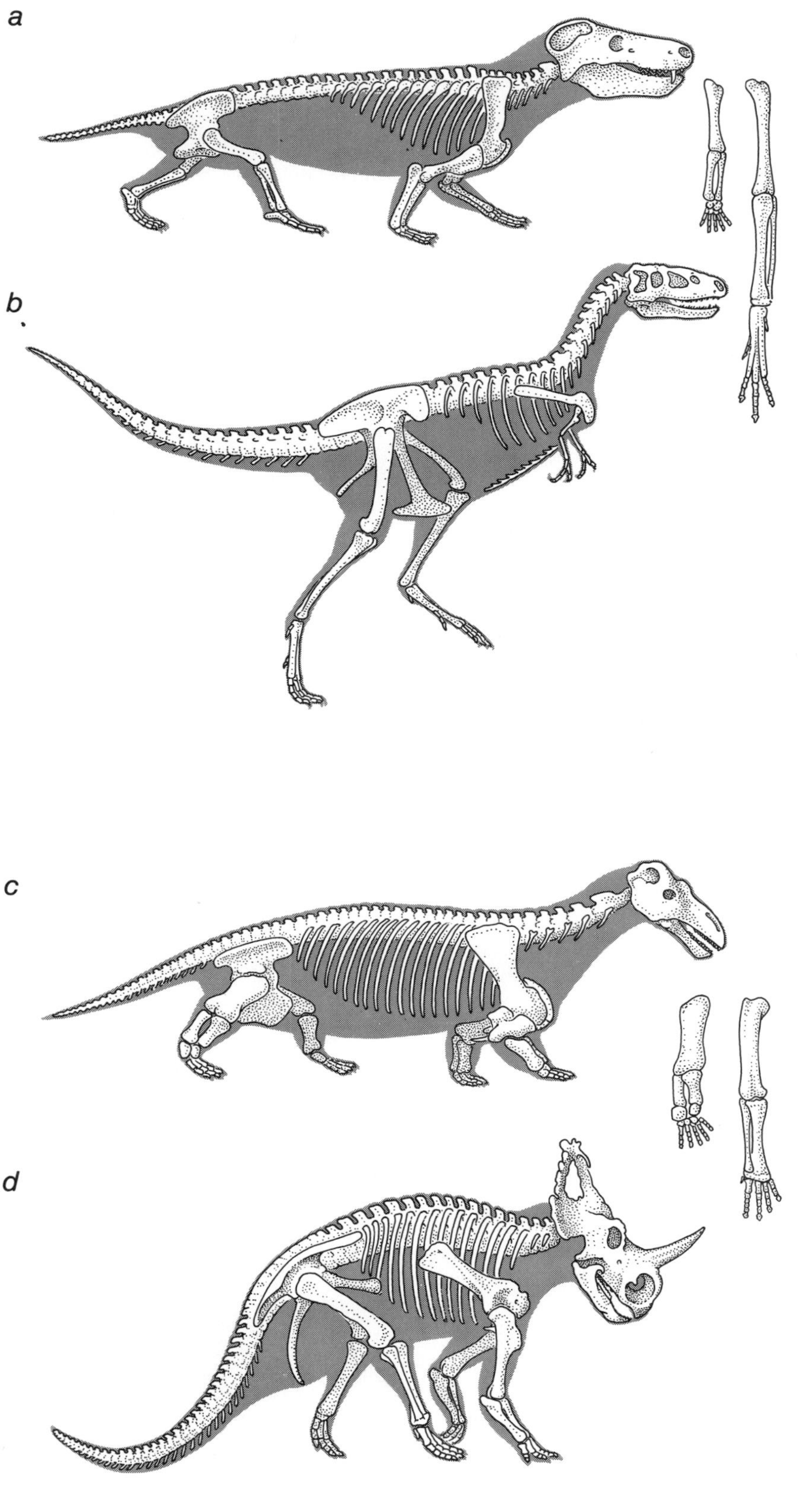

a

b

c

d

LIMB LENGTHS of dinosaurs are compared with those of two ecologically equivalent therapsids. The limbs were relatively longer in the dinosaurs and the appended muscles were larger, indicating that the dinosaurs had a larger capacity for high levels of exercise metabolism. The two top drawings represent the animals as if they were the same weight; the adult carnivorous therapsid Cynognathus (a) actually weighed 100 kilograms and the juvenile dinosaur Albertosaurus (b) 600 kilograms. Two herbivores, therapsid Struthiocephalus (c) and horned dinosaur Centrosaurus (d), weighed about 1,500 kilograms.

there.

Did some of the thecodonts have thermal insulation? Direct evidence comes from the discoveries of A. Sharov of the Academy of Sciences of the U.S.S.R. Sharov found a partial skeleton of a small thecodont and named it *Longisquama* for its long scales: strange parachutelike devices along the back that may have served to break the animal's fall when it leaped from trees. More important is the covering of long, overlapping, keeled scales that trapped an insulating layer of air next to its body [*see top illustration on page 126*]. These scales lacked the complex anatomy of real feathers, but they are a perfect ancestral stage for the insulation of birds. Feathers are usually assumed to have appeared only late in the Jurassic with the first bird, *Archaeopteryx*. The likelihood that some thecodonts had insulation is supported, however, by another of Sharov's discoveries: a pterosaur, or flying reptile, whose fossils in Jurassic lake beds still show the epidermal covering. This beast (appropriately named *Sordus pilosus*, the "hairy devil") had a dense growth of hair or hairlike feathers all over its body and limbs. Pterosaurs are descendants of Triassic thecodonts or perhaps of very primitive dinosaurs. The insulation in both *Sordus* and *Longisquama*, and the presence of big erythrosuchid thecodonts at the southern limits of Gondwanaland, strongly suggest that some endothermic thecodonts had acquired insulation by the early Triassic.

The Dinosaurs

Dinosaurs, descendants of early thecodonts, appeared first in the middle Triassic and by the end of the period had replaced thecodonts and the remaining therapsids as the dominant terrestrial vertebrates. Zonal evidence for endothermy in dinosaurs is somewhat equivocal. The Jurassic was a time of climatic optimum, when the poleward temperature gradient was the gentlest that has prevailed from the Permian until the present day. In the succeeding Cretaceous period latitudinal zoning of oceanic plankton and land plants seems, however, to have been a bit sharper. Rhinoceros-sized Cretaceous dinosaurs and big marine lizards are found in the rocks of the Canadian far north, within the Cretaceous Arctic Circle. Dale A. Russell of the National Museums of Canada points out that at these latitudes the sun would have been below the horizon for months at a time. The environment of the dinosaurs would have been far severer than the environment of the marine reptiles because of the lack of a wind-chill factor in the water and because of the ocean's temperature-buffering effect. Moreover, locomotion costs far less energy per kilometer in water than on land, so that the marine reptiles could have migrated away from the arctic winter. These considerations suggest, but do not prove, that arctic dinosaurs must have been able to cope with cold stress.

Dinosaur bone histology is less equivocal. All dinosaur species that have been investigated show fully endothermic bone, some with a blood-vessel density higher than that in living mammals. Since bone histology separates endotherms from ectotherms in the Permian and the Triassic, this evidence alone should be a strong argument for the endothermy of dinosaurs. Yet the predator-prey ratios are even more compelling. Dinosaur carnivore fossils are exceedingly rare. The predator-prey ratios for dinosaur communities in the Triassic, Jurassic and Cretaceous are usually from 1 to 3 percent, far lower even than those of therapsids and fully as low as those in large samples of fossils from advanced mammal communities in the Cenozoic. I am persuaded that all the available quantitative evidence is in favor of high heat production and a large annual energy budget in dinosaurs.

Were dinosaurs insulated? Explicit evidence comes from a surprising source: *Archaeopteryx*. As an undergraduate a decade ago I was a member of a paleontological field party led by John H. Ostrom of Yale University. Near Bridger, Mont., Ostrom found a remarkably preserved little dinosaurian carnivore, *Deinonychus*, that shed a great deal of light on carnivorous dinosaurs in general. A few years later, while looking for pterosaur fossils in European museums, Ostrom came on a specimen of *Archaeopteryx* that had been mislabeled for years as a flying reptile, and he noticed extraordinary points of resemblance between *Archaeopteryx* and carnivorous dinosaurs. After a detailed anatomical analysis Ostrom has now established beyond any reasonable doubt that the immediate ancestor of *Archaeopteryx* must have been a small dinosaur, perhaps one related to *Deinonychus*. Previously it had been thought that the ancestor of *Archaeopteryx*, and thus of birds, was a thecodont rather far removed from dinosaurs themselves.

Archaeopteryx was quite thoroughly feathered, and yet it probably could not fly: the shoulder joints were identical with those of carnivorous dinosaurs and were adapted for grasping prey, not for the peculiar arc of movement needed for wing-flapping. The feathers were probably adaptations not for powered flight or gliding but primarily for insulation. *Archaeopteryx* is so nearly identical in all known features with small carnivorous dinosaurs that it is hard to believe feathers were not present in such dinosaurs. Birds inherited their high metabolic rate and most probably their feathered insulation from dinosaurs; powered flight probably did not evolve until the first birds with flight-adapted shoulder joints appeared during the Cretaceous, long after *Archaeopteryx*.

It has been suggested a number of times that dinosaurs could have achieved a fairly constant body temperature in a warm environment by sheer bulk alone; large alligators approach this condition in the swamps of the U.S. Gulf states. This proposed thermal mechanism would not give rise to endothermic bone histology or low predator-prey ratios, however, nor would it explain arctic dinosaurs or the success of many small dinosaur species with an adult weight of between five and 50 kilograms.

Dinosaur Brains and Limbs

Large brain size and endothermy seem to be linked; most birds and mammals have a ratio of brain size to body size much larger than that of living reptiles and amphibians. The acquisition of endothermy is probably a prerequisite for the enlargement of the brain because the proper functioning of a complex central nervous system calls for the guarantee of a constant body temperature. It is not surprising that endothermy appeared before brain enlargement in the evolutionary line leading to mammals. Therapsids had small brains with reptilian organization; not until the Cenozoic did mammals attain the large brain size characteristic of most modern species. A large brain is certainly not necessary for endothermy, since the physio-

	BONE HISTOLOGY	PRESENT IN TEMPERATE ZONE	PREDATOR–PREY RATIO	LIMB LENGTH
FINBACKS, OTHER EARLY PERMIAN LAND VERTEBRATES		NO	50 PERCENT	SHORT
LATE PERMIAN CASEIDS AND BIG CAPTORHINIDS		NO	NOT APPLICABLE	SHORT
LATE PERMIAN– EARLY TRIASSIC THERAPSIDS		YES	10 PERCENT	SHORT
EARLIEST THECODONTS		?	?	SHORT
MOST LAND THECODONTS		YES UP TO 600 KILOGRAMS	10 PERCENT	SHORT
FRESHWATER THECODONTS		NO UP TO 500 KILOGRAMS	?	SHORT
DINOSAURS		YES	1–3 PERCENT	LONG
CENOZOIC MAMMALS		YES	1–5 PERCENT	LONG

PALEOBIOENERGETIC EVIDENCE is summed up here. The appropriate blocks are shaded to show whether the available data constitute evidence for ectothermy (*gray*) or endothermy (*color*) according to criteria discussed in the text of this article. Caseids and captorhinids are herbivores, so that there is no predator-prey ratio. There are early thecodonts in temperate-zone deposits, but they are small and so the evidence is not significant.

logical feedback mechanisms responsible for thermoregulation are deep within the "old" region of the brain, not in the higher learning centers. Most large dinosaurs did have relatively small brains. Russell has shown, however, that some small and medium-sized carnivorous dinosaurs had brains as large as or larger than modern birds of the same body size.

Up to this point I have concentrated on thermoregulatory heat production. Metabolism during exercise can also be read from fossils. Short bursts of intense exercise are powered by anaerobic metabolism within muscles, and the oxygen debt incurred is paid back afterward by the heart-lung system. Most modern birds and mammals have much higher

levels of maximum aerobic metabolism than living reptiles and can repay an oxygen debt much faster. Apparently this difference does not keep small ectothermic animals from moving fast: the top running speeds of small lizards equal or exceed those of small mammals. The difficulty of repaying oxygen debt increases with increasing body size, however, and the living large reptiles (crocodilians, giant lizards and turtles) have noticeably shorter limbs, less limb musculature and lower top speeds than many large mammals, such as the big cats and the hoofed herbivores.

The early Permian ectothermic dynasty was also strikingly short-limbed; evidently the physiological capacity for

high sprinting speeds in large animals had not yet evolved. Even the late therapsids, including the most advanced cynodonts, had very short limbs compared with the modern-looking running mammals that appeared early in the Cenozoic. Large dinosaurs, on the other hand, resembled modern running mammals, not therapsids, in locomotor anatomy and limb proportions. Modern, fast-running mammals utilize an anatomical trick that adds an extra limb segment to the forelimb stroke. The scapula, or shoulder blade, which is relatively immobile in most primitive vertebrates, is free to swing backward and forward and thus increase the stride length. Jane A. Peterson of Harvard has shown that

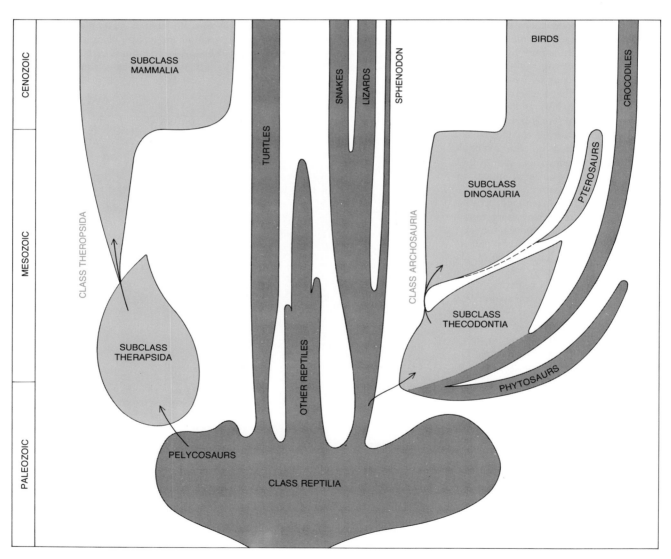

RECLASSIFICATION of land vertebrates (excluding the Amphibia) is suggested by the author on the basis of bioenergetic and anatomical evidence. The critical break comes with the development of endothermy (*color*), which is competitively superior to ectothermy (*gray*) for large land vertebrates. Therapsids were endothermic, closer in physiology to mammals than to today's reptiles. Birds almost certainly inherited their bioenergetics (as well as their joint anatomy) from dinosaurs. The new classes presented here, the Theropsida and the Archosauria, reflect energetic evolution more faithfully than the traditional groupings (*see illustration on page 118*). The width of the pathways representing the various groups is proportional to the biomass represented by their fossils.

living chameleonid lizards have also evolved scapular swinging, although its details are different from those in mammals. Quadrupedal dinosaurs evolved a chameleon-type scapula, and they must have had long strides and running speeds comparable to those of big savanna mammals today.

When the dinosaurs fell at the end of the Cretaceous, they were not a senile, moribund group that had played out its evolutionary options. Rather they were vigorous, still diversifying into new orders and producing a variety of big-brained carnivores with the highest grade of intelligence yet present on land. What caused their fall? It was not competition, because mammals did not begin to diversify until after all the dinosaur groups (except birds!) had disappeared. Some geochemical and microfossil evidence suggests a moderate drop in ocean temperature at the transition from the Cretaceous to the Cenozoic, and so cold has been suggested as the reason. But the very groups that would have been most sensitive to cold, the large crocodilians, are found as far north as Saskatchewan and as far south as Argentina before and immediately after the end of the Cretaceous. A more likely reason is the draining of shallow seas on the continents and a lull in mountain-building activity in most parts of the world, which would have produced vast stretches of monotonous topography. Such geological events decrease the variety of habitats that are available to land animals, and thus increase competition. They can also cause the collapse of intricate, highly evolved ecosystems; the larger animals seem to be the more affected. At the end of the Permian similar changes had been accompanied by catastrophic extinctions among therapsids and other land groups. Now, at the end of the Cretaceous, it was the dinosaurs that suffered a catastrophe; the mammals and birds, perhaps because they were so much smaller, found places for themselves in the changing landscape and survived.

The success of the dinosaurs, an enigma as long as they were considered "cold-blooded," can now be seen as the predictable result of the superiority of their high heat production, high aerobic exercise metabolism and insulation. They were endotherms. Yet the concept of dinosaurs as ectotherms is deeply entrenched in a century of paleontological literature. Being a reptile connotes being an ectotherm, and the dinosaurs have always been classified in the subclass Archosauria of the class Reptilia; the other land-vertebrate classes were the Mammalia and the Aves. Perhaps, then, it is time to reclassify.

Taxonomic Conclusion

What better dividing line than the invention of endothermy? There has been no more far-reaching adaptive breakthrough, and so the transition from ectothermy to endothermy can serve to separate the land vertebrates into higher taxonomic categories. For some time it has been suggested that the therapsids should be removed from the Reptilia and joined with the Mammalia; in the light of the sudden increase in heat production and the probable presence of hair in early therapsids, I fully agree. The term Theropsida has been applied to mammals and their therapsid ancestors. Let us establish a new class Theropsida, with therapsids and true mammals as two subclasses [see illustration on preceding page].

How about the class Aves? All the quantitative data from bone histology and predator-prey ratios, as well as the dinosaurian nature of Archaeopteryx, show that all the essentials of avian biology—very high heat production, very high aerobic exercise metabolism and feathery insulation—were present in the dinosaur ancestors of birds. I do not believe birds deserve to be put in a taxonomic class separate from dinosaurs. Peter Galton of the University of Bridgeport and I have suggested a more reasonable classification: putting the birds into the Dinosauria. Since bone histology suggests that most thecodonts were endothermic, the thecodonts could then be joined with the Dinosauria in a great endothermic class Archosauria, comparable to the Theropsida. The classification may seem radical at first, but it is actually a good deal neater bioenergetically than the traditional Reptilia, Aves and Mammalia. And for those of us who are fond of dinosaurs the new classification has a particularly happy implication: The dinosaurs are not extinct. The colorful and successful diversity of the living birds is a continuing expression of basic dinosaur biology.

IV

SOME MAJOR PATTERNS
IN THE HISTORY OF LIFE

SOME MAJOR PATTERNS IN THE HISTORY OF LIFE

IV

INTRODUCTION

In this section we want to look at some of the broader aspects of the fossil record, particularly during the last 600 million years of the Phanerozoic. In Section I, we saw how organisms interact with their environment—how each species occupies an adaptive niche in the economy of nature. In a static, unchanging world, we would expect life to "top out" in kinds and numbers of animals and plants. But the world is not static. In the animate world, variations generated by mutation and sexual reproduction continuously provide biological novelties better suited to coping with the environment. In the inanimate world, there is constant change owing to the ongoing processes of geological activity. Consequently, we can predict—at least in a general way—that the history of life will flow and ebb in concert with the dynamic evolution of the planet earth.

What then are some of the overall tendencies of life's history since its inception billions of years ago? First, there has been an increase in its organizational complexity, from single-celled creatures to multicellular; from simple tissues such as the outer layer, or ectoderm, of jellyfish to complex organs such as the eyes of an insect; from the sluggish nerve nets of earthworms to the elaborate central nervous system of a dog. The apotheosis of this tendency is, of course, the human organism with its conscious and thinking mentality.

Another obvious trend in the history of life is the steady expansion into more and more different kinds of environments: from marine waters into fresh, from coastal wetlands into dry upland plateaus, from the earth's surface into the trees and sky above and into the soil and caves below, and, most recently, our own tentative ventures out into space. With this expansion of environments has come a corresponding increase in diversity of organisms, for, as more and more adaptive niches are sought out and occupied, there is an increase in species. The increase in diversity has also been self-stimulating, because as more and more kinds of animals and plants evolve, owing to new physical opportunities, still other organisms will appear, to capitalize on these new biological opportunities.

A third major tendency in the history of life is the more or less continual turnover of species, due to the constant changes experienced by the earth during its history. If we think of a species as a finely tuned adjustment of organisms to the local environment, we can expect that changes in the environment will often occur faster than the organisms can adapt to it and evolve with it. Thus, if environmental change oversteps the adaptive response of a species, that species will become extinct. But the unoccupied niche will not remain unfilled for long, because, given the inherent variation of organisms, natural selection will produce another species to fill it. If we look at higher—that is, more inclusive—biological categories, such as orders, classes, and phyla, we find that this turnover is dampened. Thus, although there are no bivalved clam *species* surviving from the Cambrian, this *class* of molluscs has, indeed, survived from then. Why? Because the basic general design of the bivalves has remained adaptive to an aquatic way of life. (Consider the wheel, which has always remained useful, even though many specific variations of the wheel have come and gone.)

Although extinction is the natural outcome for virtually all species—if not the higher categories of life to which they belong—there are some few instances where species, or at least genera, have persisted for very long times. These living fossils represent broadly adapted forms that have survived in the face of environmental change. This knowledge should temper the gloom of anticipating the end of our own species. For the quintessence of what makes humans a distinct species is our very ability to cope with our environment. And because this coping is as much cultural as genetic, we can respond rapidly to change. Moreover, given our intellect, we can also presumably foresee changes in advance and prepare for them. Whether, in fact, we humans realize this potential to preserve our species remains to be seen.

"Crises in the History of Life," by Norman D. Newell, discusses both change and persistence of organisms as revealed by the fossil record. As Newell points out, wholesale disappearances of ancient floras and faunas were interpreted by earlier scientists as the result of periodic, sudden, violent catastrophes or revolutions. Although paleontologists subsequently eschewed using such possible revolutions or catastrophes as explanations for mass extinctions, the real causes eluded them, and still do today, even though we are closer to the answers.

Newell clearly states the most baffling aspect of mass extinction: the demise of large proportions of organisms that inhabit a wide range of habitats. The very variety of life becoming extinct at the end of the Permian and Cretaceous requires an equally large-scale cause. What we postulate as a factor effecting extinction for bottom-dwelling invertebrates must also affect terrestrial vertebrates. A corollary complication is that periods of major extinction for animals and plants rarely coincide.

Newell suggests that, whatever factor we invoke, it must be correlated to the restriction or destruction of habitat for the species that is becoming extinct. Loss of habitat decreases the geographic range and spatial variety for the occupying organisms. It also reduces not only the total numbers of individuals but, equally critical, the numbers of relatively isolated breeding groups. For Newell, the transgression and regression of shallow seas on and off the continents throughout geologic time would cause broad destruction of habitats, with accompanying mass extinctions. Although this article was written just before the theory of plate tectonics was formulated, Newell was alert to the role that tectonics played in shaping the ocean basins and in producing changes in sea level world wide. Finally, Newell emphasizes the role that key species play in a community, so that their extinction might, like falling dominoes, wipe out other interdependent species.

"Plate Tectonics and the History of Life in the Oceans," by James W. Valentine and Eldridge M. Moores, examines the environmental implications and biological responses to global changes wrought by continental fragmentation and reassembly. We would, of course, expect plate tectonics to have major impact on the habitats of the world. Valentine and Moores explain what those impacts are and how they influence the diversity of life. In general, they argue that when continents are large and few, climates are more seasonal or fluctuating than when the continents are small and dispersed. The shallow waters surrounding such continents would thus be highly variable environments or rather constant environments, respectively.

And, when environments are constant or stable, there is more diversity of life, because organisms can afford to specialize, thereby producing more species. Variable environments, on the other hand, favor generalists that can tolerate wide ranges of physical and chemical conditions, thus reducing the total numbers of species. In addition, food resources vary by latitude, so that continents in the tropics have steady food production year round, whereas in high latitudes food production fluctuates seasonally. Consequently, equatorial marine faunas are much more diverse than boreal faunas. Finally, continents that are fragmented will have their shallow shelves more isolated from one another than if the continents are assembled into one supercontinent with a single shelf running around it. As we noted in Section I, geographic isolation of populations leads to differentiation of species, and so there will be more diversity of life when continents are dispersed than when they are welded together.

With this line of reasoning, Valentine and Moores show how the rise and fall in the diversity of shallow marine life correlates with continental drift and plate tectonics. The initial expansion of life in the late Precambrian and early Cambrian is due, they believe, to the breakup of a large supercontinent called Pangaea. Continental fragmentation led to stability of environments and food resources as well as isolated marine shelves, all of which, in turn, brought about a great increase in life's diversity. Continental reassembly in the late Paleozoic had opposite effects and culminated in the Permian crisis among marine invertebrates. No doubt, Newell's concept of the restriction of the shallow marine habitat was also important.

The Mesozoic breakup of the two great northern and southern continents, Laurasia and Gondwanaland, led to renewed diversity, and continued separation of the continents in the Cenozoic brought about still greater increases in diversity. As plausible as this idea is—namely, that the drifting apart and coming together of continents has caused significant changes in diversity of marine organisms—not all paleontologists agree. Some interpret the rise, decline, recovery, and expansion of marine fossils throughout the Phanerozoic as simply a reflection of the corresponding abundance of marine sedimentary rocks (see Raup, 1976).

Björn Kurtén's article, "Continental Drift and Evolution," further presses home the importance of continental fragmentation in generating diversity among terrestrial vertebrates, such as reptiles and mammals. Here the argument runs as follows: Given a specific area of land surface, there will be some more or less equilibrium number of vertebrate types evolving there. Thus, in the late Mesozoic we can recognize about a dozen higher categories of reptiles, slightly more than half originating on the southern continent of Gondwanaland and less than half first appearing in the northern continent of Laurasia. The subsequent fragmentation and separation of these two supercontinents into South America, Africa, Australia, India, and Antarctica, on the one hand, and into North America and Eurasia, on the other, led to more than a doubling of the higher categories of mammals that succeeded the reptiles. Rather than attributing this increase in diversity to either something inherent in the mammalian condition or to more habitat variation of the continents, Kurtén ascribes it essentially to ecological duplication of basic adaptive types on continents isolated geographically from each other.

Strong support for Kurtén's hypothesis comes from the *loss* of diversity that resulted when North and South America were rejoined after tens of millions of years of separation during most of mammalian evolutionary history. In the late Cenozoic, the Isthmus of Panama provided a corridor for migrating mammals, both north and south. This faunal interchange of previously isolated similar adaptive types led to widespread extinctions among the less successful. Whereas there were some fifty terrestrial mammalian families in the late Pliocene, just before the reconnection, ten of these families have become extinct in the few million years since the continents were rejoined.

We thus return to the discussion and interpretation of the evolutionary significance of the geographic distributions of organisms. As we noted in Section I, these biogeographic observations strongly influenced Darwin's own thoughts. The concept of plate tectonics, dealing as it does with global processes of geological evolution, provides a new frame of reference within which paleontologists can view the flow and ebb of ancient life over the last half billion years. As that great American paleontologist, George Gaylord Simpson, put it a generation ago, "The history of organisms runs parallel with, is environmentally contained in, and continuously interacts with the physical history of the earth."

SUGGESTED FURTHER READING

Flessa, K., 1975, "Area, Continental Drift, and Mammalian Diversity," *Paleobiology*, vol. 1, pp. 189–194. Like Simberloff, 1974 (below), except the time is the Cenozoic and the organisms are terrestrial mammals. Both articles strongly reinforce the concept that geographic area and biological diversity are rather closely linked. If this is true, the shifting of landmasses and changing sea levels associated with plate tectonics must surely have had important effects on the history and evolution of life.

Raup, D. M. 1976. "Species Diversity in the Phanerozoic: Tabulation and Interpretation," *Paleobiology*, vol. 2, pp. 279–297. One of several articles in recent years by Raup that challenges the notion that life's diversity as recorded in the fossil record has any more significance than merely reflecting the vicissitudes of sediment accumulation and fossil preservation.

Simberloff, D. S. 1974. "Permo-Triassic Extinctions: Effects of Area on Biotic Equilibrium," *Journal of Geology*, vol. 82, pp. 267–274. Convincing argument that restriction of the Permian shallow seas correlates with widespread extinction among marine organisms at the end of the Paleozoic.

Valentine, J. W. 1973. *Evolutionary Paleoecology of the Marine Biosphere*, Englewood Cliffs, NJ.: Prentice-Hall. A brilliant integration of the functional and ecological interactions of organisms at various levels from the individual, population, and community levels up to the total marine biosphere. The crux of the book, however, lies in the hypothesis of how continental configurations influence stability in the marine environment and how that, in turn, controls the diversity of life therein. The last chapter "tells all" about the last three-quarter billion years.

11

Crises in the History of Life

by Norman D. Newell
February 1963

How is it that whole groups of animals have simultaneously died out? Paleontologists are returning to an earlier answer: natural catastrophe. The catastrophes they visualize, however, are not sudden but gradual

The stream of life on earth has been continuous since it originated some three or four billion years ago. Yet the fossil record of past life is not a simple chronology of uniformly evolving organisms. The record is prevailingly one of erratic, often abrupt changes in environment, varying rates of evolution, extermination and repopulation. Dissimilar biotas replace one another in a kind of relay. Mass extinction, rapid migration and consequent disruption of biological equilibrium on both a local and a world-wide scale have accompanied continual environmental changes.

The main books and chapters of earth history—the eras, periods and epochs—were dominated for tens or even hundreds of millions of years by characteristic groups of animals and plants. Then, after ages of orderly evolution and biological success, many of the groups suddenly died out. The cause of these mass extinctions is still very much in doubt and constitutes a major problem of evolutionary history.

The striking episodes of disappearance and replacement of successive biotas in the layered fossil record were termed revolutions by Baron Georges Cuvier, the great French naturalist of the late 18th and early 19th centuries. Noting that these episodes generally correspond to unconformities, that is, gaps in the strata due to erosion, Cuvier attributed them to sudden and violent catastrophes. This view grew out of his study of the sequence of strata in the region of Paris. The historic diagram on the opposite page was drawn by Cuvier nearly 150 years ago. It represents a simple alternation of fossil-bearing rocks of marine and nonmarine origin, with many erosional breaks and marked interruptions in the sequence of fossils.

The objection to Cuvier's catastro-

phism is not merely that he ascribed events in earth history to cataclysms; many normal geological processes are at times cataclysmic. The objection is that he dismissed known processes and appealed to fantasy to explain natural phenomena. He believed that "the march of nature is changed and not one of her present agents could have sufficed to have effected her ancient works." This hypothesis, like so many others about extinction, is not amenable to scientific test and is hence of limited value. In fairness to Cuvier, however, one must recall that in his day it was widely believed that the earth was only a few thousand years old. Cuvier correctly perceived that normal geological processes could not have produced the earth as we know it in such a short time.

Now that we have learned that the earth is at least five or six billion years old, the necessity for invoking Cuverian catastrophes to explain geological history would seem to have disappeared. Nevertheless, a few writers such as Immanuel Velikovsky, the author of *Worlds in Collision,* and Charles H. Hapgood, the author of *The Earth's Shifting Crust,* continue to propose imaginary catastrophes on the basis of little or no historical evidence. Although it is well established that the earth's crust has shifted and that climates have changed, these changes almost certainly were more gradual than Hapgood suggests. Most geologists, following the "uniformitarian" point of view expounded in the 18th century by James Hutton and in the 19th by Charles Lyell, are satisfied that observable natural processes are quite adequate to explain the history of the earth. They agree, however, that these processes must have varied greatly in rate.

Charles Darwin, siding with Hutton and Lyell, also rejected catastrophism as

an explanation for the abrupt changes in the fossil record. He attributed such changes to migrations of living organisms, to alterations of the local environment during the deposition of strata and to unconformities caused by erosion. Other important factors that are now given more attention than they were in Darwin's day are the mass extinction of organisms, acceleration of the rate of evolution and the thinning of strata due to extremely slow deposition.

The Record of Mass Extinctions

If we may judge from the fossil record, eventual extinction seems to be the lot of all organisms. Roughly 2,500 families of animals with an average longevity of somewhat less than 75 million years have left a fossil record. Of these, about a third are still living. Although a few families became extinct by evolving into new families, a majority dropped out of sight without descendants.

In spite of the high incidence of extinction, there has been a persistent gain in the diversity of living forms: new forms have appeared more rapidly than old forms have died out. Evidently organisms have discovered an increasing number of ecological niches to fill, and by modifying the environment they have produced ecological systems of great complexity, thereby making available still more niches. In fact, as I shall develop later, the interdependence of living organisms, involving complex chains of food supply, may provide an important key to the understanding of how relatively small changes in the environment could have triggered mass extinctions.

The fossil record of animals tells more about extinction than the fossil record of plants does. It has long been known

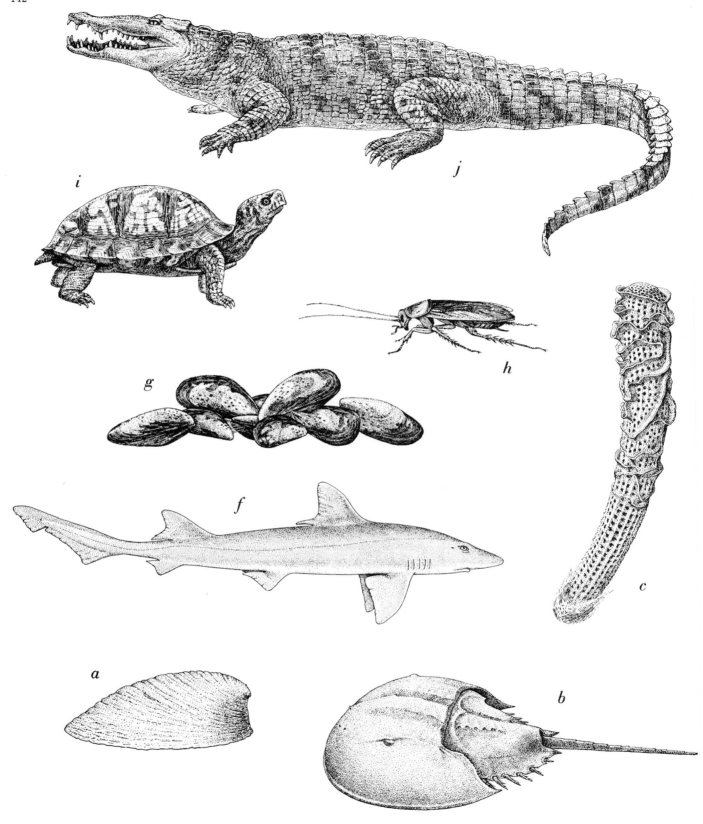

GALLERY OF HARDY ANIMALS contains living representatives of 11 groups that have weathered repeated crises in evolutionary history. Four of the groups can be traced back to the Cambrian period: the mollusk *Neopilina* (*a*), the horseshoe crab (*b*), the Venus's-flower-basket, *Euplectella* (*c*) and the brachiopod *Lingula* (*d*). One animal represents a group that goes back to the Ordovician period: the ostracode *Bairdia* (*e*). Two arose in the Devonian period: the shark (*f*) and the mussel (*g*). The cockroach

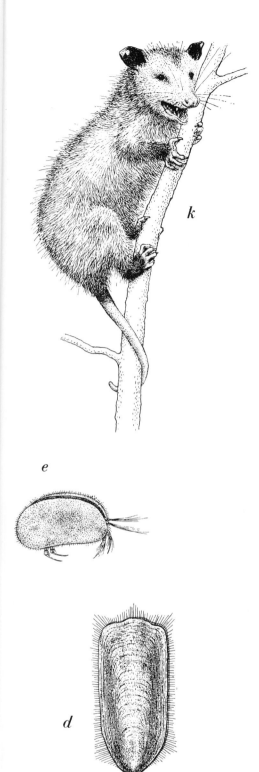

(*h*) goes back to the Pennsylvanian period. Two arose in the late Triassic: the turtle (*i*) and the crocodile (*j*). The opossum (*k*) appeared during the Cretaceous period.

that the major floral changes have not coincided with the major faunal ones. Each of the three successive principal land floras—the ferns and mosses, the gymnosperms and angiosperms—were ushered in by a short episode of rapid evolution followed by a long period of stability. The illustration on page 145 shows that once a major group of plants became established it continued for millions of years. Many groups of higher plants are seemingly immortal. Since green plants are the primary producers in the over-all ecosystem and animals are the consumers, it can hardly be doubted that the great developments in the plant kingdom affected animal evolution, but the history of this relation is not yet understood.

Successive episodes of mass extinction among animals—particularly the marine invertebrates, which are among the most abundant fossils—provide world-wide stratigraphic reference points that the paleontologist calls datums. Many of the datums have come to be adopted as boundaries of the main divisions of geologic time, but there remains some uncertainty whether the epochs of extinction constitute moments in geologic time or intervals of significant duration. In other words, did extinction occur over hundreds, thousands or millions of years? The question has been answered in many ways, but it still remains an outstanding problem.

A good example of mass extinction is provided by the abrupt disappearance of nearly two-thirds of the existing families of trilobites at the close of the Cambrian period. Before the mass extinction of these marine arthropods, which are distantly related to modern crustaceans, there were some 60 families of them. The abrupt disappearance of so many major groups of trilobites at one time has served as a convenient marker for defining the upper, or most recent, limit of the Cambrian period [*see illustration on page 146*].

Similar episodes of extinction characterize the history of every major group and most minor groups of animals that have left a good fossil record. It is striking that times of widespread extinction generally affected many quite unrelated groups in separate habitats. The parallelism of extinction between some of the aquatic and terrestrial groups is particularly remarkable [*see illustration on page 148*].

One cannot doubt that there were critical times in the history of animals. Widespread extinctions and consequent revolutionary changes in the course of animal life occurred roughly at the end of the Cambrian; Ordovician, Devonian, Permian, Triassic and Cretaceous periods. Hundreds of minor episodes of extinction occurred on a more limited scale at the level of species and genera throughout geologic time, but here we shall restrict our attention to a few of the more outstanding mass extinctions.

At or near the close of the Permian period nearly half of the known families of animals throughout the world disappeared. The German paleontologist Otto Schindewolf notes that 24 orders and superfamilies also dropped out at this point. At no other time in history, save possibly the close of the Cambrian, has the animal world been so decimated. Recovery to something like the normal variety was not achieved until late in the Triassic period, 15 or 20 million years later.

Extinctions were taking place throughout Permian time and a number of major groups dropped out well before the end of the period, but many more survived to go out together, climaxing one of the greatest of all episodes of mass extinction affecting both land and marine animals. It was in the sea, however, that the decimation of animals was particularly dramatic. One great group of animals that disappeared at this time was the fusulinids, complex protozoans that ranged from microscopic sizes to two or three inches in length. They had populated the shallow seas of the world for 80 million years; their shells, piling up on the ocean floor, had formed vast deposits of limestone. The spiny productid brachiopods, likewise plentiful in the late Paleozoic seas, also vanished without descendants. These and many other groups dropped suddenly from a state of dominance to one of oblivion.

By the close of the Permian period 75 per cent of amphibian families and more than 80 per cent of the reptile families had also disappeared. The main suborders of these animals nonetheless survived the Permian to carry over into the Triassic.

The mass extinction on land and sea at the close of the Triassic period was almost equally significant. Primitive reptiles and amphibians that had dominated the land dropped out and were replaced by the early dinosaurs that had appeared and become widespread before the close of the period. It is tempting to conclude that competition with the more successful dinosaurs was an important factor in the disappearance of these early land animals, but what bearing could this have had on the equally impressive and

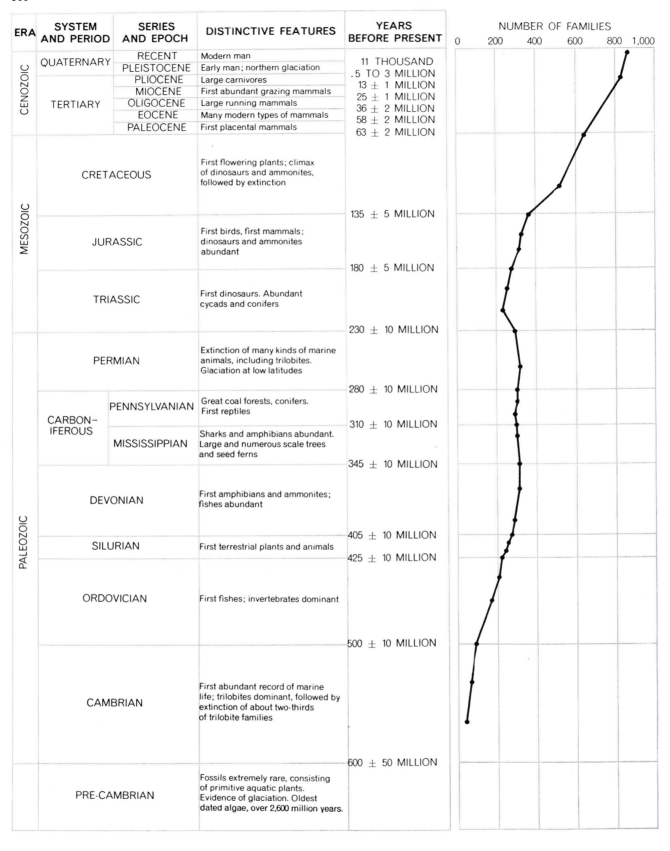

ERA	SYSTEM AND PERIOD		SERIES AND EPOCH	DISTINCTIVE FEATURES	YEARS BEFORE PRESENT
CENOZOIC	QUATERNARY		RECENT	Modern man	11 THOUSAND
			PLEISTOCENE	Early man; northern glaciation	.5 TO 3 MILLION
	TERTIARY		PLIOCENE	Large carnivores	13 ± 1 MILLION
			MIOCENE	First abundant grazing mammals	25 ± 1 MILLION
			OLIGOCENE	Large running mammals	36 ± 2 MILLION
			EOCENE	Many modern types of mammals	58 ± 2 MILLION
			PALEOCENE	First placental mammals	63 ± 2 MILLION
MESOZOIC	CRETACEOUS			First flowering plants; climax of dinosaurs and ammonites, followed by extinction	135 ± 5 MILLION
	JURASSIC			First birds, first mammals; dinosaurs and ammonites abundant	180 ± 5 MILLION
	TRIASSIC			First dinosaurs. Abundant cycads and conifers	230 ± 10 MILLION
PALEOZOIC	PERMIAN			Extinction of many kinds of marine animals, including trilobites. Glaciation at low latitudes	280 ± 10 MILLION
	CARBON-IFEROUS	PENNSYLVANIAN		Great coal forests, conifers. First reptiles	310 ± 10 MILLION
		MISSISSIPPIAN		Sharks and amphibians abundant. Large and numerous scale trees and seed ferns	345 ± 10 MILLION
	DEVONIAN			First amphibians and ammonites; fishes abundant	405 ± 10 MILLION
	SILURIAN			First terrestrial plants and animals	425 ± 10 MILLION
	ORDOVICIAN			First fishes; invertebrates dominant	500 ± 10 MILLION
	CAMBRIAN			First abundant record of marine life; trilobites dominant, followed by extinction of about two-thirds of trilobite families	600 ± 50 MILLION
	PRE-CAMBRIAN			Fossils extremely rare, consisting of primitive aquatic plants. Evidence of glaciation. Oldest dated algae, over 2,600 million years.	

NUMBER OF FAMILIES

0 200 400 600 800 1,000

GEOLOGICAL AGES can be dated by comparing relative amounts of radioactive elements remaining in samples of rock obtained from different stratigraphic levels. The expanding curve at the right indicates how the number of major families of fossil animals increased through geologic time. The sharp decline after the Permian reflects the most dramatic of several mass extinctions.

simultaneous decline in the sea of the ammonite mollusks? Late in the Triassic there were still 25 families of widely ranging ammonites. All but one became extinct at the end of the period and that one gave rise to the scores of families of Jurassic and Cretaceous time.

The late Cretaceous extinctions eliminated about a quarter of all the known families of animals, but as usual the plants were little affected. The beginning of a decline in several groups is discernible near the middle of the period, some 30 million years before the mass extinction at the close of the Cretaceous. The significant point is that many characteristic groups—dinosaurs, marine reptiles, flying reptiles, ammonites, bottom-dwelling aquatic mollusks and certain kinds of extinct marine plankton—were represented by several world-wide families until the close of the period. Schindewolf has cited 16 superfamilies and orders that now became extinct. Many world-wide genera of invertebrates and most of the known species of the youngest Cretaceous period drop out near or at the boundary between the Cretaceous and the overlying Paleocene rocks. On the other hand, many families of bottom-dwelling sea organisms, fishes and nautiloid cephalopods survived with only minor evolutionary modifications. This is also true of primitive mammals, turtles, crocodiles and most of the plants of the time.

In general the groups that survived each of the great episodes of mass extinction were conservative in their evolution. As a result they were probably able to withstand greater changes in environment than could those groups that disappeared, thus conforming to the well-known principle of "survival of the unspecialized," recognized by Darwin. But there were many exceptions and it does not follow that the groups that disappeared became extinct simply because they were highly specialized. Many were no more specialized than some groups that survived.

The Cretaceous period was remarkable for a uniform and world-wide distribution of many hundreds of distinctive groups of animals and plants, which was probably a direct result of low-lying lands, widespread seas, surprisingly uniform climate and an abundance of migration routes. Just at the top of the Cretaceous sequence the characteristic fauna is abruptly replaced by another, which is distinguished not so much by radically new kinds of animals as by the elimination of innumerable major groups that had characterized the late Cre-

taceous. The geological record is somewhat obscure at the close of the Cretaceous, but most investigators agree that there was a widespread break in sedimentation, indicating a brief but general withdrawal of shallow seas from the area of the continents.

Extinctions in the Human Epoch

At the close of the Tertiary period, which immediately preceded the Quaternary in which we live, new land connections were formed between North America and neighboring continents. The horse and camel, which had evolved in North America through Tertiary time, quickly crossed into Siberia and spread throughout Eurasia and Africa. Crossing the newly formed Isthmus of Panama at about the same time, many North American animals entered South America. From Asia the mammoth, bison, bear and large deer entered North America, while from the south came ground sloths and other mammals that had originated and evolved in South America. Widespread migration and concurrent episodes of mass extinction appear to mark the close of the Pliocene (some two or three million years ago) and the middle of the Pleistocene in both North America and

Eurasia. Another mass extinction, particularly notable in North America, occurred at the very close of the last extensive glaciation, but this time it apparently was not outstandingly marked by intercontinental migrations. Surprisingly, none of the extinctions coincided with glacial advances.

It is characteristic of the fossil record that immigrant faunas tend to replace the old native faunas. In some cases newly arrived or newly evolved families replaced old families quite rapidly, in less than a few million years. In other cases the replacement has been a protracted process, spreading over tens of millions or even hundreds of millions of years. We cannot, of course, know the exact nature of competition between bygone groups, but when they occupied the same habitat and were broadly overlapping in their ecological requirements, it can be assumed that they were in fact competitors for essential resources. The selective advantage of one competing stock over another may be so slight that a vast amount of time is required to decide the outcome.

At the time of the maximum extent of the continental glaciers some 11,000 years ago the ice-free land areas of the Northern Hemisphere supported a rich

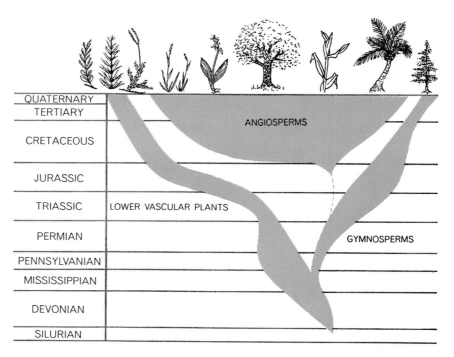

HISTORY OF LAND PLANTS shows the spectacular rise of angiosperms in the last 135 million years. The bands are roughly proportional to the number of genera of plants in each group. Angiosperms are flowering plants, a group that includes all the common trees (except conifers), grasses and vegetables. Lower vascular plants include club mosses, quillworts and horsetails. The most familiar gymnosperms (naked-seed plants) are the conifers, or evergreens. The diagram is based on one prepared by Erling Dorf of Princeton University.

and varied fauna of large mammals comparable to that which now occupies Africa south of the Sahara. Many of the species of bears, horses, elks, beavers and elephants were larger than any of their relatives living today. As recently as 8,000 years ago the horse, elephant and camel families roamed all the continents but Australia and Antarctica. Since that time these and many other families have retreated into small regions confined to one or two continents.

In North America a few species dropped out at the height of the last glaciation, but the tempo of extinction stepped up rapidly between about 12,000 and 6,000 years ago, with a maximum rate around 8,000 years ago, when the climate had become milder and the glaciers were shrinking [see illustration on page 152]. A comparable, but possibly more gradual, loss of large mammals occurred at about the same time in Asia and Australia, but not in Africa. Many of the large herbivores and carnivores had been virtually world-wide through a great range in climate, only to become extinct within a few hundred years. Other organisms were generally unaffected by this episode of extinction.

On the basis of a limited series of radiocarbon dates Paul S. Martin of the University of Arizona has concluded that now extinct large mammals of North America began to disappear first in Alaska and Mexico, followed by those in the Great Plains. Somewhat questionable datings suggest that the last survivors may have lived in Florida only 2,000 to 4,000 years ago. Quite recently, therefore, roughly three-quarters of the North American herbivores disappeared, and most of the ecological niches that were vacated have not been filled by other species.

Glaciation evidently was not a significant agent in these extinctions. In the first place, they were concentrated during the final melting and retreat of the continental glaciers after the entire biota had successfully weathered a number of glacial and interglacial cycles. Second, the glacial climate certainly did not reach low latitudes, except in mountainous areas, and it is probable that the climate over large parts of the tropics was not very different from that of today.

Studies of fossil pollen and spores in many parts of the world show that the melting of the continental glaciers was accompanied by a change from a rainy climate to a somewhat drier one with higher mean temperatures. As a result of these changes forests in many parts of the world retreated and were replaced by deserts and steppes. The changes, however, probably were not universal or severe enough to result in the elimination of any major habitat.

A number of investigators have proposed that the large mammals may have been hunted out of existence by prehistoric man, who may have used fire as a weapon. They point out that the mass extinctions coincided with the rapid growth of agriculture. Before this stage in human history a decrease in game supply would have been matched by a decrease in human populations, since man could not have destroyed a major food source without destroying himself.

In Africa and Eurasia, where man had lived in association with game animals throughout the Pleistocene, extinctions were not so conspicuously concentrated in the last part of the epoch. There was ample opportunity in the Old World for animals to become adapted to man through hundreds of thousands of years of coexistence. In the Americas and

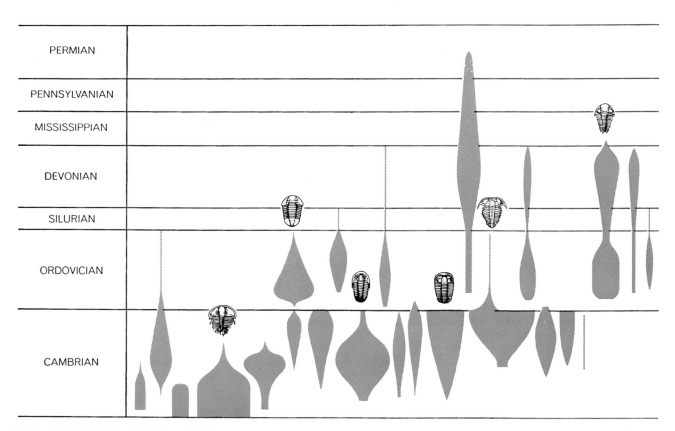

MASS EXTINCTION OF TRILOBITES, primitive arthropods, occurred at the close of the Cambrian period about 500 million years ago. During the Cambrian period hundreds of kinds of trilobites populated the shallow seas of the world. The chart depicts 15 superfamilies of Cambrian trilobites; the width of the shapes is roughly proportional to the number of members in each superfamily. Final extinction took place in the Permian. The chart is based on the work of H. B. Whittington of Harvard University.

Australia, where man was a comparative newcomer, the animals may have proved easy prey for the hunter.

We shall probably never know exactly what happened to the large mammals of the late Pleistocene, but their demise did coincide closely with the expansion of ancient man and with an abrupt change from a cool and moist to a warm and dry climate over much of the world. Possibly both of these factors contributed to this episode of mass extinction. We can only guess.

The Modern Crisis

Geological history cannot be observed but must be deduced from studies of stratigraphic sequences of rocks and fossils interpreted in the context of processes now operating on earth. It is helpful, therefore, to analyze some recent extinctions to find clues to the general causes of extinction.

We are now witnessing the disastrous effects on organic nature of the explosive spread of the human species and the concurrent development of an efficient technology of destruction. The human demand for space increases, hunting techniques are improved, new poisons are used and remote areas that had long served as havens for wildlife are now easily penetrated by hunter, fisherman, lumberman and farmer.

Studies of recent mammal extinctions show that man has been either directly or indirectly responsible for the disappearance, or near disappearance, of more than 450 species of animals. Without man's intervention there would have been few, if any, extinctions of birds or mammals within the past 2,000 years. The heaviest toll has been taken in the West Indies and the islands of the Pacific and Indian oceans, where about 70 species of birds have become extinct in the past few hundred years. On the continents the birds have fared somewhat better. In the same period five species of birds have disappeared from North America, three from Australia and one from Asia. Conservationists fear, however, that more North American birds will become extinct in the next 50 years than have in the past 5,000 years.

The savannas of Africa were remarkable until recently for a wealth of large mammals comparable only to the rich Tertiary and Pleistocene faunas of North America. In South Africa stock farming, road building, the fencing of grazing lands and indiscriminate hunting had wiped out the wild populations of large grazing mammals by the beginning of the 20th century. The depletion of ani-

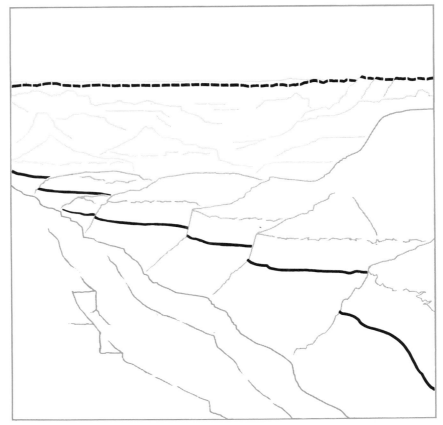

PALEONTOLOGICAL BOUNDARIES are clearly visible in this photograph of the Grand Canyon. The diagram below identifies the stratigraphic boundary between the Cambrian and Ordovician periods (*solid line*) and the top of the Permian rocks (*broken line*). These are world-wide paleontological division points, easily identified by marine fossils.

148

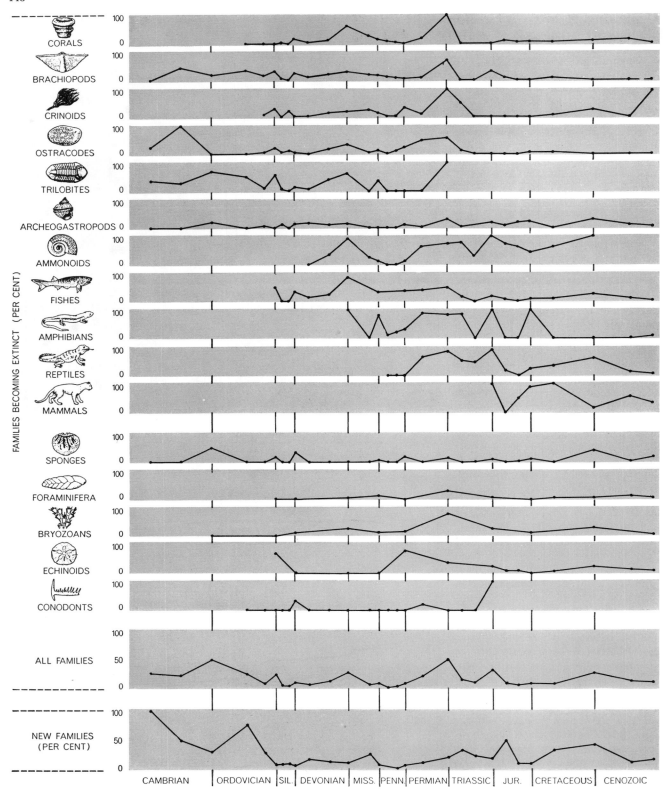

FAMILIES BECOMING EXTINCT (PER CENT)

CORALS

BRACHIOPODS

CRINOIDS

OSTRACODES

TRILOBITES

ARCHEOGASTROPODS

AMMONOIDS

FISHES

AMPHIBIANS

REPTILES

MAMMALS

SPONGES

FORAMINIFERA

BRYOZOANS

ECHINOIDS

CONODONTS

ALL FAMILIES

NEW FAMILIES (PER CENT)

CAMBRIAN | ORDOVICIAN | SIL. | DEVONIAN | MISS. | PENN. | PERMIAN | TRIASSIC | JUR. | CRETACEOUS | CENOZOIC

RECORD OF ANIMAL EXTINCTIONS makes it quite clear that the history of animals has been punctuated by repeated crises. The top panel of curves plots the ups and downs of 11 groups of animals from Cambrian times to the present. Massive extinctions took place at the close of the Ordovician, Devonian and Permian periods. The second panel shows the history of five other groups for which the evidence is less complete. (Curves are extrapolated between dots.) The next to bottom curve depicts the sum of extinctions for all the fossil groups plotted above (plus bivalves and caenogastropods). The bottom curve shows the per cent of new families in the main fossil groups. It indicates that periods of extinction were usually followed by an upsurge in evolutionary activity.

mals has now spread to Equatorial Africa as a result of poaching in and around the game reserves and the practice of eradicating game as a method of controlling human and animal epidemics. Within the past two decades it has become possible to travel for hundreds of miles across African grasslands without seeing any of the large mammals for which the continent is noted. To make matters worse, the great reserves that were set aside for the preservation of African wildlife are now threatened by political upheavals.

As a factor in extinction, man's predatory habits are supplemented by his destruction of habitats. Deforestation, cultivation, land drainage, water pollution, wholesale use of insecticides, the building of roads and fences—all are causing fragmentation and reduction in range of wild populations with resulting loss of environmental and genetic resources. These changes eventually are fatal to populations just able to maintain themselves under normal conditions. A few species have been able to take advantage of the new environments created by man, but for the most part the changes have been damaging.

Reduction of geographic range is prejudicial to a species in somewhat the same way as overpopulation. It places an increasing demand on diminishing environmental resources. Furthermore, the gene pool suffers loss of variability by reduction in the number of local breeding groups. These are deleterious changes, which can be disastrous to species that have narrow tolerances for one or more environmental factors. No organism is stronger than the weakest link in its ecological chain.

Man's direct attack on the organic world is reinforced by a host of competing and pathogenic organisms that he intentionally or unwittingly introduces to relatively defenseless native communities. Charles S. Elton of the University of Oxford has documented scores of examples of the catastrophic effects on established communities of man-sponsored invasions by pathogenic and other organisms. The scale of these ecological disturbances is world-wide; indeed, there are few unmodified faunas and floras now surviving.

The ill-advised introduction of predators such as foxes, cats, dogs, mongooses and rats into island communities has been particularly disastrous; many extinctions can be traced directly to this cause. Grazing and browsing domestic animals have destroyed or modified vegetation patterns. The introduction of

European mammals into Australia has been a primary factor in the rapid decimation of the native marsupials, which cannot compete successfully with placental mammals.

An illustration of invasion by a pathogenic organism is provided by an epidemic that in half a century has nearly wiped out the American sweet chestnut tree. The fungus infection responsible for this tragedy was accidentally introduced from China on nursery plants. The European chestnut, also susceptible to the fungus, is now suffering rapid decline, but the Chinese chestnut, which evolved in association with the blight, is comparatively immune.

Another example is provided by the marine eelgrass *Zostera*, which gives food and shelter to a host of invertebrates and fishes and forms a protective blanket over muddy bottoms. It is the

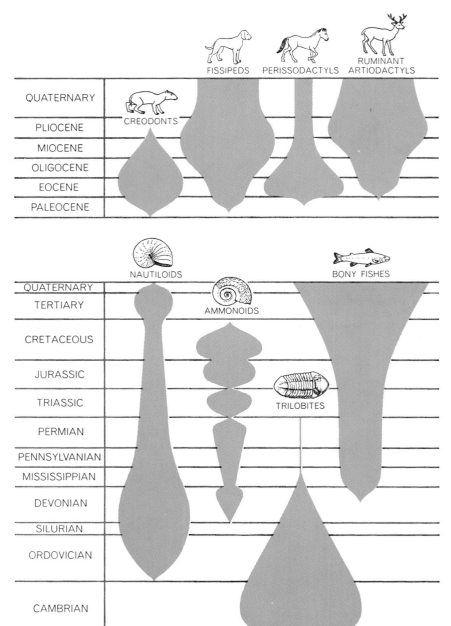

ECOLOGICAL REPLACEMENT appears to be a characteristic feature of evolution. The top diagram shows the breadth of family representation among four main groups of mammals over the last 60-odd million years. The bottom diagram shows a similar waxing and waning among four groups of marine swimmers, dating back to the earliest fossil records. The ammonoid group suffered near extinction twice before finally expiring. The diagrams are based on the work of George Gaylord Simpson of Harvard University and the author.

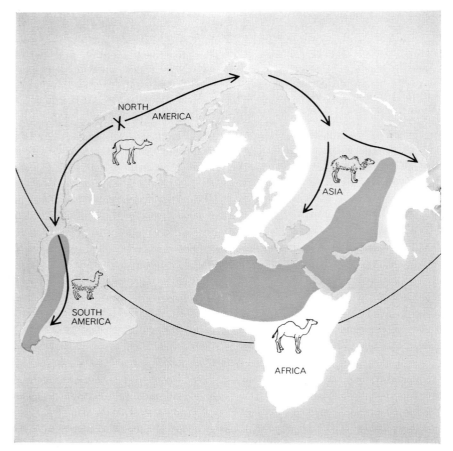

DISPERSAL OF CAMEL FAMILY from its origin (*X*) took place during Pleistocene times. Area in light color shows the maximum distribution of the family; dark color shows present distribution. This map is based on one in *Life: An Introduction to Biology*, by Simpson, C. S. Pittendrigh and L. H. Tiffany, published by Harcourt, Brace and Company.

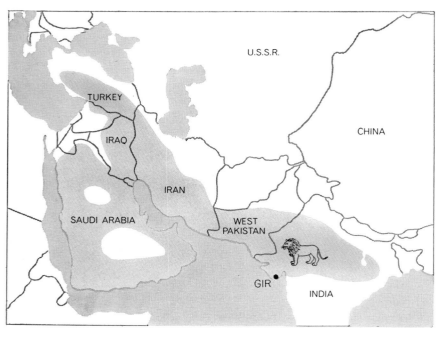

DISTRIBUTION OF ASIATIC LION has contracted dramatically just since 1800, when it roamed over large areas (*shown in color*) of the Middle East, Pakistan and India. Today the Asiatic lion is found wild only in Gir, a small game preserve in western India.

most characteristic member of a distinctive community that includes many plant and animal species. In the 1930's the eelgrass was attacked by a virus and was almost wiped out along the Atlantic shores of North America and Europe. Many animals and plants not directly attacked nevertheless disappeared for a time and the community was greatly altered. Resistant strains of *Zostera* fortunately escaped destruction and have slowly repopulated much of the former area. Eelgrass is a key member of a complex ecological community, and one can see that if it had not survived, many dependent organisms would have been placed in jeopardy and some might have been destroyed.

This cursory glance at recent extinctions indicates that excessive predation, destruction of habitat and invasion of established communities by man and his domestic animals have been primary causes of extinctions within historical time. The resulting disturbances of community equilibrium and shock waves of readjustment have produced ecological explosions with far-reaching effects.

The Causes of Mass Extinctions

It is now generally understood that organisms must be adapted to their environment in order to survive. As environmental changes gradually pass the limits of tolerance of a species, that species must evolve to cope with the new conditions or it will die. This is established by experiment and observation. Extinction, therefore, is not simply a result of environmental change but is also a consequence of failure of the evolutionary process to keep pace with changing conditions in the physical and biological environment. Extinction is an evolutionary as well as an ecological problem.

There has been much speculation about the causes of mass extinction; hypotheses have ranged from worldwide cataclysms to some kind of exhaustion of the germ plasm—a sort of evolutionary fatigue. Geology does not provide support for the postulated cataclysms and biology has failed to discover any compelling evidence that evolution is an effect of biological drive, or that extinction is a result of its failure. Hypotheses of extinction based on supposed racial old age or overspecialization, so popular among paleontologists a few generations ago and still echoed occasionally, have been generally abandoned for lack of evidence.

Of the many hypotheses advanced to explain mass extinctions, most are un-

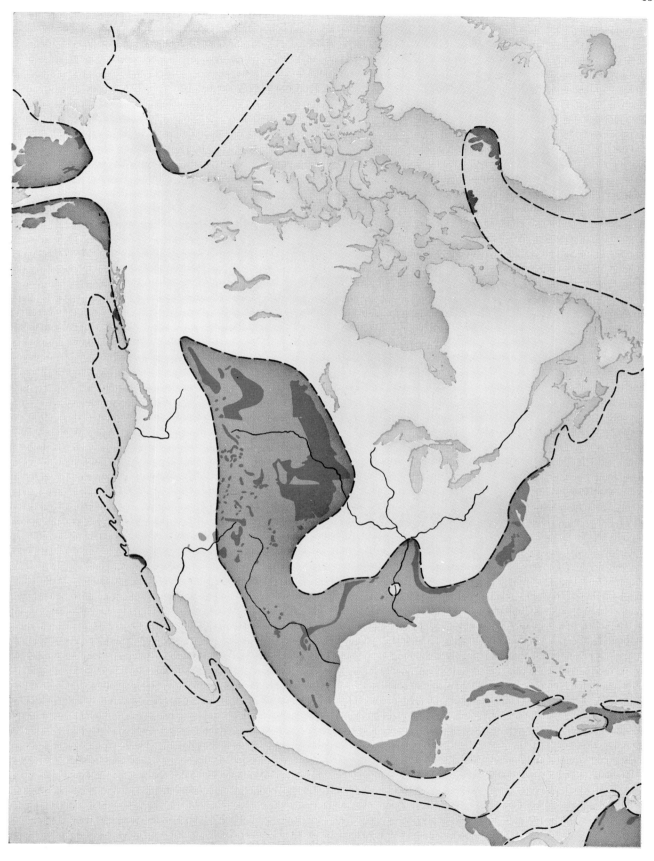

LATE CRETACEOUS SEA covered large portions of Central and North America (*dark gray*). Fossil-bearing rocks laid down at that time, and now visible at the surface of the earth, are shown in dark color. The approximate outline of North America in the Cretaceous period is represented by the broken line. The map is based on the work of the late Charles Schuchert of Yale University.

MASTODON

COLUMBIAN MAMMOTH

DIRE WOLF

CAMEL

HORSE

GIANT ARMADILLO

BISON OCCIDENTALIS

BISON ANTIQUUS

GIANT GROUND SLOTH

GROUND SLOTH

EXTINCT TAPIR

WOOLLY MAMMOTH

IMPERIAL MAMMOTH

DWARF ELEPHANT

EXTINCT PECCARY

SABER-TOOTHED CAT

14,000   13,000   12,000   11,000   10,000   9,000   8,000   7,000   6,000

YEARS BEFORE PRESENT

satisfactory because they lack testable corollaries and are designed to explain only one episode of extinction. For example, the extinction of the dinosaurs at the end of the Cretaceous period has been attributed to a great increase in atmospheric oxygen and alternatively to the explosive evolution of pathogenic fungi, both thought to be by-products of the dramatic spread of the flowering plants during late Cretaceous time.

The possibility that pathogenic fungi may have helped to destroy the dinosaurs was a recent suggestion of my own. I was aware, of course, that it would not be a very useful suggestion unless a way could be found to test it. I was also aware that disease is one of the most popular hypotheses for explaining mass extinctions. Unfortunately for such hypotheses, pathogenic organisms normally attack only one species or at most a few related species. This has been interpreted as an indication of a long antecedent history of coadaptation during which parasite and host have become mutually adjusted. According to this theory parasites that produce pathological reactions are not well adapted to the host. On first contact the pathogenic organism might destroy large numbers of the host species; it is even possible that extinction of a species might follow a pandemic, but there is no record that this has happened in historical times to any numerous and cosmopolitan group of species.

It is well to keep in mind that living populations studied by biologists generally are large, successful groups in which the normal range of variation provides tolerance for all the usual exigencies, and some unusual ones. It is for this reason that the eelgrass was not extinguished by the epidemic of the 1930's and that the human race was not eliminated by the influenza pandemic following World War I. Although a succession of closely spaced disasters of various kinds might have brought about extinction, the particular virus strains responsible for these diseases did not directly attack associated species.

Another suggestion, more ingenious

ICE-AGE MAMMALS provided North America with a fauna of large herbivores comparable to that existing in certain parts of Africa today. Most of them survived a series of glacial periods only to become extinct about 8,000 years ago, when the last glaciers were shrinking. The chart is based on a study by Jim J. Hester of the Museum of New Mexico in Santa Fe.

than most, is that mass extinctions were caused by bursts of high-energy radiation from a nearby supernova. Presumably the radiation could have had a dramatic impact on living organisms without altering the climate in a way that would show up in the geological record. This hypothesis, however, fails to account for the patterns of extinction actually observed. It would appear that radiation would affect terrestrial organisms more than aquatic organisms, yet there were times when most of the extinctions were in the sea. Land plants, which would be more exposed to the radiation and are more sensitive to it, were little affected by the changes that led to the animal extinctions at the close of the Permian and Cretaceous periods.

Another imaginative suggestion has been made recently by M. J. Salmi of the Geological Survey of Finland and Preston E. Cloud, Jr., of the University of Minnesota. They have pointed out that excessive amounts or deficiencies of certain metallic trace elements, such as copper and cobalt, are deleterious to organisms and may have caused past extinctions. This interesting hypothesis, as applied to marine organisms, depends on the questionable assumption that deficiencies of these substances have occurred in the ocean, or that a lethal concentration of metallic ions might have diffused throughout the oceans of the world more rapidly than the substances could be concentrated and removed from circulation by organisms and various common chemical sequestering agents. To account for the disappearance of land animals it is necessary to postulate further that the harmful elements were broadcast in quantity widely over the earth, perhaps as a result of a great volcanic eruption. This is not inconceivable; there have probably been significant variations of trace elements in time and place. But it seems unlikely that such variations sufficed to produce worldwide biological effects.

Perhaps the most popular of all hypotheses to explain mass extinctions is that they resulted from sharp changes in climate. There is no question that large-scale climatic changes have taken place many times in the past. During much of geologic time shallow seas covered large areas of the continents; climates were consequently milder and less differentiated than they are now. There were also several brief episodes of continental glaciation at low latitudes, but it appears that mass extinctions did not coincide with ice ages.

It is noteworthy that fossil plants,

which are good indicators of past climatic conditions, do not reveal catastrophic changes in climate at the close of the Permian, Triassic and Cretaceous periods, or at other times coincident with mass extinctions in the animal kingdom. On theoretical grounds it seems improbable that any major climatic zone of the past has disappeared from the earth. For example, climates not unlike those of the Cretaceous period probably have existed continuously at low latitudes until the present time. On the other hand, it is certain that there have been great changes in distribution of climatic zones. Severe shrinkage of a given climatic belt might adversely affect many of the contained species. Climatic changes almost certainly have contributed to animal extinctions by destruction of local habitats and by inducing wholesale migrations, but the times of greatest extinction commonly do not clearly correspond to times of great climatic stress.

Finally, we must consider the evidence that so greatly impressed Cuvier and many geologists. They were struck by the frequent association between the last occurrence of extinct animals and unconformities, or erosional breaks, in the geological record. Cuvier himself believed that the unconformities and the mass extinctions went hand in hand, that both were products of geologic revolutions, such as might be caused by paroxysms of mountain building. The idea still influences some modern thought on the subject.

It is evident that mountains do strongly influence the environment. They can alter the climate, soils, water supply and vegetation over adjacent areas, but it is doubtful that the mountains of past ages played dominant roles in the evolutionary history of marine and lowland organisms, which constitute most of the fossil record. Most damaging to the hypothesis that crustal upheavals played a major role in extinctions is the fact that the great crises in the history of life did not correspond closely in time with the origins of the great mountain systems. Actually the most dramatic episodes of mass extinction took place during times of general crustal quiet in the continental areas. Evidently other factors were involved.

Fluctuations of Sea Level

If mass extinctions were not brought about by changes in atmospheric oxygen, by disease, by cosmic radiation, by trace-element poisoning, by climatic changes or by violent upheavals of the earth's crust, where is one to look for a satisfactory—and testable—hypothesis?

The explanation I have come to favor, and which has found acceptance among many students of the paleontological record, rests on fluctuations of sea level. Evidence has been accumulating to show an intimate relation between many fossil zones and major advances and retreats of the seas across the continents. It is clear that diastrophism, or reshaping, of the ocean basins can produce universal changes in sea level. The evidence of long continued sinking of the sea floor under Pacific atolls and guyots (flat-topped submarine mountains) and the present high stand of the continents indicate that the Pacific basin has been subsiding differentially with respect to the land at least since Cretaceous time.

During much of Paleozoic and Mesozoic time, spanning some 540 million years, the land surfaces were much lower than they are today. An appreciable rise in sea level was sufficient to flood large areas; a drop of a few feet caused equally large areas to emerge, producing major environmental changes. At least 30 major and hundreds of minor oscillations of sea level have occurred in the past 600 million years of geologic time.

Repeated expansion and contraction of many habitats in response to alternate flooding and draining of vast areas of the continents unquestionably created profound ecological disturbances among offshore and lowland communities, and repercussions of these changes probably extended to communities deep inland and far out to sea. Intermittent draining of the continents, such as occurred at the close of many of the geologic epochs and periods, greatly reduced or eliminated the shallow inland seas that provided most of the fossil record of marine life. Many organisms adapted to the special estuarine conditions of these seas evidently could not survive along the more exposed ocean margins during times of emergence and they had disappeared when the seas returned to the continents. There is now considerable evidence that evolutionary diversification was greatest during times of maximum flooding of the continents, when the number of habitats was relatively large. Conversely, extinction and natural selection were most intense during major withdrawals of the sea.

It is well known that the sea-level oscillations of the Pleistocene epoch caused by waxing and waning of the continental glaciers did not produce numerous extinctions among shallow-water marine communities, but the situation was quite unlike that which prevailed during much of geological history. By Pleistocene times the continents stood high above sea level and the warm interior seas had long since disappeared. As a result the Pleistocene oscillations did not produce vast geographic and climatic changes. Furthermore, they were of short duration compared with major sea-level oscillations of earlier times.

Importance of Key Species

It might be argued that nothing less than the complete destruction of a habitat would be required to eliminate a world-wide community of organisms. This, however, may not be necessary. After thousands of years of mutual accommodation, the various organisms of a biological community acquire a high order of compatibility until a nearly steady state is achieved. Each species plays its own role in the life of the community, supplying shelter, food, chemical conditioners or some other resource in kind and amount needed by its neighbors. Consequently any changes involving evolution or extinction of species, or the successful entrance of new elements into the community, will affect the associated organisms in varying degrees and result in a wave of adjustments.

The strength of the bonds of interdependence, of course, varies with species, but the health and welfare of a community commonly depend on a comparatively small number of key species low in the community pyramid; the extinction of any of these is sure to affect adversely many others. Reduction and fragmentation of some major habitats, accompanied by moderate changes in climate and resulting shrinkage of populations, may have resulted in extinction of key species not necessarily represented in the fossil record. Disappearance of any species low in the pyramid of community organization, as, for example, a primary food plant, could lead directly to the extinction of many ecologically dependent species higher in the scale. Because of this interdependence of organisms a wave of extinction originating in a shrinking coastal habitat might extend to more distant habitats of the continental interior and to the waters of the open sea.

This theory, in its essence long favored by geologists but still to be fully developed, provides an explanation of the common, although not invariable, parallelism between times of widespread emergence of the continents from the seas and episodes of mass extinction that closed many of the chapters of geological history.

12

Plate Tectonics and the History of Life in the Oceans

by James W. Valentine and Eldridge M. Moores
April 1974

*The breakup of the ancient supercontinent of Pangaea
triggered a long-term evolutionary trend that has led
to the unprecedented variety of the present biosphere*

During the 1960's a conceptual revolution swept the earth sciences. The new world view fundamentally altered long established notions about the permanency of the continents and the ocean basins and provided fresh perceptions of the underlying causes and significance of many major features of the earth's mantle and crust. As a consequence of this revolution it is now generally accepted that the continents have greatly altered their geographic position, their pattern of dispersal and even their size and number. These processes of continental drift, fragmentation and assembly have been going on for at least 700 million years and perhaps for more than two billion years.

Changes of such magnitude in the relative configuration of the continents and the oceans must have had far-reaching effects on the environment, repatterning the world's climate and influencing the composition and distribution of life in the biosphere. These more or less continual changes in the environment must also have had profound effects on the course of evolution and accordingly on the history of life.

Natural selection, the chief mechanism by which evolution proceeds, is a very complex process. Although it is constrained by the machinery of inheritance, natural selection is chiefly an ecological process based on the relation between organisms and their environment. For any species certain heritable variations are favored because they are particularly well suited to survive and to reproduce in their prevailing environment. To answer the question of why any given group of organisms has evolved, then, one needs to understand two main factors. First, it is necessary to know what the ancestral organisms were that formed the "raw material" on which selection worked. And second, one must have some idea of the sequence of environmental conditions that led the ancestral stock to evolve along a particular pathway to a descendant group. Given these factors, one can then infer the organism-environment interactions that gave rise to the evolutionary events. The study of the relations between ancient organisms and their environment is called paleoecology.

The new ideas of continental drift that came into prominence in the 1960's revolve around the theory of plate tectonics. According to this theory, new sea floor and underlying mantle are currently being added to the crust of the earth at spreading centers under deep-sea ridges and in small ocean basins at rates of up to 10 centimeters per year. The sea floor spreads laterally away from these centers and eventually sinks into the earth's interior at subduction zones, which are marked by deep-sea trenches. Volcanoes are created by the consumption process and flank the trenches. The lithosphere, or rocky outer shell of the earth, therefore comprises several major plates that are generated at spreading centers and consumed at subduction zones. Most lithospheric plates bear one continent or more, which passively move with the plate on which they rest. Because the continents are too light to sink into the trenches they remain on the surface. Continents can fragment at new ridges, however, and hence oceans may appear across them. Conversely, continents can be welded together when they collide at the site of a trench. Thus continents may be assembled into supercontinents, fragmented into small continents and generally moved about the earth's surface as passive riders on plates. In tens or hundreds of millions of years entire oceans may be created or destroyed, and the number, size and dispersal pattern of continents may be vastly altered.

The record of such continental fragmentation and reassembly is evident as deformed regions in the earth's mountain belts, particularly those mountain belts that contain the rock formations known as ophiolites. These formations are characterized by a certain sequence of rocks consisting (from bottom to top) of ultramafic rock (a magnesium-rich rock composed mostly of olivine), gabbro (a coarse-grained basaltic rock), volcanic rocks and sedimentary rocks. The major ophiolite belts of the earth are believed to represent preserved fragments of vanished ocean basins [*see illustration on pages 156 and 157*]. The existence of such a belt within a continent (for example the Uralian belt in the U.S.S.R.) is evidence for the former presence there of an ocean basin separating two continental fragments that at some time in the past collided with each other and were welded into the single larger continent. The timing of such events as the opening of ocean basins, the dispersal of continents and the closing of oceans by continental collisions can accordingly be "read" from the geology of a given mountain system.

Of course, the biological environment is constantly being altered as well. For example, the changes in continental configuration will greatly affect the ocean currents, the temperature, the nature of seasonal fluctuations, the distribution of nutrients, the patterns of productivity and many other factors of fundamental importance to living organisms. Therefore evolutionary trends in marine animals must have varied through geologic time in response to the major environmental changes, as natural selection acted to adapt organisms to the new conditions.

It should in principle be possible to detect these changes in the fossil record. Indeed, paleontologists have long recognized that vast changes in the composition, distribution and diversity of marine life are well documented by the fossil record. Now for the first time, however, it is possible to reconstruct the sequence of environmental changes based on the theory of plate tectonics, to determine their environmental consequences and to attempt to correlate them with the sequence of faunal changes that is seen in the fossil record. Such a thorough reconstruction ultimately may explain many of the enigmatic faunal changes known for many years. Even at this early stage paleontologists have succeeded in shedding much new light on a number of major extinctions and diversifications of the past.

As a first step toward understanding the relation between plate tectonics and the history of life it is helpful to investigate the relations that exist today between marine life, the present pattern of continental drift and plate-tectonic theory. The vast majority of marine species (about 90 percent) live on the continental shelves or on shallow-water portions of islands or subsurface "rises" at depths of less than about 200 meters

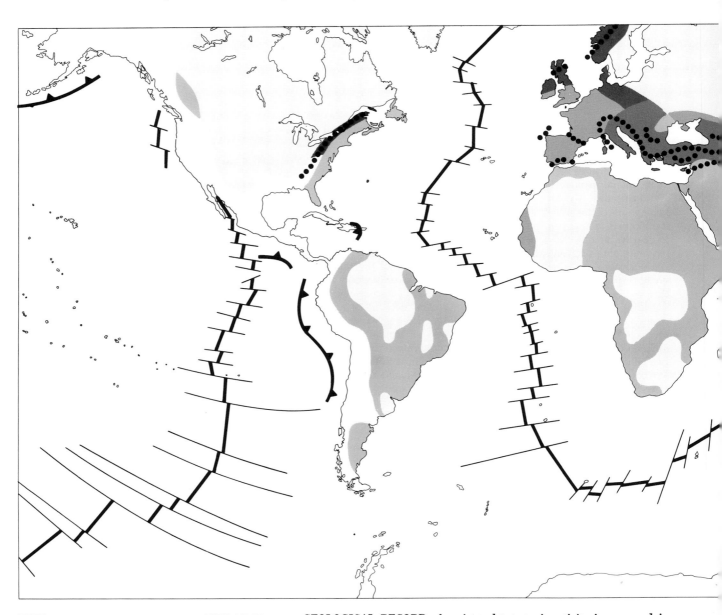

PRECAMBRIAN

PAN-AFRICAN-BAIKALIAN

CALEDONIAN

APPALACHIAN-HERCYNIAN

URALIAN

CORDILLERAN-TETHYAN

GEOLOGICAL RECORD of ancient plate-tectonic activity is preserved in certain deformed mountain belts (color), particularly those that contain the characteristic rock sequences known as ophiolites (black dots). The Pan-African–Baikalian belt, for example, is made up of rocks dating from 873 to 450 million years ago and may represent the assembly of all or nearly all the landmasses near the beginning of Phanerozoic time. This supercontinent may then have fragmented into four or more smaller continents, sometime just before and during the Cambrian period. The Caledonian mountain system may represent the collision of two continents at about late Silurian or early De-

(660 feet); most of the fossil record also consists of these faunas. Therefore it is the pattern of shallow-water sea-floor animal life that is of particular interest here.

The richest shallow-water faunas are found today at low latitudes in the Tropics, where communities are packed with vast numbers of highly specialized species. Proceeding to higher latitudes, diversity gradually falls; in the Arctic or Antarctic regions less than a tenth as many animals are living as in the Tropics, when comparable regions are considered [see illustration, pages 158 and 159]. The diversity gradient correlates well with a gradient in the stability of food supplies; as the seasons become more pronounced, fluctuations in primary productivity become greater. Although this strong latitudinal gradient dominates the earth's overall diversity pattern, there are important longitudinal diversity trends as well. In regions of similar latitude, for example, diversity is lower where there are sharp seasonal changes (such as variations in the surface-current pattern or in the upwelling of cold water) that affect the nutrient supply by causing large fluctuations in productivity.

At any given latitude, therefore, diversity is highest off the shores of small islands or small continents in large oceans, where fluctuations in nutrient supplies are least affected by the seasonal effects of landmasses, whereas diversity is lowest off large continents, particularly when they face small oceans, where shallow-water seasonal variations are greatest. In short, whereas latitudinal diversity increases generally from high latitudes to low, longitudinal diversity increases generally with distance from large continental landmasses. In both of these trends the increase in diversity is correlated with increasing stability of food resources. The resource-stability pattern depends largely on the shape of the continents and should also be sensitive to the extent of inland seas and to the presence of coastal mountains. Seas lying on continental platforms are particularly important: not only do extensive shallow seas provide much habitat area for shallow-water faunas but also such seas tend to damp seasonal climatic changes and to have an ameliorating influence on the local environment.

Today shallow marine faunas are highly provincial, that is, the species living in different oceans or on opposite sides of the same ocean tend to be quite different. Even along continuous coastlines there are major changes in species composition from place to place that generally correspond to climatic changes. The deep-sea floor, generated at oceanic ridges, forms a significant barrier to the dispersal of shallow-water organisms, and latitudinal climatic changes clearly form other barriers. The present dominantly north-south series of ridges forms a pattern of longitudinally alternating oceans and continents, thereby creating a series of barriers to shallow-water marine organisms. The steep latitudinal climatic gradient, on the other hand, creates chains of provinces along north-south coastlines. As a result the marine faunas today are partitioned into more than 30 provinces, among which there is in general only a low percentage of common species [see illustration on pages 160 and 161]. It is estimated that the shallow-water marine fauna represents more than 10 times as many species today as would be present in a world with only a single province, even a highly diverse one.

The volcanic arcs that appear over subduction zones form fairly continuous

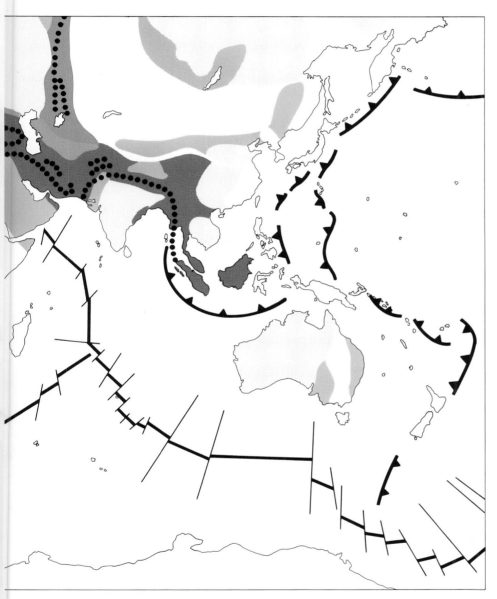

vonian time (approximately 400 million years ago). The Appalachian-Hercynian system may represent a two-continent collision during the late Carboniferous period (300 million years ago). The Uralian mountains may represent a similar collision at about Permo-Triassic time (220 million years ago). The Cordilleran-Tethyan system represents regions of Mesozoic mountain-building and includes the continental collisions that resulted in the Alpine-Himalayan mountain system. The ophiolite belts shown are the preserved remnants of ocean floor exposed in the mountain systems in question. Spreading ridges such as the Mid-Atlantic Ridge are indicated by heavy lines cut by lighter lines, which correspond to transform faults. Subduction zones are marked by heavy black curved lines with triangles.

island chains and provide excellent dispersal routes. When long island chains are arranged in an east-west pattern so as to lie within the same climatic zone, they are inhabited by wide-ranging faunas that are highly diverse for their latitude. Indeed, the widest ranging marine province, and also by far the most diverse, is the Indo-Pacific province, which is based on island arcs in its central regions. The faunal life of this province spills from these arcs onto tropical continental shelves in the west (India and East Africa) and also onto tropical intraplate volcanoes (the Polynesian and Micronesian islands) that are reasonably close to them. This vast tropical biota is cut off from the western American mainland by the East Pacific Barrier, a zoogeographic obstruction formed by a spreading ridge.

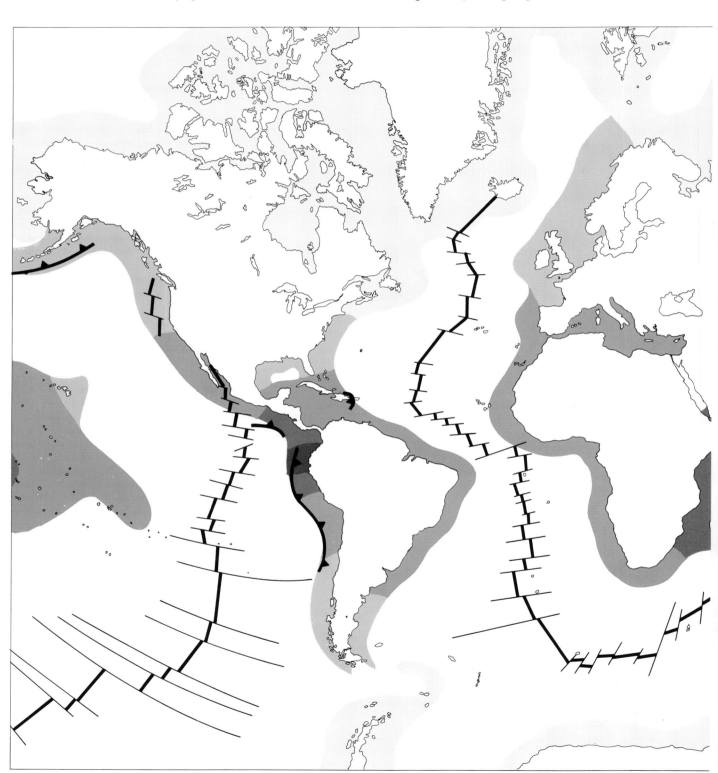

RELATIVE DIVERSITY of shallow-water, bottom-dwelling species in the present oceans is suggested by the colored patterns in this world map. The diversity classes are not based on absolute counts but are inferred from the diversity patterns of the best-

Since current patterns of marine provinciality and diversity fit closely with the present oceanic and continental geography and the resulting environmental patterns, one would expect ancient provinces and ancient diversity patterns also to fit past geographies. One

of the best-established of ancient geographies is the one that existed near the beginning of the Triassic period, about 225 million years ago. The continents were then assembled into a single supercontinent named Pangaea, which must have had a continuous shallow-water

margin running all the way around it, with no major physical barriers to the dispersal of shallow-water marine animals [see illustration on page 162]. Therefore provinciality must have been low compared with today, and it must have been attributable entirely to climatic effects. It is likely that the marine climate was quite mild and that even in high latitudes water temperatures were much warmer than they are today. As a result climatic provinciality must have been greatly reduced also. Furthermore, the seas at that time were largely confined to the ocean basins and did not extend significantly over the continental shelves. Thus the habitat area for shallow-water marine organisms was greatly reduced, first by the diminution of coastline that accompanies the creation of a supercontinent from smaller continents, and second by the general withdrawal of seas from continental platforms. The reduced habitat area would make for low species diversity. Finally, the extreme emergence of such a supercontinent would provide unstable nearshore conditions, with the result that food resources would have been very unstable compared with those of today. All these factors tend to reduce species diversity; hence one would expect to find that Triassic biotas were widespread and were made up of comparatively few species. That is precisely what the fossil record indicates.

Prior to the Triassic period, during the late Paleozoic, diversity appears to have been much higher [see top illustration on page 164]. It was sharply reduced again near the close of the Permian period during a vast wave of extinction that on balance is the most severe known to have been suffered by the marine fauna. The late Paleozoic species that were the more elaborately adapted specialists became virtually extinct, whereas the surviving descendants tended to have simple skeletons. A high proportion of these survivors appear to have been detritus feeders or suspension feeders that harvested the water layers just above the sea floor. These successful types seem to be ecologically similar to the populations found today in unstable environments, for instance in high latitudes; the unsuccessful specialists, on the other hand, seem ecologically similar to the populations found in stable environments, for instance in the Tropics. Thus the extinctions appear to have been caused by the reduced potential for diversity of the shallow seas, a trend associated with less provinciality, less habitat area and less stable environmental conditions.

In the period following the great extinction, as Pangaea broke up and the

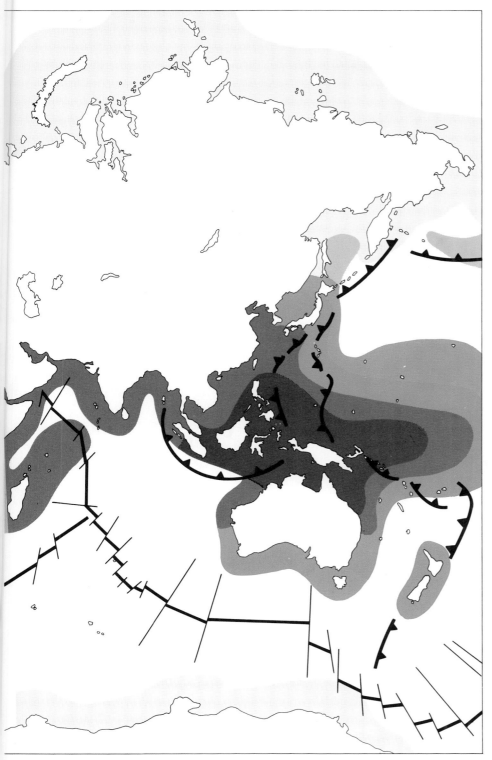

known skeletonized groups, chiefly the bivalves, gastropods, echinoids and corals. The highest class (darkest color) is about 20 times as diverse as the lowest (lightest color).

resulting continents themselves gradually fragmented and migrated to their present positions, provinciality increased, communities in stabilized regions became filled with numerous specialized animals and the overall diversity of species in the world ocean rose to un-

precedented heights, even though occasional waves of extinctions interrupted this long-range trend.

There is another time in the past besides the early Triassic period when low provinciality and low diversity were

coupled with the presence of a high proportion of detritus feeders and near-bottom suspension feeders. That is in the late Precambrian and Cambrian periods, when a widespread, soft-bodied fauna of low diversity gave way to a slightly provincialized, skeletonized fauna of some-

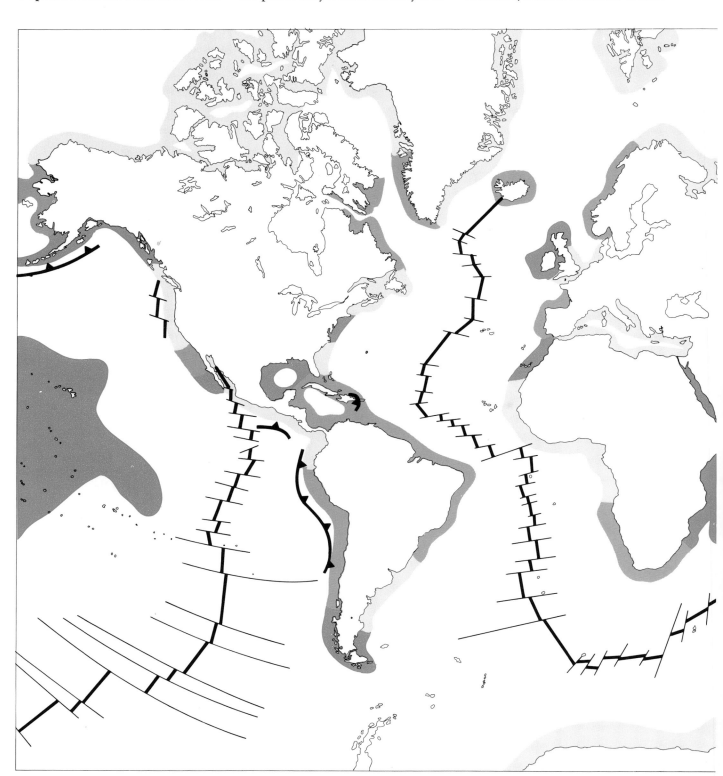

PRINCIPAL SHALLOW-WATER MARINE PROVINCES at present are indicated by the colored areas. The dominant north-

south chains of provinces along the continental coastlines are created by the present high latitudinal gradient in ocean temperature

what higher diversity. It seems likely that the late Precambrian environment was quite unstable and that there may well have been a supercontinent in existence, or at least that the continents then were collected into a more compact assemblage than at present. In the late Precambrian period one finds the first unequivocal records of invertebrate life, including burrowing forms that were probably coelomic, or hollow-bodied, worms. In the Cambrian four continents may have existed although they were not arranged in the present pattern. During the Cambrian a skeletonized fauna appears that is at first almost entirely surface-dwelling and that includes chiefly detritus-feeding and suspension-feeding forms, probably with some browsers.

It seems possible, therefore, that the late Precambrian species were adapted to highly unstable conditions and became diversified chiefly as a bottom-living, detritus-feeding assemblage. The coelomic body cavity, evidently a primitive adaptation for burrowing, was developed and diversified into a variety of forms, perhaps as many as five basic ones: highly segmented worms that lived under the ocean floor and were detritus feeders; slightly segmented worms that lived attached to the ocean floor and were suspension feeders; slightly segmented worms that lived attached to the ocean floor and were detritus feeders; "pseudosegmented" worms that lived on the ocean floor and were detritus feeders or browsers, and nonsegmented worms that lived under the ocean floor and fed by means of an "introvert." In addition to these coelomates there were a number of coelenterate stocks (such as corals, sea anemones and jellyfishes) and probably also flatworms and other noncoelomate worms.

From the chiefly wormlike coelomate stocks higher forms of animal life have originated; many of them appear in the Cambrian period, when they evidently first became organized into the groups that characterize them today. Animals with skeletons appeared in the fossil record at that time. Presumably the invasion of the sea-floor surface by coelomates and the origin of numerous skeletonized species accompanied a general amelioration of environmental conditions as the continents became dispersed; the skeletons themselves can be viewed as adaptations required for worms to lead various modes of life on the surface of the sea floor rather than under it. The sudden appearance of skeletons in the fossil record therefore is associated with a generalized elaboration of the bottom-dwelling members of the marine ecosystem. Later, free-swimming and underground lineages developed from the skeletonized ocean-floor dwellers, with the result that skeletons became general in all marine environments.

The correlation of major events in the history of life with major environmental changes inferred from plate-tectonic processes is certainly striking. Even though details of the interpretation are still provisional, it seems certain that further work on this relation will prove fruitful. Indeed, the ability of geologists

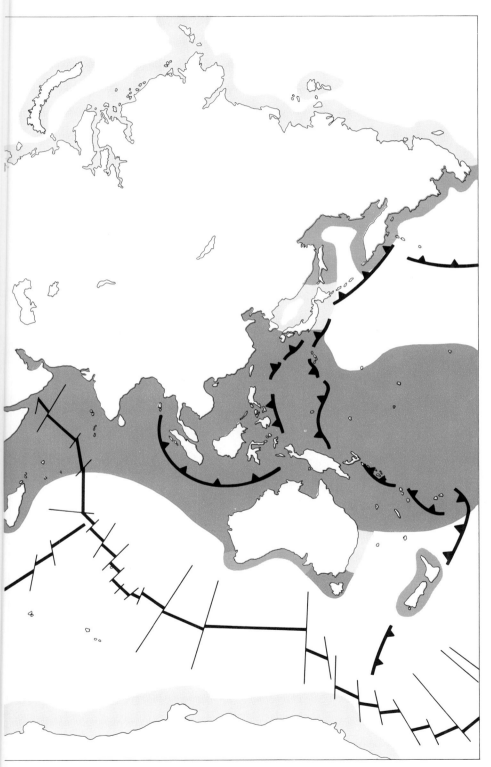

and by the undersea barriers formed by spreading ridges. The vast Indo-Pacific province (*darkest color*) spills out onto scattered islands as indicated. There are 31 provinces shown.

to determine past continental geographies should provide the basis for reconstructing the historical sequence of global environmental conditions for the first time. That sequence can then be compared with the sequence of organisms revealed in the fossil record. The following tentative account of such a comparison, on the broadest scale and without detail, will indicate the kind of history that is emerging; it is based on the examples reviewed above and on similar considerations.

Before about 700 million years ago bottom-dwelling, multicellular animals had developed that somewhat resembled flatworms. As yet no fossil evidence for their evolutionary pathways exists, but evidence from embryology and comparative anatomy suggests that they arose from swimming forms, possibly larval jellyfish, which in turn evolved from primitive single-celled animals.

Approximately 700 million years ago, perhaps in response to the onset of fluctuating environmental conditions brought about by continental clustering, a true coelomic body cavity was evolved to act as a hydrostatic skeleton in roundworms; this adaptation allowed burrowing in soft sea floors and led to the diversification of a host of worm architectures as that mode of life was explored. Burrows of this type are still preserved in some late Precambrian rocks. As the environment later became more stable, several of the worm lineages evolved more varied modes of life. The changes in body plan necessary to adapt to such

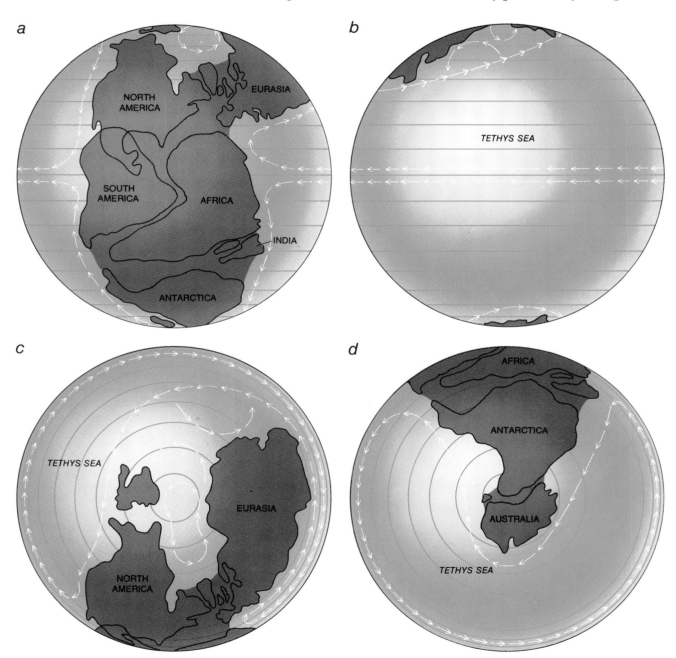

ANCIENT OCEAN CURRENTS in the vicinity of Pangaea, the single "supercontinent" that is believed to have existed near the beginning of the Triassic period some 225 million years ago, are indicated here in two equatorial views (a, b) and two polar views (c, d). Owing to a combination of geographic and environmental factors, including the predominantly warm-water currents shown, one would expect the continuous shallow-water margin that surrounded Pangaea to have been populated by comparatively few but widespread species. Such low species diversity combined with low provinciality is precisely what the fossil record indicates.

a life commonly involved the development of a skeleton. There were evidently three or four main types of worms that are represented by skeletonized descendants today. One type was highly segmented like earthworms, and presumably burrowed incessantly for detrital food; these were represented in the Cambrian period by the trilobites and related species. A second type was segmented into two or three coelomic compartments and burrowed weakly for domicile, afterward filtering suspended food from the seawater just above the ocean floor; these evolved into such forms as brachiopods and bryozoans. A third type consisted of long-bodied creepers with a series of internal organs but without true segmentation; from these the classes of mollusks (such as snails, clams and cephalopods) have descended. Probably a fourth type consisted of unsegmented burrowers that fed on surface detritus and gave rise to the modern sipunculid worms. These may also have given rise to the echinoderms (which include the sea cucumber and the spiny sea urchin), and eventually to the chordates and to man. Although the lines of descent are still uncertain among these primitive and poorly known groups, the adaptive steps are becoming clearer.

The major Cambrian radiation of the underground species into sea-floor surface habitats established the basic evolutionary lineages and occupied the major marine environments. Further evolutionary episodes tended to modify these basic animals into more elaborate structures. After the Cambrian period shallow-water marine animals became more highly specialized and richer in species, suggesting a continued trend toward resource stabilization. Suspension feeders proliferated and exploited higher parts of the water column, and predators also became more diversified. This trend seems to have reached a peak (or perhaps a plateau) in the Devonian period, some 375 million years ago. The characteristic Paleozoic fauna was finally swept away during the reduction in diversity that accompanied the great Permian-Triassic extinctions. Thus the rise of the Paleozoic fauna accompanied an amelioration in environmental conditions and increased provinciality, whereas the decline of the fauna accompanied a reestablishment of severe, unstable conditions and decreased provinciality. The subsequent breakup and dispersal of the continents has led to the present biosphere.

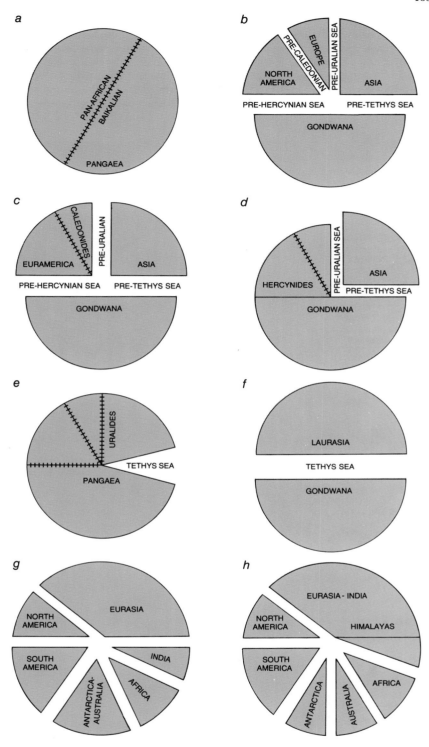

SIMPLIFIED DIAGRAMS are employed to suggest the relative configuration of the continents and the oceans during the past 700 million years. The late Precambrian supercontinent (*a*), which probably existed some 700 million years ago, may have been formed from previously separate continents. The Cambrian world (*b*) of about 570 million years ago consisted of four continents. The Devonian period (*c*) of about 390 million years ago was distinguished by three continents following the collapse of the pre-Caledonian Ocean and the collision of ancient Europe and North America. In the late Carboniferous period (*d*), about 300 million years ago, Euramerica became welded to Gondwana along the Hercynian belt. In the late Permian period (*e*), about 225 million years ago, Asia was welded to the remaining continents along the Uralian belt to form Pangaea. In early Mesozoic time (*f*), about 190 million years ago, Laurasia and Gondwana were more or less separate. In the late Cretaceous period (*g*), about 70 million years ago, Gondwana was highly fragmented and Laurasia partially so. The present continental pattern (*h*) shows India welded to Eurasia.

Today we live in a highly diverse world, probably harboring as many species as have ever lived at any time, associated in a rich variety of communities and a large number of provinces, probably the richest and largest ever to have existed at one time. We have been furnished with an enviably diverse and interesting biosphere; it would be a tragedy if we were to so perturb the environment as to return the biosphere to a low-diversity state, with the concomitant extinction of vast arrays of species. Of course, natural processes might eventually recoup the lost diversity, if we waited patiently for perhaps a few tens of millions of years. Alternatively we can work to preserve the environment in its present state and therefore to preserve the richness and variety of nature.

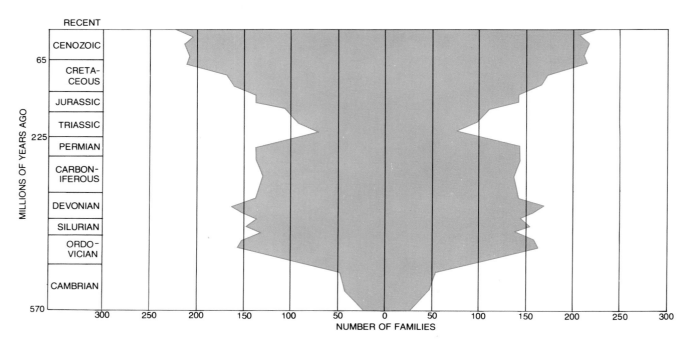

FLUCTUATIONS in the number of families, and hence in the level of diversity, of well-skeletonized invertebrates living on the world's continental shelves during the past 570 million years are plotted by geologic epoch in this graph. Time proceeds upward.

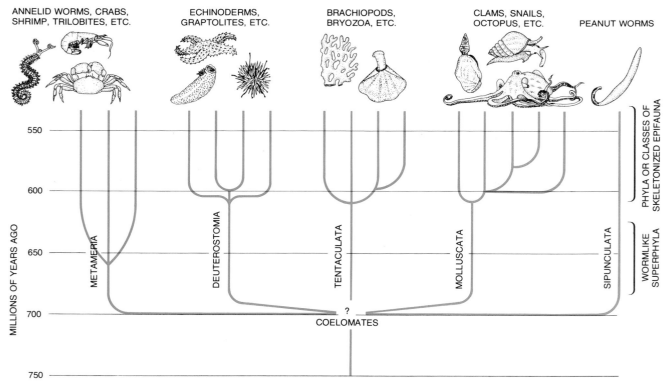

PHYLOGENETIC MODEL of the evolution of coelomate, or hollow-bodied, marine organisms is based on inferred adaptive pathways. The late Precambrian lineages were chiefly worms, which gave rise to epifaunal (bottom-dwelling), skeletonized phyla during the Cambrian period. The organisms depicted in the drawings at top are modern descendants of the major Cambrian lineages.

Continental Drift and Evolution

13

by Björn Kurtén
March 1969

*The breakup of ancient supercontinents would have
had major effects on the evolution of living organisms.
Does it explain the difference in the diversification of
reptiles and mammals?*

The history of life on the earth, as it is revealed in the fossil record, is characterized by intervals in which organisms of one type multiplied and diversified with extraordinary exuberance. One such interval is the age of reptiles, which lasted 200 million years and gave rise to some 20 reptilian orders, or major groups of reptiles. The age of reptiles was followed by our own age of mammals, which has lasted for 65 million years and has given rise to some 30 mammalian orders.

The difference between the number of reptilian orders and the number of mammalian ones is intriguing. How is it that the mammals diversified into half again as many orders as the reptiles in a third of the time? The answer may lie in the concept of continental drift, which has recently attracted so much attention from geologists and geophysicists [see "The Confirmation of Continental Drift," by Patrick M. Hurley; SCIENTIFIC AMERICAN Offprint 874]. It now seems that for most of the age of reptiles the continents were collected in two supercontinents, one in the Northern Hemisphere and one in the Southern. Early in the age of mammals the two supercontinents had apparently broken up into the continents of today but the present connections between some of the continents had not yet formed. Clearly such events would have had a profound effect on the evolution of living organisms.

The world of living organisms is a world of specialists. Each animal or plant has its special ecological role. Among the mammals of North America, for instance, there are grass-eating prairie animals such as the pronghorn antelope, browsing woodland animals such as the deer, flesh-eating animals specializing in large game, such as the mountain lion, or in small game, such as the fox, and so on. Each order of mammals

comprises a number of species related to one another by common descent, sharing the same broad kind of specialization and having a certain physical resemblance to one another. The order Carnivora, for example, consists of a number of related forms (weasels, bears, dogs, cats, hyenas and so on), most of which are flesh-eaters. There are a few exceptions (the aardwolf is an insect-eating hyena and the giant panda lives on bamboo shoots), but these are recognized as late specializations.

Radiation and Convergence

In spite of being highly diverse, all the orders of mammals have a common origin. They arose from a single ancestral species that lived at some unknown time in the Mesozoic era, which is roughly synonymous with the age of reptiles. The American paleontologist Henry Fairfield Osborn named the evolution of such a diversified host from a single ancestral type "adaptive radiation." By adapting to different ways of life—walking, climbing, swimming, flying, plant-eating, flesh-eating and so on—the descendant forms come to diverge more and more from one another. Adaptive radiation is not restricted to mammals; in fact we can trace the process within every major division of the plant and animal kingdoms.

The opposite phenomenon, in which stocks that were originally very different gradually come to resemble one another through adaptation to the same kind of life, is termed convergence. This too seems to be quite common among mammals. There is a tendency to duplication—indeed multiplication—of orders performing the same function. Perhaps the most remarkable instance is found among the mammals that have specialized in large-scale predation on termites

and ants in the Tropics. This ecological niche is filled in South America by the ant bear *Myrmecophaga* and its related forms, all belonging to the order Edentata. In Asia and Africa the same role is played by mammals of the order Pholidota: the pangolins, or scaly anteaters. In Africa a third order has established itself in this business: the Tubulidentata, or aardvarks. Finally, in Australia there is the spiny anteater, which is in the order Monotremata. Thus we have members of four different orders living the same kind of life.

One can cite many other examples. There are, for instance, several living and extinct orders of hoofed herbivores. There are two living orders (the Rodentia, or rodents, and the Lagomorpha, or rabbits and hares) whose chisel-like incisor teeth are specialized for gnawing. Some extinct orders specialized in the same way, and an early primate, an ice-age ungulate and a living marsupial have also intruded into the "rodent niche" [see top illustration on page 169]. This kind of duplication, or near-duplication, is an essential ingredient in the richness of the mammalian life that unfolded during the Cenozoic era, or the age of mammals. Of the 30 or so orders of land-dwelling mammals that appeared during this period almost two-thirds are still extant.

The Reptiles of the Cretaceous

The 65 million years of the Cenozoic are divided into two periods: the long Tertiary and the brief Quaternary, which includes the present [see illustration on page 167]. The 200-million-year age of reptiles embraces the three periods of the Mesozoic era (the Triassic, the Jurassic and the Cretaceous) and the final period (the Permian) of the preceding era. It is instructive to compare the number

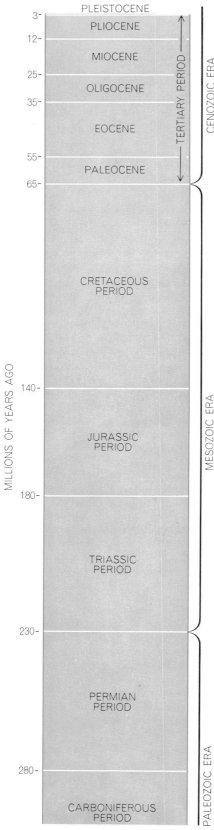

SIX PERIODS of earth history were occupied by the age of reptiles and the age of
mammals. The reptiles' rise began 280 million years ago, in the final period of the
Paleozoic era. Mammals replaced reptiles as
dominant land animals 65 million years ago.

of reptilian orders that flourished during
some Mesozoic interval about as long as
the Cenozoic era with the number of
mammalian orders in the Cenozoic. The
Cretaceous period is a good candidate.
Some 75 million years in duration, it is
only slightly longer than the age of mammals. Moreover, the Cretaceous was the
culmination of reptilian life and its fossil
record on most continents is good. In the
Cretaceous the following orders of land
reptiles were extant:

Order Crocodilia: crocodiles, alligators and the like. Their ecological role
was amphibious predation; their size,
medium to large.

Order Saurischia: saurischian dinosaurs. These were of two basic types:
bipedal upland predators (Theropoda)
and very large amphibious herbivores
(Sauropoda).

Order Ornithischia: ornithischian dinosaurs. Here there were three basic
types: bipedal herbivores (Ornithopoda),
heavily armored quadrupedal herbivores
(Stegosauria and Ankylosauria) and
horned herbivores (Ceratopsia).

Order Pterosauria: flying reptiles.

Order Chelonia: turtles and tortoises.

Order Squamata: The two basic types
were lizards (Lacertilia) and snakes (Ser-
pentes). Both had the same principal
ecological role: small to medium-sized
predator.

Order Choristodera (or suborder in
the order Eosuchia): champsosaurs.
These were amphibious predators.

One or two other reptilian orders may
be represented by rare forms. Even if
we include them, we get only eight or
nine orders of land reptiles in Cretaceous
times. One could maintain that an order
of reptiles ranks somewhat higher than
an order of mammals; some reptilian orders include two or even three basic
adaptive types. Even if these types are
kept separate, however, the total rises
only to 12 or 13. Furthermore, there
seems to be only one clear-cut case of
ecological duplication: both the crocodilians and the champsosaurs are sizable
amphibious predators. (The turtles cannot be considered duplicates of the armored dinosaurs. For one thing, they
were very much smaller.) A total of
somewhere between seven and 13 orders
over a period of 75 million years seems
a sluggish record compared with the
mammalian achievement of perhaps 30
orders in 65 million years. What light
can paleogeography shed on this matter?

The Mesozoic Continents

The two supercontinents of the age of
reptiles have been named Laurasia (after

Laurentian and Eurasia) and Gondwanaland (after a characteristic geological
formation, the Gondwana). Between
them lay the Tethys Sea (named for the
wife of Oceanus in Greek myth, who
was mother of the seas). Laurasia, the
northern supercontinent, consisted of
what would later be North America,
Greenland and Eurasia north of the Alps
and the Himalayas. Gondwanaland, the
southern one, consisted of the future
South America, Africa, India, Australia
and Antarctica. The supercontinents
may have begun to split up as early as
the Triassic period, but the rifts between
them did not become effective barriers
to the movement of land animals until
well into the Cretaceous, when the age
of reptiles was nearing its end.

When the mammals began to diversify in the late Cretaceous and early Tertiary, the separation of the continents appears to have been at an extreme. The
old ties were sundered and no new ones
had formed. The land areas were further
fragmented by a high sea level; the waters flooded the continental margins and
formed great inland seas, some of which
completely partitioned the continents.
For example, South America was cut in
two by water in the region that later
became the Amazon basin, and Eurasia
was split by the joining of the Tethys
Sea and the Arctic Ocean. In these circumstances each chip of former supercontinent became the nucleus for an
adaptive radiation of its own, each fos-
tering a local version of a balanced
fauna. There were at least eight such
nuclei at the beginning of the age of
mammals. Obviously such a situation is
quite different from the one in the age
of reptiles, when there were only two
separate land masses.

Where the Reptiles Originated

The fossil record contains certain clues
to some of the reptilian orders' probable
areas of origin. The immense distance in
time and the utterly different geography,
however, make definite inferences haz-
ardous. Let us see what can be said
about the orders of Cretaceous reptiles
(most of which, of course, arose long before the Cretaceous):

Crocodilia. The earliest fossil crocodilians appear in Middle Triassic formations in a Gondwanaland continent
(South America). The first crocodilians
in Laurasia are found in Upper Triassic
formations. Thus a Gondwanaland origin is suggested.

Saurischia. The first of these dinosaurs appear on both supercontinents in
the Middle Triassic, but they are more

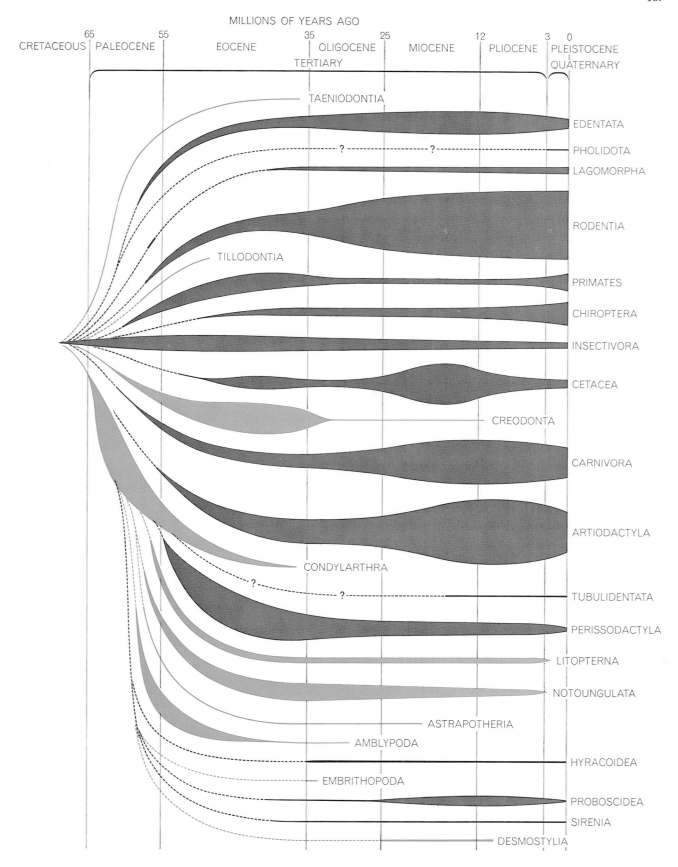

MILLIONS OF YEARS AGO

65		55			35	25		12		3	0
CRETACEOUS	PALEOCENE		EOCENE		OLIGOCENE	MIOCENE		PLIOCENE		PLEISTOCENE	
					TERTIARY					QUATERNARY	

TAENIODONTIA

EDENTATA

PHOLIDOTA

LAGOMORPHA

RODENTIA

TILLODONTIA

PRIMATES

CHIROPTERA

INSECTIVORA

CETACEA

CREODONTA

CARNIVORA

ARTIODACTYLA

CONDYLARTHRA

TUBULIDENTATA

PERISSODACTYLA

LITOPTERNA

NOTOUNGULATA

ASTRAPOTHERIA

AMBLYPODA

HYRACOIDEA

EMBRITHOPODA

PROBOSCIDEA

SIRENIA

DESMOSTYLIA

ADAPTIVE RADIATION of the mammals has been traced from its starting point late in the Mesozoic era by Alfred S. Romer of Harvard University. Records for 25 extinct and extant orders of placental mammals are shown here. The lines increase and decrease in width in proportion to the abundance of each order. Extinct orders are shown in color; broken lines mean that no fossil record exists during the indicated interval and question marks imply doubt about the suggested ancestral relation between some orders.

varied in the south. A Gondwanaland origin is very tentatively suggested.

Ornithischia. These dinosaurs appear in the Upper Triassic of South Africa (Gondwanaland) and invade Laurasia somewhat later. A Gondwanaland origin is indicated.

Pterosauria. The oldest fossils of flying reptiles come from the early Jurassic of Europe. They represent highly specialized forms, however, and their antecedents are unknown. No conclusion seems possible.

Chelonia. Turtles are found in Triassic formations in Laurasia. None are found in Gondwanaland before Cretaceous times. This suggests a Laurasian origin. On the other hand, a possible forerunner of turtles appears in the Permian of South Africa. If the Permian form was in fact ancestral, a Gondwanaland origin would be indicated. In any case, the order's main center of evolution certainly lay in the northern supercontinent.

Squamata. Early lizards are found in the late Triassic of the north, which may suggest a Laurasian origin. Unfortunately the lizards in question are aberrant gliding animals. They must have had a long history, of which we know nothing at present.

Choristodera. The crocodile-like champsosaurs are found only in North America and Europe, and so presumably originated in Laurasia.

The indications are, then, that three orders of reptiles—the crocodilians and the two orders of dinosaurs—may have originated in Gondwanaland. Three others—the turtles, the lizards and snakes and the champsosaurs—may have originated in Laurasia. The total number of basic adaptive types in the Gondwanaland group is six; the Laurasia group has four. The Gondwanaland radiation may well have been slightly richer than the Laurasian because it seems that the southern supercontinent was somewhat larger and had a slightly more varied climate. Laurasian climates seem to have been tropical to temperate. Southern parts of Gondwanaland were heavily glaciated late in the era preceding the Mesozoic, and its northern shores (facing the Tethys Sea) had a fully tropical climate.

Although some groups of reptiles, such as the champsosaurs, were confined to one or another of the supercontinents, most of the reptilian orders sooner or later spread into both of them. This means that there must have been ways for land animals to cross the Tethys Sea. The Tethys was narrow in the west and wide in the east. Presumably whatever land connection there was—a true land bridge or island stepping-stones—was located in the western part of the sea. In any case, migration along such routes meant that there was little local differentiation among the reptiles of the Mesozoic era. It was over an essentially uniform reptilian world that the sun finally set at the end of the age of reptiles.

Early Mammals of Laurasia

The conditions of mammalian evolution were radically different. In early and middle Cretaceous times the connections between continents were evidently close enough for primitive mammals to spread into all corners of the habitable world. As the continents drifted farther apart, however, populations of these primitive forms were gradually isolated from one another. This was particularly the case, as we shall see, with the mammals that inhabited the daughter continents of Gondwanaland. Among the Laurasian continents North America was drifting away from Europe, but at the beginning of the age of mammals the distance was not great and there is good evidence that some land connection remained well into the early Tertiary. North American and European mammals were practically identical as late as early Eocene times. Furthermore, throughout the Cenozoic era there was a connection between Alaska and Siberia, at least intermittently, across the Bering Strait. On the other hand, the inland sea extending from the Tethys to the Arctic Ocean formed a complete barrier to direct migration between Europe and Asia in the early Tertiary. Migrations could take place only by way of North America.

In this way the three daughter continents of ancient Laurasia formed three semi-isolated nuclear areas. Many orders of mammals arose in these Laurasian nuclei, among them seven orders that are now extinct but that covered a wide spectrum of specialized types, including primitive hoofed herbivores, carnivores, insectivores and gnawers. The orders of mammals that seem to have arisen in the northern daughter continents and that are extant today are:

Insectivora: moles, hedgehogs, shrews and the like. The earliest fossil insectivores are found in the late Cretaceous of North America and Asia.

Chiroptera: bats. The earliest-known bat comes from the early Eocene of North America. At a slightly later date bats were also common in Europe.

Primates: prosimians (for example, tarsiers and lemurs), monkeys, apes, man. Early primates have recently been found in the late Cretaceous of North America. In the early Tertiary they are common in Europe as well.

Carnivora: cats, dogs, bears, weasels and the like. The first true carnivores appear in the Paleocene of North America.

Perissodactyla: horses, tapirs and other odd-toed ungulates. The earliest forms appear at the beginning of the Eocene in the Northern Hemisphere.

Artiodactyla: cattle, deer, pigs and other even-toed ungulates. Like the odd-toed ungulates, they appear in the early Eocene of the Northern Hemisphere.

Rodentia: rats, mice, squirrels, beavers and the like. The first rodents appear in the Paleocene of North America.

Lagomorpha: hares and rabbits. This order makes its first appearance in the Eocene of the Northern Hemisphere.

Pholidota: pangolins. The earliest come from Europe in the middle Tertiary.

The fact that a given order of mammals is found in older fossil deposits in North America than in Europe or Asia does not necessarily mean that the order arose in the New World. It may simply reflect the fact that we know much more about the early mammals of North America than we do about those of Eurasia. All we can really say is that a total of 16 extant or extinct orders of mammals probably arose in the Northern Hemisphere.

Early Mammals of South America

The fragmentation of Gondwanaland seems to have started earlier than that of Laurasia. The rifting certainly had a much more radical effect. Looking at South America first, we note that at the beginning of the Tertiary this continent was tenuously connected to North America but that for the rest of the period it was completely isolated. The evidence for the tenuous early linkage is the presence in the early Tertiary beds of North America of mammalian fossils representing two predominantly South American orders: the Edentata (the order that includes today's ant bears, sloths and armadillos) and the Notoungulata (an order of extinct hoofed herbivores).

Four other orders of mammals are exclusively South American: the Paucituberculata (opossum rats and other small South American marsupials), the Pyrotheria (extinct elephant-like animals), the Litopterna (extinct hoofed herbivores, including some forms resembling

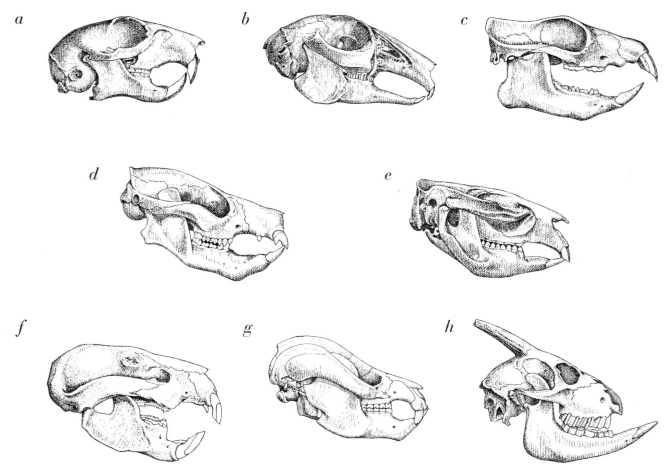

CHISEL-LIKE INCISORS, specialized for gnawing, appear in animals belonging to several extinct and extant orders in addition to the rodents, represented by a squirrel (*a*), and the lagomorphs, represented by a hare (*b*), which are today's main specialists in this ecological role. Representatives of other orders with chisel-like incisor teeth are an early tillodont, *Trogosus* (*c*), an early primate, *Plesiadapis* (*d*), a living marsupial, the wombat (*e*), one of the extinct multituberculate mammals, *Taeniolabis* (*f*), a mammal-like reptile of the Triassic, *Bienotherium* (*g*), and a Pleistocene cave goat, *Myotragus* (*h*), whose incisor teeth are in the lower jaw only.

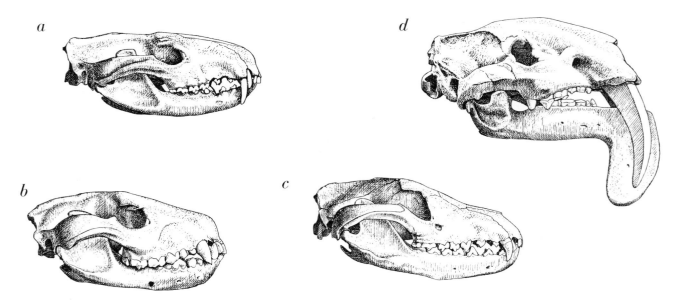

CARNIVOROUS MARSUPIALS, living and extinct, fill an ecological niche more commonly occupied by the placental carnivores today. Illustrated are the skulls of two living forms, the Australian "cat," *Dasyurus* (*a*), and the Tasmanian devil, *Sarcophilus* (*b*). The Tasmanian "wolf," *Thylacinus* (*c*), has not been seen for many years and may be extinct. A tiger-sized predator of South America, *Thylacosmilus* (*d*) became extinct in Pliocene times, long before the placental sabertooth of the Pleistocene, *Smilodon*, appeared.

horses and camels) and the Astrapotheria (extinct large hoofed herbivores of very peculiar appearance). Thus a total of six orders, extinct or extant, probably originated in South America. Still another order, perhaps of even more ancient origin, is the Marsupicarnivora. The order is so widely distributed, with species found in South America, North America, Europe and Australia, that its place of origin is quite uncertain. It includes, in addition to the extinct marsupial carnivores of South America, the opossums of the New World and the native "cats" and "wolves" of the Australian area.

The most important barrier isolating South America from North America in the Tertiary period was the Bolívar Trench. This arm of the sea cut across the extreme northwest corner of the continent. In the late Tertiary the bottom of the Bolívar Trench was lifted above sea level and became a mountainous land area. A similar arm of sea, to which I have already referred, extended across the continent in the region that is now the Amazon basin. This further enhanced the isolation of the southern part of South America.

Africa's role as a center of adaptive radiation is problematical because practically nothing is known of its native mammals before the end of the Eocene. We do know, however, that much of the continent was flooded by marginal seas, and that in the early Tertiary, Africa was cut up into two or three large islands. Still, there must have been a land route to Eurasia even in the Eocene; some of the African mammals of the following epoch (the Oligocene) are clearly immigrants from the north or northeast. Nonetheless, the majority of African mammals are of local origin. They include the following orders:

Proboscidea: the mastodons and elephants.

Hyracoidea: the conies and their extinct relatives.

Embrithopoda: an extinct order of very large mammals.

Tubulidentata: the aardvarks.

In addition the order Sirenia, consisting of the aquatic dugongs and manatees, is evidently related to the Proboscidea and hence presumably also originated in Africa. The same may be true of another order of aquatic mammals, the extinct Desmostylia, which also seems to be related to the elephants. The one snag in this interpretation is that desmostylian fossils are found only in the North Pacific, which seems rather a long way from Africa. Nonetheless, once they were waterborne, early desmostylians might have crossed the Atlantic, which was then only a narrow sea, navigated the Bolívar Trench and, rather like Cortes (but stouter), found themselves in the Pacific.

Early Mammals of Africa

Thus there are certainly four, and possibly six, mammalian orders for which an African origin can be postulated. Here it should be noted that Africa had an impressive array of primates in the Oligocene. This suggests that the order Primates had a comparatively long history in Africa before that time. Even though the order as such does not have its roots in Africa, it is possible that the higher primates—the Old World monkeys, the apes and the ancestors of man —may have originated there. Most of the fossil primates found in the Oligocene

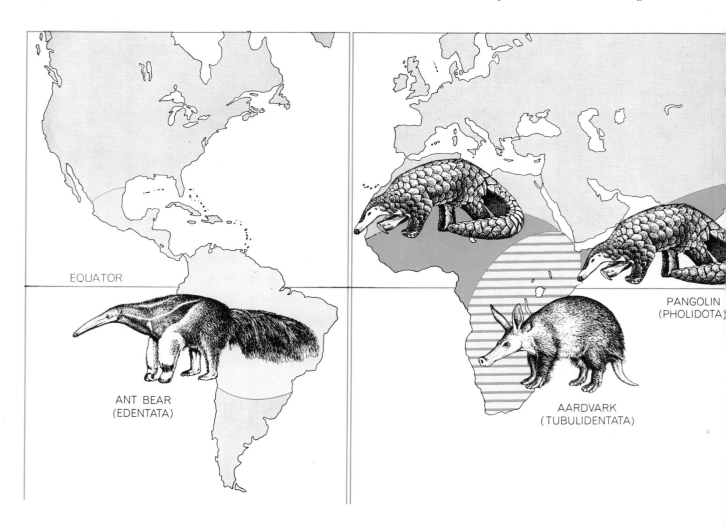

EQUATOR

ANT BEAR
(EDENTATA)

PANGOLIN
(PHOLIDOTA)

AARDVARK
(TUBULIDENTATA)

formations of Africa are primitive apes or monkeys, but there is at least one form (*Propliopithecus*) whose dentition looks like a miniature blueprint of a set of human teeth.

The Rest of Gondwanaland

We know little or nothing of the zoogeographic roles played by India and Antarctica in the early Tertiary. Mammalian fossils from the early Tertiary are also absent in Australia. It may be assumed, however, that the orders of mammals now limited to Australia probably originated there. These include two orders of marsupials: the Peramelina, comprised of several bandicoot genera, and the Diprotodonta, in which are found the kangaroos, wombats, phalangers and a number of extinct forms. In addition the order Monotremata, a very primitive group of mammals that includes the spiny anteater and the platypus, is likely to be of Australian origin. This gives us a total of three orders probably founded in Australia.

Summing up, we find that the three Laurasian continents produced a total of 16 orders of mammals, an average of five or six orders per continent. As for Gondwanaland, South America produced six orders, Africa four to six and Australia three. The fact that Australia is a small continent probably accounts for the lower number of orders founded there. Otherwise the distribution—the average of five or six orders per subdivision—is remarkably uniform for both the Laurasian and Gondwanaland supercontinents. The mammalian record should be compared with the data on Cretaceous reptiles, which show that the two supercontinents produced a total of 12 or 13 orders (or adaptively distinct suborders). A regularity is suggested, as if a single nucleus of radiation would tend in a given time to produce and support a given amount of basic zoological variation.

As the Tertiary period continued new land connections were gradually formed, replacing those sundered when the old supercontinents broke up. Africa made its landfall with Eurasia in the Oligocene and Miocene epochs. Laurasian orders of mammals spread into Africa and crowded out some of the local forms, but at the same time some African mammals (notably the mastodons and elephants) went forth to conquer almost the entire world. In the Western Hemisphere the draining and uplifting of the Bolívar Trench was followed by intense intermigration and competition among the mammals of the two Americas. In the process much of the typical South American mammal population was exterminated, but a few forms pressed successfully into North America to become part of the continent's spectacular ice-age wildlife.

India, a fragment of Gondwanaland that finally became part of Asia, must have made a contribution to the land fauna of that continent but just what it was cannot be said at present. Of all the drifting Noah's arks of mammalian evolution only two—Antarctica and Australia—persist in isolation to this day. The unknown mammals of Antarctica have long been extinct, killed by the ice that engulfed their world. Australia is therefore the only island continent that still retains much of its pristine mammalian fauna. [*see illustration on pages 172–173*].

If the fragmentation of the continents at the beginning of the age of mammals promoted variety, the amalgamation in the latter half of the age of mammals has promoted efficiency by means of a large-scale test of the survival of the fittest. There is a concomitant loss of variety; 13 orders of land mammals have become extinct in the course of the Cenozoic. Most of the extinct orders are island-continent productions, which suggests that a system of semi-isolated provinces, such as the daughter continents of Laurasia, tends to produce a more efficient brood than the completely isolated nuclei of the Southern Hemisphere. Not all the Gondwanaland orders were inferior, however; the edentates were moderately successful and the proboscidians spectacularly so.

As far as land mammals are concerned, the world's major zoogeographic provinces are at present four in number: the Holarctic-Indian, which consists of North America and Eurasia and also northern Africa; the Neotropical, made up of Central America and South America; the Ethiopian, consisting of Africa south of the Sahara, and the Australian. This represents a reduction from seven provinces with about 30 orders of mammals to four provinces with about 18 orders. The reduction in variety is proportional to the reduction in the number of provinces.

In conclusion it is interesting to note that we ourselves, as a subgroup within the order Primates, probably owe our origin to a radiation within one of Gondwanaland's island continents. I have noted that an Oligocene primate of Africa may have been close to the line of human evolution. By Miocene times there were definite hominids in Africa, identified by various authorities as members of the genus *Ramapithecus* or the genus *Kenyapithecus*. Apparently these early hominids spread into Asia and Europe toward the end of the Miocene. The cycle of continental fragmentation and amalgamation thus seems to have played an important part in the origin of man as well as of the other land mammals.

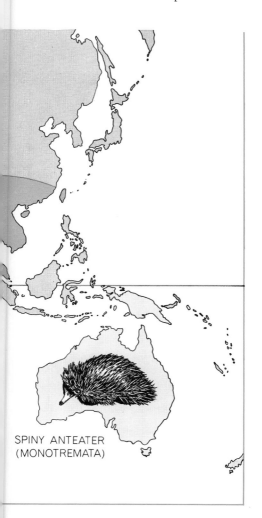

SPINY ANTEATER
(MONOTREMATA)

FOUR ANT-EATING MAMMALS have become adapted to the same kind of life although each is a member of a different mammalian order. Their similar appearance provides an example of an evolutionary process known as convergence. The ant bears of the New World Tropics are in the order Edentata. The aardvark of Africa is the only species in the order Tubulidentata. Pangolins, found both in Asia and in Africa, are members of the order Pholidota. The spiny anteater of Australia, a very primitive mammal, is in the order Monotremata.

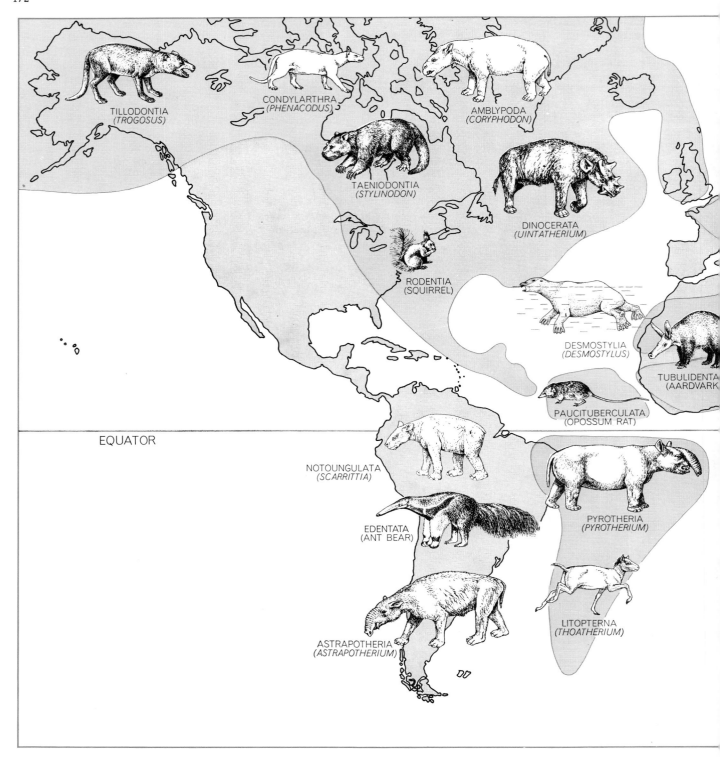

CONTINENTAL DRIFT affected the evolution of the mammals by fragmenting the two supercontinents early in the Cenozoic era. In the north, Europe and Asia, although separated by a sea, remained connected with North America during part of the era. The

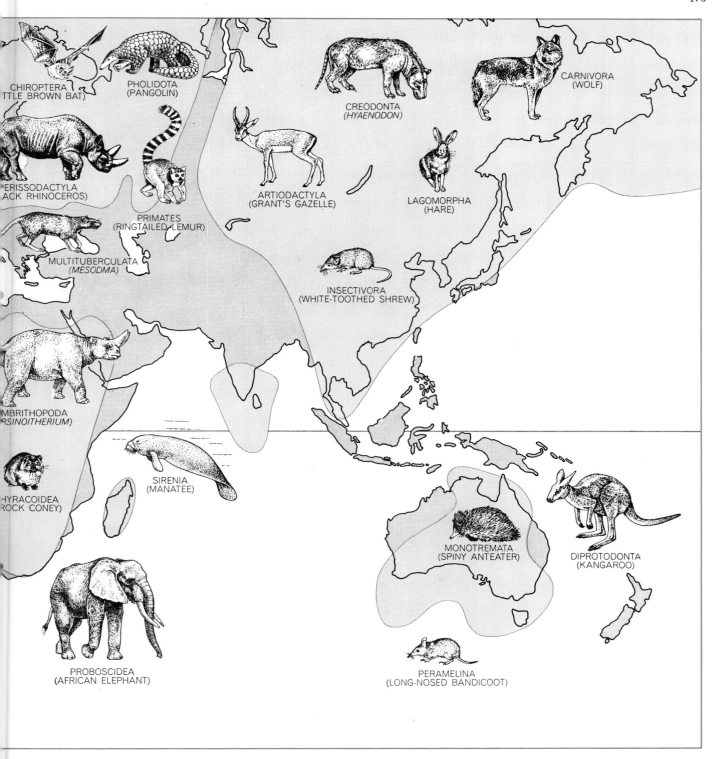

CHIROPTERA
(LITTLE BROWN BAT)

PHOLIDOTA
(PANGOLIN)

CREODONTA
(*HYAENODON*)

CARNIVORA
(WOLF)

PERISSODACTYLA
(BLACK RHINOCEROS)

PRIMATES
(RINGTAILED LEMUR)

ARTIODACTYLA
(GRANT'S GAZELLE)

LAGOMORPHA
(HARE)

MULTITUBERCULATA
(*MESODMA*)

INSECTIVORA
(WHITE-TOOTHED SHREW)

EMBRITHOPODA
(*ARSINOITHERIUM*)

SIRENIA
(MANATEE)

HYRACOIDEA
(ROCK CONEY)

MONOTREMATA
(SPINY ANTEATER)

DIPROTODONTA
(KANGAROO)

PROBOSCIDEA
(AFRICAN ELEPHANT)

PERAMELINA
(LONG-NOSED BANDICOOT)

free migration that resulted prevents certainty regarding the place of origin of many orders of mammals that evolved in the north.

The far wider rifting of Gondwanaland allowed the evolution of unique groups of mammals in South America, Africa and Australia.

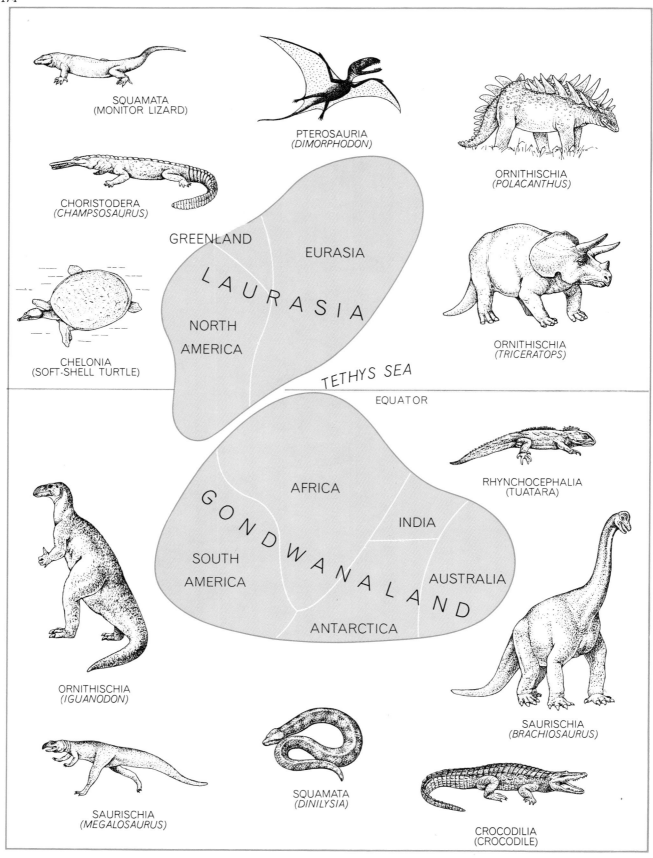

SQUAMATA
(MONITOR LIZARD)

PTEROSAURIA
(DIMORPHODON)

ORNITHISCHIA
(POLACANTHUS)

CHORISTODERA
(CHAMPSOSAURUS)

GREENLAND

EURASIA

L A U R A S I A

NORTH
AMERICA

ORNITHISCHIA
(TRICERATOPS)

CHELONIA
(SOFT-SHELL TURTLE)

TETHYS SEA

EQUATOR

AFRICA

RHYNCHOCEPHALIA
(TUATARA)

G O N D W A N A L A N D

INDIA

SOUTH
AMERICA

AUSTRALIA

ANTARCTICA

ORNITHISCHIA
(IGUANODON)

SAURISCHIA
(BRACHIOSAURUS)

SAURISCHIA
(MEGALOSAURUS)

SQUAMATA
(DINILYSIA)

CROCODILIA
(CROCODILE)

TWO SUPERCONTINENTS of the Mesozoic era were Laurasia in the north and Gondwanaland in the south. The 12 major types of reptiles, represented by typical species, are those whose fossil remains are found in Cretaceous formations. Most of the orders inhabited both supercontinents; migrations were probably by way of a land bridge in the west, where the Tethys Sea was narrowest.

V

HUMAN EVOLUTION

HUMAN EVOLUTION   V

INTRODUCTION

Only in the concluding pages of *On the Origin of Species* does Darwin allude to human evolution by suggesting that owing to the ideas presented therein, "Light will be thrown on man and his history." (By the sixth edition, Darwin had added the modifier "much" to *light*.) Although Darwin skirted the issue, readers of *Origin* clearly saw the implications of his thesis for humans. And, of course, much of the opposition to Darwinism then, as well as now, comes out of those implications. Darwin directly addressed the issue a dozen years later in his *Descent of Man* (1871) where he argues a common ancestry for apes and humans based on their similarities in anatomy and behavior. As convincing as his argument was—namely, if the lower animals, like apes, have evolved through the ages by descent with modification owing to natural selection and if humans share many common biological traits with apes, then it is logical to suppose that humans too have evolved—what Darwin sorely lacked to clinch his argument was fossil evidence demonstrating the various stages leading up to modern humans. During Darwin's lifetime, the only human fossils known were those of our own species, *Homo sapiens*, such as Neanderthal Man from Germany and Cro-Magnon Man from France. (One might well claim that no amount of fossil evidence for human evolution will suffice for those unalterably opposed to the whole idea in the first place.)

Ernst Haeckel, a nineteenth century champion of Darwin in Germany, further speculated on human origins and even named the hypothetical ape-to-human ancestor, *Pithecanthropus* (literally "ape-man"), and he suggested its fossils would be found in Asia where the roots of civilization seemed to be. In 1891 Eugene Dubois, a young doctor in the Dutch army stationed in Java who was greatly influenced by Haeckel, discovered fossils that he named *Pithecanthropus* and which today are considered to belong to *Homo erectus*, the species immediately preceding *Homo sapiens*. In 1918 other examples of *H. erectus* were found in China by Davidson Black, so-called Peking Man. These discoveries vindicated, temporarily, Haeckel's notion of Asia being the central locale of early human origins. (See the accompanying chart that lists important hominid discoveries.)

In 1924 Raymond Dart discovered in some South African cave deposits the Taung child's skull, which would eventually be recognized as *Australopithecus africanus*, a slender species of the australopithecine genus ancestral to that of *Homo*. In 1938, also in South Africa, Robert Broom discovered the "Paranthropus" skull that became the type specimen for *A. robustus*, a robust species of autralopithecine. These two South African ape-like forms were small-brained and yet did have some affinities with *Homo*, especially in their dentition. However, one of the major difficulties with the South African discoveries was that they occurred in isolated cave deposits, which

IMPORTANT HOMINOID DISCOVERIES

DATE OF DISCOVERY	FOSSIL	PLACE	AGE	CURRENT INTERPRETATION
1856	"Neanderthal"	Germany	Late Pleistocene	*H. sapiens neanderthalensis*
1868	"Cro-Magnon"	France	Late Pleistocene	*H. sapiens sapiens*
1891–93	"Java Man"	Java	Early Pleistocene	*H. erectus*
1912	"Piltdown Man"	England	—	Big-Brained Hoax!!
1918–29	"Peking Man"	China	Middle Pleistocene	*H. erectus*
1924	"Taung Child"	South Africa	Pleistocene	Gracile Australopithecine (*A. africanus*)
1938	"Paranthropus"	South Africa	Pleistocene	Robust Australopithecine (*A. robustus*)
1959	"Zinj"	Olduvai	Early Pleistocene	Robust Australopithecine (*A. boisei*)
1960	"Handy Man"	Olduvai	Early Pleistocene	*H. habilis*
1972	"1470" Skull	East Turkana	Early Pleistocene	*H. habilis*
1973–77	"Lucy" *et al.*	Laetoli; Hadar	Pliocene	*A. afarensis*
1977	Hominid Tracks	Laetoli	Pliocene	Bipedalism

made it impossible to place them precisely within the geologic time scale. Although "Pleistocene," they were not at all well dated.

The Piltdown hoax, perpetrated in 1912 in England, also helped obscure the real significance of the australopithecine discoveries. The combination of a big brained modern cranium with a doctored chimp's jaw came out of, and further reinforced, the then-prevailing opinion that enlarged brains characterized early hominids. Not until the Piltdown forgery was exposed in 1953 and additional finds of australopithecines were made did the South African fossils assume their rightful place in early human evolutionary history.

The next very important find was made by Mary Leakey in 1959 in Olduvai Gorge, northern Tanzania, East Africa. Initially called "Zinjanthropus" or nut-cracker man because of the massive jaws and teeth, it was later recognized as another robust australopithecine. Unlike earlier South African finds, however, this fossil had very good stratigraphic control in that the associated, nonhominid vertebrates indicated an early Pleistocene age. Subsequently, the lavas that were interbedded with the fossil-bearing sediments were reliably dated, radiometrically, at 1.75 million years, thereby indicating for the first time that early human origins went considerably further back than previously supposed.

Just a year later, in 1960, Mary and Louis Leakey turned up still more hominid fossils that seemed to link the somewhat older australopithecines with younger *H. erectus*; they called this *H. habilis*, or "handy man," because of the associated artifacts found with the bones. Then in 1972, Richard Leakey, the son of Mary and Louis, working in northern Kenya discovered a fragmented skull (referred to as "1470" from its museum catalogue number) that had a fairly large brain case almost 75 percent larger than the australopithecines. Although the initial radiometric dating of the beds in which 1470 occurred was somewhat ambiguous, a confirmed age of 1.8 million years subsequently demonstrated for the first time that there was a relatively large-brained hominid living contemporaneously with the small-brained australopithecines.

Most recently, in the early and middle 1970s, expeditions to Hadar in Ethiopia and to Laetoli in northern Tanzania turned up a number of specimens, including the famous Lucy skeleton, in rocks that varied in age from 2.6 to 3.8 million years. These fossils have been recently described as the most primitive australopithecine species, *Australopithecus afarensis*. Finally, in the late 1970s Mary Leakey and her colleagues at Laetoli discovered a trackway extending some 25 meters in volcanic ashbeds dated at 3.6 to 3.8 million years that records the presence of a fully erect, bipedal hominid that far back in time.

This short review of hominid discoveries suggests several important points. First, that for a long time human evolution was discussed and speculated upon without much of an actual fossil record. What little there was was rather scrappy and highly disputed—what people have referred to as "bones of contention." Second, that evidence for human evolution consequently had to be based on the comparative morphology with living primates and on what philosophers call "arguments of consistency"; that is, that the notion of human evolution was fully consistent with evolution seen in other animal groups for which there *was* a good fossil record. Third, that the time scale over which hominids have evolved has been increasingly extended backward, now approaching some 4 million years. And finally, the geographic focus of fossil hominid discovery has shifted from Europe, to Asia, and now to sub-Saharan Africa.

"The Evolution of Man" (and obviously, Woman) by Sherwood L. Washburn summarizes the current broad understanding we have of our human origins. As Washburn points out, were it not that we are searching for our own immediate ancestors, the fossil record of human evolution would be considered quite good. However, because much of that evolution, especially in its latter stages, involves speech and language, manufacture and manipulation of objects, elaborate social interactions, and complex reproductive behavior, there is little or no opportunity for such phenomena to be directly preserved in the stratigraphic record. Hence, what we can infer from bones, teeth, and stony artifacts does indeed tell us a great deal about our morphological and early cultural evolution, yet much is missing to explain how we got from there to where we are today. Nevertheless, certain conclusions are obvious. We share a common ancestry with the African apes; our initial adaptation was upright, bipedal locomotion; the forelimbs released from locomotor functions became adapted to handling objects and eventually making objects; there was a subsequent rapid evolution of the brain, reinforcing these tool-using adaptations, and undoubtedly also related to speech, language, and interpersonal communication.

What the fossil record does not tell us unequivocally is from what stock of ancestral apes, presumably in the late Miocene some five to ten million years ago, the australopithecines (who gave rise to the line *Homo*) arose.

Alan Walker and Richard E. F. Leakey, in "The Hominids of East Turkana," describe the stratigraphy, sites, and fossils found along the eastern shores of Lake Turkana in northern Kenya. They thus home in on a part—and a very crucial part—of the overall evolutionary picture presented by Washburn. They argue that rather than the simple evolutionary ladder whose bottom rung of *Australopithecus* leads to *Homo erectus* and then to *H. sapiens* at the top, there were two, possibly three, species of hominids living contemporaneously in East Africa some one to two million years ago: slender or gracile australopithecines, robust australopithecines, and a pre-erectus stage *Homo*. In short, a bush rather than a ladder for our metaphor.

There are a couple of intriguing implications of co-existing hominid species. First, how did they individually interact with the environment; that is, what were the respective ecologic niches? Second, given that the australopithecines became extinct and the *Homo* line continued, leading to our own extant species, did the success of the latter require the demise of the former? Or perhaps their different evolutionary success was related to factors that had little or nothing to do with the other species? Archaeologists are quick to emphasize that artifacts are found associated with each line, so that we cannot automatically assume that the rise of culture in one line gave it ultimate advantage over the other line.

"The Neanderthals" by Erik Trinkaus and William W. Howells discusses still another early misconception regarding human evolution: that they were either freak examples of our race or subhuman representatives off on some evolutionary tangent. Despite the fact that Neanderthals were the first fossils found that seemed related to modern humans, it has taken more than a century to rehabilitate them as *bona fide* members of our species that antedated modern varieties of humans. They seem to be a stage of evolution between *H. erectus* and modern *H. sapiens*. However, it is still uncertain whether they actually gave rise to modern humans or whether they were a parallel line, or subspecies, that became extinct while our subspecies survived. Or perhaps, Neanderthal populations were absorbed into the modern race by interbreeding. Whatever the mode of evolution, the anatomical form of our subspecies apparently was superior to that of the Neanderthals. As the authors point out we can now only guess what made our physique evolutionarily successful. Was it climate, ecology, or cultural advances that made the bulkier Neanderthal inefficient compared to ourselves?

These readings should make it evident to the unprejudiced that human evolution is as well documented as that of other organisms represented in the fossil record. However, given the fact of our evolution, there still remains much rich detail to be explored. Disagreements among specialists about those details should in no way obscure the fact that we humans are as much a part of nature as any other species of plant or animal. And as such, we are subject to the same laws, principles, and forces of nature as they. Yet, it cannot be denied that we are also apart from nature owing to our self-consciousness, our awareness, our sense of time that allows us to live in the past, present, and future, and our culture upon which we depend so much for our survival but which is not transmitted genetically as is the case with other organisms. Surely the rise of humans and humanity must be as big an evolutionary milestone as the origin of photosynthesis, the advent of the coelom in metazoans, or the invasion of dry land by vertebrates when they invented the amniote egg. As with these other evolutionary milestones, the earth has been irreversibly altered by our arrival.

SUGGESTED FURTHER READING

Johanson, D.C., and A. E. Maitland. 1981. *Lucy: The Beginnings of Human Evolution*. New York: Simon and Schuster. A frank look at how science is really done, with all its false hopes, starts, and ideas, yet eventually yielding something close to the truth. Worthwhile and fun to read.

Johanson, D.C., and T.D. White. 1979. "A Systematic Assessment of Early African Hominids," *Science*, vol. 203, pp. 321–330. Technical but readable description of the most primitive australopithecine, *A. afarensis*, and its place in the overall scheme of human evolution.

Tanner, N. 1981. *On Becoming Human*. New York: Cambridge University Press. Up-to-date discussion of human origins and evolution. Emphasizes woman-the-gatherer as well as man-the-hunter, redressing the usual gender bias in this subject. Avoids a piecemeal approach by integrating fossils with primate behavior in a cultural perspective.

Zihlman, A. 1981. *Human Evolution Coloring Book*. New York: Harper and Row. A painless way to learn the anatomy, geography, stratigraphy, and concepts involved in tracing human evolutionary history over the last ten million years.

14   The Evolution of Man

by Sherwood L. Washburn
September 1978

A wealth of new fossil evidence indicates that manlike creatures had already branched off from the other primates by four million years ago. Homo sapiens himself arose only some 100,000 years ago

Perhaps the most significant single fact about human evolution has a paradoxical quality: the brain with which man now begins to understand his own lengthy biological past developed under conditions that have long ceased to exist. That brain evolved both in size and in neurological complexity over some millions of years, during most of which time our ancestors lived under a daily obligation to act and react on the basis of exceedingly limited information. What is more, much of the information was wrong.

Consider what that meant. Before the information being fed to man's evolving brain began to be refined by an advancing technology our ancestors lived in a world that seemed to them small and flat and that they could assess only in very personal terms. Sharing the world with them were divine spirits, ghosts and monsters. Yet the brain that developed these concepts was the same brain that today deals with the subtleties of modern mathematics and physics. And it is this same technological progress that allows us to recognize human evolution.

One relevant example of the paradox is the extraordinary expansion over the past two centuries of man's perception of time. As Ernst Mayr points out in the introductory article of this *Scientific American* book, at the beginning of the 18th century the accepted view was that the time that had elapsed between the creation of the earth and the present was no more than a few thousand years. At the end of the 19th century the perceived interval had been enlarged a thousandfold and stood at about 40 million years. With the discovery that the slow and constant decay of certain radioactive isotopes constitutes a clock it became necessary to enlarge the interval another hundredfold, so that today we reckon the age of the earth to be about 4.6 billion years.

The human mind cannot literally comprehend such an interval; it is as ineffable as the trillions and quadrillions of dollars that are juggled in world economics. Man's common sense perceives time as being short: a rhythm of birth, growth and death. To this sense of biological time can be added a sense of social time: a less tangible interval of three to five generations that is important to the actors in the drama of human society. Longer intervals do not have the same emotional impact. The real time scale of the universe that has been developed by science can be regarded as having liberated the perception of time from the limitations of the human mind.

The modern perception of time is of course only one in a series of mental emancipations that have led man to a deeper understanding of his own evolutionary history. In what follows I shall review information bearing on human evolution that has been gathered by workers in several fields of study, and I shall also assess the contribution of each field to our overall grasp of the subject today. No one can be an expert in all these fields, and this article might better be regarded as a personal evaluation than as an objective summary.

Few things defy common sense more than the concept that the continents of the earth are constantly in motion and that their positions have shifted drastically on a time scale short compared with the age of the earth. One consequence of yesterday's common sense is that all traditional theories of human evolution have assumed that the positions of the continents are fixed. To be sure, "land bridges" between continents and shallow seas that invaded continents were postulated, but the positions of the great continental plates were not affected. Although it has been nearly 70 years since Alfred Wegener proposed that continental drift was a reality, it is only over the past 20 years, as the mechanism of plate tectonics has been developed and the movement of continents has actually been measured, that the concept of moving continents has become accepted and even respectable.

The combined data from radioactive-isotope dating and plate tectonics have fundamentally changed the background of human evolutionary studies. For example, everyone used to think that the monkeys of the New World had evolved directly from the primitive prosimians that had once flourished in North America. (The prosimians are the least advanced of the order Primates, which includes the apes and man. All living prosimians are confined to the Old World.) We now know, however, that 35 to 40 million years ago Africa was as close to South America as North America was. Some of the primates that were ancestral to the New World monkeys might just as easily have been accidentally rafted (perhaps on a tree felled by a flood) to South America from Africa as from North America. The propinquity of the three continents, which was unthinkable before continental drift was accepted, does not prove that the ancestral stock of the New World monkeys emigrated from Africa; nevertheless, it does present that entirely new and important possibility.

Another example of the effect of plate tectonics on human evolutionary hypotheses has to do with recent and repeated assertions that Africa was the focus of human origins. What the history of continental drift indicates is that there were broad connections between Africa and Eurasia from the time of their collision some 18 million years ago (when some ancestral elephants left Africa and spread over Eurasia) until the flooding of the Mediterranean basin five or six million years ago. It happens that

IMMUNOLOGICAL DISTANCES between selected mammals are indicated by the separation, as measured along the horizontal axis, between the branches of this "divergence tree." For example, the monotremes (primitive egg-laying mammals) are removed from the marsupials by a distance (in arbitrary units) of only 1.5 but are removed from the chimpanzee by a distance of nearly 17. The distance between man and the Old World monkeys is a little more than 3, between man and the Asiatic gibbons 2, and between man and the gorillas and the chimpanzees of Africa less than 1. The data are from Morris Goodman of Wayne State University.

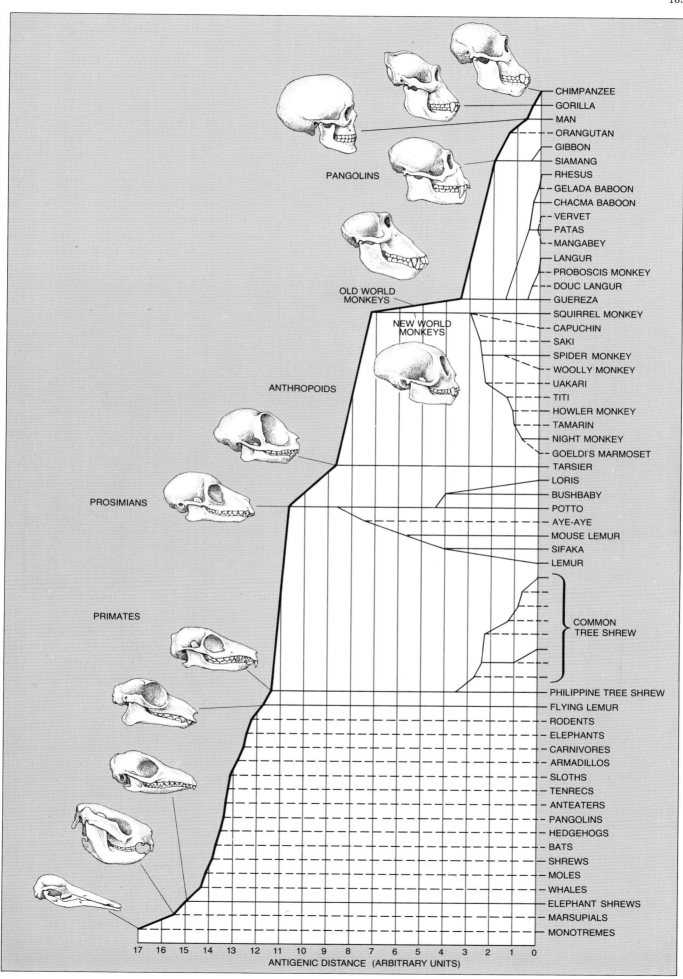

PANGOLINS

OLD WORLD
MONKEYS

NEW WORLD
MONKEYS

ANTHROPOIDS

PROSIMIANS

PRIMATES

CHIMPANZEE
GORILLA
MAN
ORANGUTAN
GIBBON
SIAMANG
RHESUS
GELADA BABOON
CHACMA BABOON
VERVET
PATAS
MANGABEY
LANGUR
PROBOSCIS MONKEY
DOUC LANGUR
GUEREZA
SQUIRREL MONKEY
CAPUCHIN
SAKI
SPIDER MONKEY
WOOLLY MONKEY
UAKARI
TITI
HOWLER MONKEY
TAMARIN
NIGHT MONKEY
GOELDI'S MARMOSET
TARSIER
LORIS
BUSHBABY
POTTO
AYE-AYE
MOUSE LEMUR
SIFAKA
LEMUR

COMMON
TREE SHREW

PHILIPPINE TREE SHREW
FLYING LEMUR
RODENTS
ELEPHANTS
CARNIVORES
ARMADILLOS
SLOTHS
TENRECS
ANTEATERS
PANGOLINS
HEDGEHOGS
BATS
SHREWS
MOLES
WHALES
ELEPHANT SHREWS
MARSUPIALS
MONOTREMES

17 16 15 14 13 12 11 10 9 8 7 6 5 4 3 2 1 0
ANTIGENIC DISTANCE (ARBITRARY UNITS)

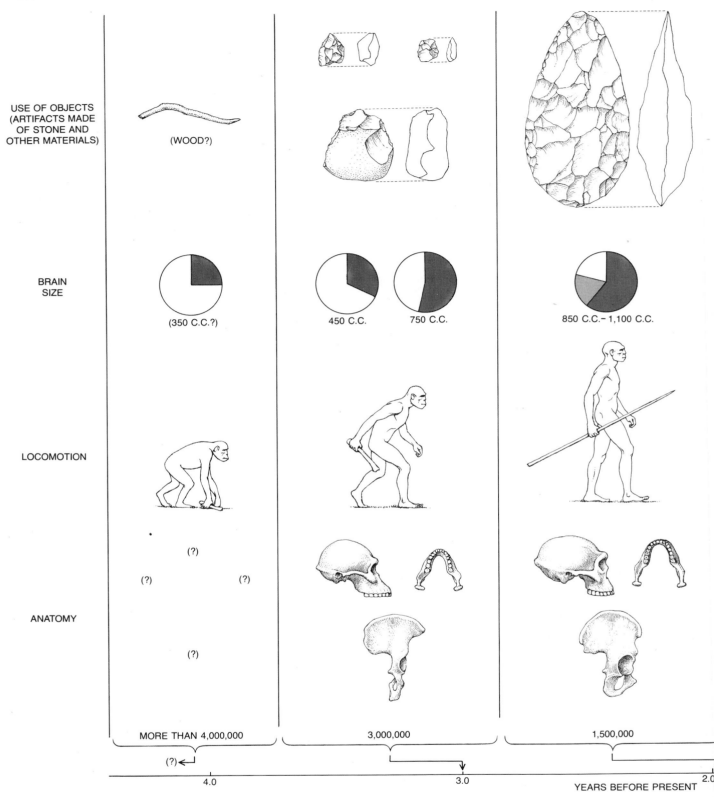

USE OF OBJECTS
(ARTIFACTS MADE
OF STONE AND
OTHER MATERIALS)

(WOOD?)

BRAIN
SIZE

(350 C.C.?)   450 C.C.   750 C.C.   850 C.C.– 1,100 C.C.

LOCOMOTION

ANATOMY

(?)

(?)   (?)

(?)

MORE THAN 4,000,000   3,000,000   1,500,000

(?)

4.0   3.0   2.0

YEARS BEFORE PRESENT

HUMAN EVOLUTION, projected over a possible span of 10 million years, begins at a slow pace when a still undiscovered hominid branches off from the hominoid stock ancestral to man, the chimpanzee and the gorilla at some time more than four million years ago (far left). It is assumed that the ancestral hominid had a small brain and walked on its knuckles. This mode of locomotion enables a quadruped to move about while holding objects in its hands, leading to the further assumption that the hominid outdid living chimpanzees in manipulating sticks and other objects. By four million years ago the African fossil record reveals the presence of an advanced hominid: *Australopithecus.* **This subhuman had a pelvis that allowed an upright posture and a bipedal gait. The size of the brain had increased to some 450 cubic centimeters. Stone tools soon appear in the archaeological record; they are simple implements made from pebbles and cobbles. The tools may have been made by a second hominid group, chiefly notable for having a much larger brain: 750 c.c. Next, about 1.5 million years ago, the first true man,** *Homo erectus,* **appeared. Still primitive with respect to the morphology of its cranium and jaw,** *H. erectus* **had an essentially modern pelvis and a striding gait. Its brain size approaches the modern average in a number of instances.**

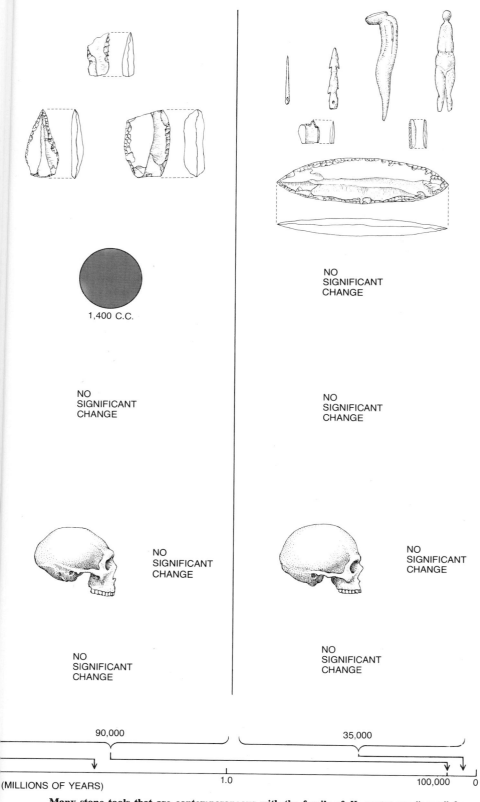

1,400 C.C.

NO
SIGNIFICANT
CHANGE

NO
SIGNIFICANT
CHANGE

NO
SIGNIFICANT
CHANGE

NO
SIGNIFICANT
CHANGE

NO
SIGNIFICANT
CHANGE

NO
SIGNIFICANT
CHANGE

NO
SIGNIFICANT
CHANGE

90,000   35,000

(MILLIONS OF YEARS)   1.0   100,000   0

Many stone tools that are contemporaneous with the fossils of *H. erectus* are "cores" from which flakes have been removed on two sides; they are representative of the Acheulian tool industry. Not until some 100,000 years ago did *Homo sapiens* appear, in the form of Neanderthal man. The shape of Neanderthal's skull is not quite modern but the size of its brain is. Most of the tools found at Neanderthal sites represent the Mousterian industry; they are made from flakes of flint rather than cores. Only 40,000 years ago modern man, *Homo sapiens sapiens*, arrived on the scene. His skull is less robust than that of Neanderthal and his brain is slightly smaller. Many of his stone tools are slender blades; some, known as laurel-leaf points, appear to be ceremonial rather than utilitarian. Among his bone artifacts are needles, harpoon heads, awls and statuettes. About 10,000 years ago man's transition from hunting to farming began.

the fossil remains of *Ramapithecus*, the Miocene-Pliocene ape commonly believed to be an ancestor of the hominid line, that is, the line of man and his extinct close relatives, are found from India and Pakistan through the Near East and the Balkans to Africa. The continuity in the distribution of these fossils suggests that the geography of Eurasia and Africa then was substantially different from what it is today. Moreover, the list of identical Indian and African faunas can be extended far beyond this single extinct ape: both regions harbor macaque monkeys, lions, leopards, cheetahs, jackals, wild dogs and hyenas. The possibility that man originated solely in Africa therefore seems less likely than it once did. In other words, the longer man's ancestors existed as intelligent, upright-walking, tool-using hunters, the less likely it is that their distribution was confined to any one continent.

Comparative anatomy is a field of study considerably older than plate tectonics. Its roots are in the 19th century, and it has been the discipline most concerned with the similarities and differences between man and his fellow primates. Its basic assumption has been that a sufficiently large quantity of information will inevitably lead to a correct conclusion; it gives little attention to questions of how anatomical data are related to evolutionary theory or to phylogeny. For example, the shape of one human tooth, the lower first premolar, has been cited as proof that man never went through an apelike stage in the course of his evolution. As late as 1972 this datum was offered as evidence that man and his ancestors had been separated from the other primates for at least 35 million years. Since then late Pliocene hominid fossils have been found that are about 3.7 million years old, and their lower first premolars show moderately apelike characteristics. It was not the description of the tooth that was wrong but the conclusions drawn from it (not to mention the belief that it is reasonable to base major phylogenetic determinations on the anatomy of a single tooth).

Comparative anatomy nonetheless makes many valuable connections. For example, the bones of the human arm are much like those of an ape's arm but are very different from the comparable bones of a monkey. Monkey arm bones are very similar to those of many other primates and indeed to those of many other mammals; their form is basic to quadrupedal locomotion. In contrast, the form of human and ape arm bones is basic to the motions of climbing. This finding is a significant one, but it can lead to two quite opposite conclusions: (1) man and apes are related or (2) man and apes have followed a parallel course of evolution, that is, the structures of the arm evolved in the same way even though the two evolutionary lines had long since separated. Deciding between the two alternatives is made all the more difficult by the fact that the comparison is being made between two living animals, each of which has evolved to an unknown degree since its divergence from a common ancestry. Fortunately powerful new analytical tools have come into existence that are a great help in resolving such dilemmas. To these I shall return, under the heading of molecular anthropology; for the moment it is necessary only to say that when the patterns of primate biology being compared are functional ones, the fit between the conclusions drawn from comparative anatomy and those drawn from molecular anthropology is quite close.

Until a few decades ago the fossil record of the primates was poor and that of the hominids, including man, was even poorer. For example, when Sir Arthur Keith undertook to array the existing hominid fossils along their probable lines of descent some 50 years ago, he had to deal with only three genera in the Miocene epoch, and he was able to spread the five (at that time) hominid genera across a later Pliocene-to-Recent time interval of less than half a million years. (Between the Pliocene and the Recent epochs was the Pleistocene; to it and the Recent combined was allotted 200,000 years.) The five genera were *Homo erectus* (then represented only by specimens from Java named *Pithecanthropus*), Neanderthal man, Piltdown man (then still accepted as a valid genus named *Eoanthropus*), Rhodesian man (*Homo rhodesiensis*, a form no longer considered a distinct species) and finally the genus and species *Homo sapiens* (from which, as can be seen, Keith excluded the Neanderthals). Keith had Java man branching off from the main human stem in Miocene times and indicated the extinction of the line at the start of the Pleistocene. The Neanderthals, today classified as *Homo sapiens neanderthalensis,* he saw as branching off in the mid-Pliocene, shortly before the appearance of Rhodesian man and well before that of Piltdown man; he had all three genera extinct in Pleistocene times.

Keith's arrangement was marvelous in its simplicity: each fossil that had to be accounted for stood at the end of its own evolutionary branch, and the time of branching was deduced from its anatomy. (Piltdown man was a problem: its genuinely modern cranium put this faked specimen's branching point higher up on the tree than Rhodesian man's, but its genuinely nonhuman jawbone demanded that the branching be put in the Pliocene.) This kind of typological thinking died slowly with the discovery of many new hominid fossils and the rise of radioactive-isotope dating, which have extended the duration of the Pleistocene from about 200,000 years to about two million. Theodosius Dobzhansky's 1944 paper "On Species and Races of Living and Fossil Man" ushered in the new era and ended nearly a century of analysis that had been primarily typological.

Today there are hundreds of primate fossils. Many of them are accurately dated and more are being discovered every year. It is no longer even practical to list the individual specimens, which was the custom until a few years ago. Problems still remain with respect to the hominid fossil record, perhaps in part because human beings are obsessively curious about the details of their own ancestry. If any animal other than the human one were involved, the hominid fossil record over the past four million years would be considered adequate and even generous.

How is the evidence of those four million years to be read? To begin with, it can now be said with some certainty that hominids have walked upright for at least three million years. That is the age of a pelvis of the early hominid *Australopithecus* recently unearthed in the Afar region of Ethiopia by Donald C. Johanson of Case Western Reserve University. Prior to Johanson's find the best evidence of upright walking was a younger *Australopithecus* pelvis uncovered at Sterkfontein in South Africa. The two fossils are nearly identical. The inference is inescapable that bipedal locomotion is not just another human anatomical adaptation but the most fundamental one. The early bipeds all had small brains (average: 450 cubic centimeters).

Not much later, perhaps about 2.5 million years ago, the bipeds were making stone tools and hunting animals for food. By about two million years ago hominid craniums with a larger capacity appeared; by 1.5 million years ago *Homo erectus* was on the scene, the brains had doubled in size and the stone tools now include bifaces, tools that have been flaked on both sides. These bifaces belong to the core-tool industry known as the Acheulian. (The characteristic form was first recognized at a French Paleolithic site, St.-Acheul.) From about two million to one million years ago another kind of early biped was also present; its robust anatomy identifies it as a separate species of *Australopithecus*. It is readily distinguishable from the less robust bipeds by its massive jaw and molar teeth that are very large compared with the incisors.

This summary of the hominid fossil record is undoubtedly oversimplified, but I think the evidence supports the main outline. What difficulties there are arise mainly from the fragmentary nature of many of the fossils. For example, Johanson has found one skeleton in the Afar region complete enough to allow reconstruction of that hominid's general proportions. The reconstruction shows that it had relatively long arms, a fact that could not be determined from the hundreds of previously discovered fragmentary remains of *Australopithecus*.

Dating also causes problems. For ex-

CLOSER VIEW of primate divergence is afforded by this schematic structure. The distance between man and the chimpanzee has a value of 1; this places both man and chimpanzee at a remove of 4 from the orangutan, and the orangutan and the Old World monkeys at a remove of 7 from the ancestor that both have in common with the New World monkeys. All anthropoids are at a remove of 7 from the ancestor they have in common with all prosimians (less advanced primates such as the lemurs) and at a remove of 11 from the most primitive primates.

ample, there are no radioactive-isotope dates for the hominid fossils found in South Africa. There is disagreement among specialists about the date of a particularly important marker layer of volcanic tuff in the East Turkana region of Kenya, where many important hominid fossils are currently being found. My response (or my bias) is to try to see the general order and to add complications only when they are absolutely inescapable.

The first conclusion I draw from this simplified picture is that upright walking evolved millions of years before a large brain, stone tools or other characteristics we think of as being human. If one accepts this conclusion, the problem of tracing human origins is primarily one of unearthing fossil evidence for that complex locomotor adaptation. How much time the adaptation required and what its intermediate stages may have been cannot be determined as long as the fossil leg bones are missing. The adaptation may have begun at any time from five to 10 million years ago. Fossil-bearing deposits of that age exist, and so all that is needed to clarify this aspect of human evolution is money for the search and a bit of luck.

The second conclusion I draw from my simplified outline is that stone tools and hunting long antedate the appearance of a large brain. Excavating in East Turkana, Glynn Isaac of the University of California at Berkeley and his colleagues uncovered a scatter of crude stone tools including both flakes and the cores that had yielded the flakes, and together with the tools were bits of animal bone. Unfortunately the creatures that deposited this material, which may be as much as 2.5 million years old, left no evidence of their own anatomy.

The East Turkana tools are very early but it is most unlikely that they are the earliest. For example, at Olduvai in neighboring Tanzania many of the stone tools from Bed I are unworked stones. They could not have been identified as tools except for the fact that they were found in a layer of volcanic ash otherwise free of stones, so that someone must have brought them there from somewhere else. In the absence of some similar circumstance the earliest stone tools are likely to go unrecognized.

My third conclusion stems both from the fossil record and from what is known about the anatomy of the human brain. As I have indicated, in my view large brains follow long after stone tools. Tools that are hard to make, such as those of the Acheulian industry, follow the earlier simple tools only after at least a million years have passed. It looks as if the successful way of life the earlier tools made possible acted in some kind of feedback relation with the evolution of the brain. What can be seen in the cortex of the human cerebrum

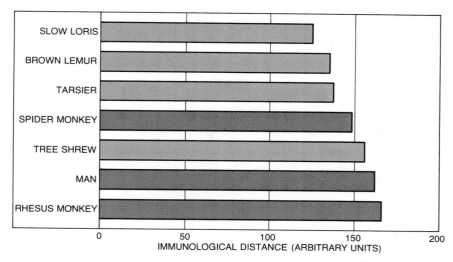

RATE OF EVOLUTION appears to be independent of the number of generations per unit time according to immunological-distance data. Bars show the distances, calculated by Vincent M. Sarich of the University of California at Berkeley, separating the carnivores from various primates. Man and the rhesus monkey were the most distant, respectively 162 and 166 units removed, although each human generation is five times longer than a rhesus generation. Four prosimians (color), also far shorter-lived than man, were even less distant from the carnivores.

mirrors this evolutionary success. Just as the proportions of the human hand, with its large and muscular thumb, reflect a selection for success in the use of tools, so does the anatomy of the human brain reflect a selection for success in manual skills.

Here a further point arises that is often forgotten. The only direct evidence for the importance of increasing brain size comes from the archaeological record. The hypothesis of a correlation between tool use and a large brain argues that the archaeological progression (from no stone tools to simple stone tools to tools of increasing refinement) is correlated with the doubling of hominid brain size. If the hypothesis is correct, the brain should not only have grown in size but also have increased in complexity. The fact remains that the fossil record contains no clues bearing on this neurological advance. Nevertheless, increasing brain size does seem to be correlated with the increasing complexity of stone tools over hundreds of thousands of years in a way that is not evident during the past 100,000 years of human evolution.

Students of the fossil record rely almost entirely on description, just as students of comparative anatomy do. When a fossil is discovered, the first requirement is to determine its geological context, its associations and its probable age. Once the fossil has been brought to the laboratory it is described and compared with similar fossils, and conclusions are drawn on the basis of the comparisons. The anatomical structures that are compared are complex ones and the work is arduous, so that a great deal of analysis remains to be completed on fossils first discovered many years ago. The method has other limitations. For example, teeth are traditionally compared in isolation one at a time. In nature, of course, upper and lower teeth interact where they meet. Comparisons that take this functional factor into account, as the anatomist W. E. Le Gros Clark has shown, give results quite different from those yielded by the traditional tooth-by-tooth method.

Comparative methods are further complicated by the simple fact that a face is full of teeth. The form of the face is related both to the teeth and to the chewing muscles; functional patterns of this kind are not well described by linear measurements. What is more, the descriptive tradition even sets limits on what is observed. For example, the lower jaw of the robust species of Australopithecus from East Africa has a very large ascending ramus, the part of the lower jaw that projects upward to hinge with the skull. At the top of the ascending ramus is what is called the mandibular condyle. As this species' lower jaw opened and its mandibular condyle moved forward, the teeth of its upper and lower jaws must have moved farther apart than they do in any other primate. When one considers all that has been written about the possible diet of Australopithecus and about this hominid's teeth, it is surprising to find a fact as fundamental as the size of its bite is not discussed.

The same jaw provides another example of the weakness of such descriptive systems. In the robust species of Australopithecus the inside of the ascending ramus has features unlike those found in any other primate. This fact is not mentioned in the formal descriptions because it is not traditional to study the inside of the ramus. I could give a number of other examples, but the point at issue is the same in all of them: there are no clearly defined rules that state how fossils should be compared or how anatomy should be understood.

Having sketched what the traditional disciplines have to suggest on the subject of man's evolution, we can now turn to the suggestions that stem from work in two relatively new disciplines: molecular anthropology and the observation of primate behavior in the wild. The first of these disciplines actually has a longer history than plate tectonics: whereas Wegener first proposed his theory in 1912, George H. F. Nuttall demonstrated that the biochemical classification of animals was a possibility in 1904. Nuttall's method was immunological. If blood serum from an animal is injected into an experimental animal, the experimental animal will manufacture antibodies against proteins in the

foreign serum. If serum from the experimental animal is added to serum from a third animal, the antibodies will combine with similar proteins in that serum to form a precipitate. The stronger the precipitation reaction, the closer the relation of the first animal to the third.

Nuttall's method was successfully applied in a number of investigations, but he attracted no more disciples than Wegener did. Not until the past decade, when findings based on immunological methods were seen to agree with those based on the similarity of amino acid sequences in proteins and the similarity of nucleotide sequences in DNA, did the concept of molecular taxonomy gain acceptance. As with the radioactive-isotope methods for determining absolute dates, the new molecular methods are objective and quantitative; they yield the same results when the tests are conducted by different workers.

The capacity of molecular taxonomy to define the relations among primates is perhaps the most important development in the study of human evolution over the past several decades. The great strength of the method is of course its objectivity. For example, data from the fossil record and from comparative anatomy have been cited to demonstrate that man's closest relative is variously the tarsier, certain monkeys, certain extinct apes, the chimpanzee or the gorilla, and to suggest that the time separating man from the last ancestor he shares with each of these candidates is variously from 50 million to four million years.

What do the data of molecular taxonomy show? The primary finding is that the molecular tests indicate little "distance" between man and the African apes. For example, when the distance separating the Old World monkeys from the New World monkeys is given a value of 1 and other distances are expressed as fractions of that value, then the distance between man and the Old World monkeys, as Vincent M. Sarich of the University of California at Berkeley has shown, is more than half a unit (.53 to .61). The distance between man and the Asian great ape the orangutan is about a quarter of a unit (.25 to .33) and the distance between man and the chimpanzee is about an eighth of a unit (.12 to .15).

The short distance between man and the African apes can be compared with similar distances among other related mammals. The relationship is about as close as that between horses and zebras and closer than that between dogs and foxes. Mary-Claire King and Allan C. Wilson of the University of California at Berkeley estimate (on the basis of comparisons between human and chimpanzee polypeptides, or protein chains) that man and the chimpanzee share more than 99 percent of their genetic material.

It might be thought such a wealth of new information about primate relationships would have been welcomed by students of human evolution. This has not been the case. The problem is that whereas the molecular data prove that man and the African apes are very closely related, the data appear to measure relationship and not time. It may be, however, that they do both. The overall picture seems clear: animals that are phylogenetically distant relatives are separated by large molecular distances and those that are close relatives are separated by small distances. This suggests that time and molecular distance are correlated. Unfortunately when it comes to the primates, the molecular distance between the New World and Old World monkeys is much too small to fit in with conventional phylogeny. What is worse, the distance between man and the African apes is startlingly less than convention demands. I suspect

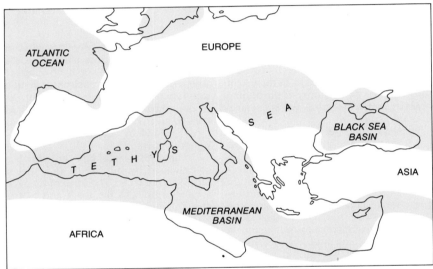

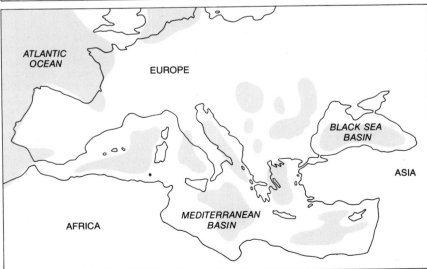

SEPARATION OF ASIA AND AFRICA, now joined by a narrow land bridge, was absolute some 20 million years ago (top), when the Tethys Sea reached from the Atlantic to the Persian Gulf. Later and until five million years ago (bottom) the Tethys was reduced to a network of lakes; the Old World primates were thus free at the time to move between the two continents.

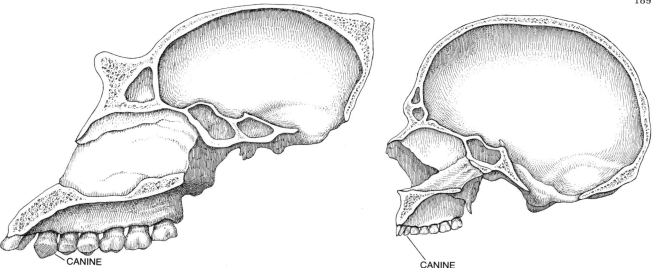

CANINE   CANINE

SKULLS OF MAN AND THE GORILLA, seen here in centerline section, have in common upper canine teeth that are much reduced in size. The gorilla is a female; male gorillas bare their very large canines when threatening to fight. The suggestion by Charles Darwin that man's use of weapons relieved him of the need for large canines seems to be supported by fossil evidence: the oldest human canines known are quite small compared with the canines of male African apes. This implies man's use of weapons for hundreds of millenniums.

that if molecular anthropology had shown man and apes to be very far apart, the concept of a correlation between genetic difference and time would have been accepted without debate.

The validity of a molecular clock is being argued at present, but I believe the problems will be worked out over the next few years. The chemical techniques are being improved and the fund of relevant information is being enlarged by work in many laboratories. Meanwhile the fossil evidence makes it highly unlikely that the ape and human lines separated less than five million years ago, and the molecular evidence makes it highly unlikely that they separated more than 10 million years ago. I have friends and colleagues who violently attack both dates, and they may be right! I am impressed by the degree of emotion that still surrounds the study of human evolution.

Studies of monkeys and apes under natural conditions have increased in number over the past few years. It is noteworthy in this connection that the heyday of evolutionary speculation was in the 19th century and that almost all the primate field studies began after 1960. The brutish, stooped Neanderthal and the monogamous chimpanzee have both proved to be products of the 19th-century imagination. Perhaps the most pertinent revision of preconceived views has to do with locomotion. All the traditional theories of human origins carefully considered how it was that a tree-dwelling ancestor became a ground-dwelling upright walker. The field studies have shown that our closest primate relatives, the African apes, are primarily ground dwellers. Moreover, their locomotor patterns suggest that the ancestor we share in common with them, howev-

er long ago, was also a ground dweller.

In the quadrupedal locomotion of most primates the hand and the foot are both placed flat on the ground; the animal cannot carry anything in them and move at the same time. Gorillas and chimpanzees (and the men who play some of the forward positions in American football), however, have developed a form of locomotion called knuckle walking that enables the apes (if not the football players) to walk normally as they carry objects between their fingers and their palm. If knuckle walking is an ancient trait, it neatly gets around the problem of how the handling and using of objects could have become a common habit. Of all living mammals except man the knuckle-walking chimpanzee is the most habitual user of objects. As Jane Goodall and her colleagues at the Gombe Stream Research Centre in Tanzania add to their observations year after year, the record of the number of objects handled by chimpanzees and the number of ways they are employed steadily increases. The chimpanzees use sticks for bluffing and attack, for poking, teasing and exploring. They use twigs and blades of grass to collect termites and ants. They use leaves to clean themselves. They use stones to crack nuts and also throw stones with moderate accuracy.

Our incredulity dies hard. When Peking man (now classified as *Homo erectus*) was first found, he was declared to be far too primitive to have made the stone tools found in association with his remains. The next unjustified victim of incredulity was *Australopithecus;* surely, the consensus had it, no one with such a small brain could have made tools. Even today many believe only one form of early biped, the form ancestral to man, could have made tools. Chimpan-

zee behavior is therefore enlightening: it shows that a typical ape is able to use objects in far greater variety and with greater effectiveness than anyone had suspected. There is no longer any reason not to suppose all the early bipeds also used objects, and probably used them far more than chimpanzees do, from a time far earlier than the time when stone tools first appear in the archaeological record.

Darwin suggested that the reason men had small canine teeth and the gorilla had huge ones was that man's possession of weapons had eliminated the need for fangs. It is clear that the large canine of the male gorilla has nothing to do with efficient chewing; the canine of the female gorilla is small but she is as well nourished as the male. Is the male canine part of an adaptation for bluffing and fighting? Before such an anatomical feature could have been reduced in the course of evolution its offensive function would have had to have been transferred to some other structure or mechanism. On this view the evolution of small human canine teeth would, as Darwin surmised, have depended on the use of weapons. The chimpanzee field studies, with their evidence for the frequent use of objects, support Darwin's interpretation. Sticks are seldom fossilized and unworked stones can rarely be proved to be artifacts, but teeth are the commonest of all hominid fossils and the earliest bipeds already had small canines. They had probably been using tools for many hundreds of thousands of years.

Not all behavioral information comes from studies in the field. Consider speech. The nonhuman primates cannot learn to speak even though great efforts have been made in the laboratory to teach them to do so. The recent remarkable successes in teaching apes how to communicate by symbols have been achieved in ways other than verbal ones. There is a lesson here, since human beings learn to speak with the greatest of ease.

The sounds made by monkeys primarily convey emotions and are controlled by brain systems more primitive than the cerebral cortex; removal of the cortex does not affect the production of sounds. In man the cortex of the dominant side of the brain is very important in speech. Speech is of course the form of behavior that more than any other differentiates man from other animals. Yet in spite of many ingenious attempts at investigation the origins of human speech remain a mystery. There is no clue to its presence or absence to be found in the fossils.

The archaeological record, however, does offer clues. What we see in the last 40,000 years of prehistory may have been triggered by the development of speech as we know it today. This is to say that although man was surely not mute for most of his development, an increased capacity for verbal communication may have been the ability that led to the extraordinary spread of modern man, *Homo sapiens sapiens*.

For most of the past million years the progress of human evolution, both biological and technological, was slow. Traditions of stone-tool manufacture, as reflected in the rise of successive stone-tool industries, persisted for hundreds of thousands of years with little change. Then came the great acceleration of about 40,000 years ago. Men who were anatomically modern now dominated the scene. Primitive forms of man disappeared; there are not enough fossils to make it possible to decide whether the disappearance was by evolution, hybridization or extinction. Then, in far less than 1 percent of the

time that bipeds are present in the fossil record, came a technological revolution. Its fruits included entirely new and complex tools and weapons, the construction of shelters, the invention of boats, the addition of fish and shellfish to the human diet, deep-water voyages (to Australia, for example), the peopling of the Arctic, the migration to the Americas and the proliferation of a lively variety of arts and a wide range of personal adornment.

The rate of change continued to accelerate. Agriculture and animal husbandry appeared at roughly the same time around the globe. Technological progress, the mastery of new materials (such as metals) and new energy sources (such as wind and water power) led in an amazingly short time to the Industrial Revolution and the world of today. The acceleration of human history cannot be better illustrated than by comparing the changes of the past 10,000 years with those of the previous four million.

Language, that marriage of speech and cognitive abilities, may well have been the critical new factor that provided a biological base for the acceleration of history. Just as upright walking and toolmaking were the unique adaptation of the earlier phases of human evolution, so was the physiological capacity for speech the biological base for the later stages. Without this remarkably effective mode of communication man's technological advance would perforce have been slow and limited. Given an open system of communication rapid change becomes possible and social systems can grow in complexity. Human social systems are all mediated by language; perhaps this is why there are no forms of behavior among the nonhuman primates that correspond to religion, politics or even economics.

If all this seems too pat, I should remind the reader that some of the

oldest and most troublesome questions about human evolution remain unanswered. Looking to the future, I expect that molecular biology will determine the relationships between man and the other living primates and the times of their mutual divergence more accurately than any other discipline can. But there will still be other major problems, particularly in determining the rates of evolution. As in the past, the present proponents of various hypotheses may be wrong on the very points on which they are surest they are right.

At this stage, then, it is probably wise to entertain more than one hypothesis and to state opinions in terms of the odds in favor of their being right rather than presenting them as conclusions. On this basis I would guess from the present evidence that the odds are 100 to one (in favor) that man and the African apes do in fact form a closely related group. I would also guess that a very recent separation of man and the apes, say five to six million years ago, was not nearly as probable; there my odds are only two to one in favor.

Perhaps by presenting opinions in this way we might demonstrate that all views of human evolution are built on seeming facts that vary widely in their degree of reliability. For example, if it is accepted that man is particularly close in his relationship to the African apes, it does not necessarily follow that man and the apes separated in Africa. At the time when the ape and human lines separated there were apes in the Near East and India; man may be descended from the apes of those areas. Perhaps the reason there are no longer any apes in India is that those apes evolved into men. Both the African and the non-African theories of the evolution of the earliest upright walkers are reasonable; only the discovery of more fossils will determine which theory is correct.

The Hominids of East Turkana

by Alan Walker and Richard E. F. Leakey
August 1978

This region on the shore of Lake Turkana in northeastern Kenya is a treasure trove of fossils of early members of the genus Homo and their close relatives dating back 1.5 million years and more

The continent of Africa is rich in the fossilized remains of extinct mammals, and one of the richest repositories of such remains is located in Kenya, near the border with Ethiopia. The first European to explore the area was a 19th-century geographer, Count Samuel Teleki, who reached the forbidding eastern shore of an unmapped brackish lake there early in 1888. Exercising the explorer's prerogative, he named the 2,500-square-mile body of water after the Austro-Hungarian emperor Franz Josef's son and heir, the archduke Rudolf (who within a year had committed suicide in the notorious Mayerling episode). Teleki's party evidently passed a major landmark on the shore of the lake, the Koobi Fora promontory, during the last week in March. They took notes on the local geology and even collected a few fossil shells, but they missed the mammal remains.

Teleki's name has faded into history, and even the name òf the lake has now been changed by the government of Kenya from Lake Rudolf to Lake Turkana. Yet today the Koobi Fora area is world-famous among students of early man. Its mammalian fossils include the partial remains of some 150 individual hominids, early relatives of modern man. They represent the most abundant and varied assemblage of early hominid fossils found so far anywhere in the world.

The fossil beds of East Turkana (formerly East Rudolf) might have been found at any time after Teleki's reconnaissance. It was not until 1967, however, that the deposits came to notice. At that time an international group was authorized by the government of Ethiopia to study the geology of a remote southern corner of the country: the valley of the Omo River, a tributary of Lake Turkana. Erosion in the area has exposed sedimentary strata extending backward in time from the Pleistocene to the Pliocene, that is, from about one million to about four million years ago.

Supplies going to the Omo camps were flown over the East Turkana area; on one such trip one of us (Leakey) noticed that part of the terrain consisted of sedimentary beds that had been dissected by streams and that appeared to be potentially fossil-bearing. A brief survey afterward by helicopter revealed that the exposed sediments contained not only mammalian fossils but also stone tools. This reconnaissance was followed up in 1968 by an expedition to the vast, hot and inhospitable area. Out of a total area of several thousand square kilometers the expedition located some 800 square kilometers of fossil-bearing sediments, mainly in the vicinity of Koobi Fora, Ileret and to the south at Allia Bay. The expedition also found the fossil remains of many kinds of mammals, most of them beautifully preserved. Only in the category of hominids were the finds disappointing: the total was only three jaws, all of them badly weathered. Nevertheless, the overall richness of the fossil deposits made it clear that further prospecting would be worthwhile.

The large task of establishing the geological context of both the fossils and the stone tools discovered in East Turkana began in 1969. The work that season was highlighted by the excavation of the first stone tools to be found in stratified sequences there and by the discovery of two skulls of early hominids. It was with these discoveries that the enormous importance of the area for the study of human evolution began to be recognized.

A formal organization was established: the Koobi Fora Research Project. The project operates under the joint leadership of one of us (Leakey) and Glynn Isaac of the University of California at Berkeley. In the years since it was founded the project has brought together workers from many countries who represent many different disciplines: geology, geophysics, paleontology, anatomy, archaeology, ecology and taphonomy. (Taphonomy is a new discipline concerned with the study of the processes that convert living plant and animal communities into collections of fossils.) The interaction of specialists has become a particular strength of the project as the workers have become increasingly aware of the particular outlook (and also the limitations) of fields other than their own.

The project area extends from the Kenya-Ethiopia border on the north to a point south of Allia Bay where the land surface is of volcanic origin. The western boundary is the lakeshore and the eastern is marked by another volcanic outcropping. The promontory of Koobi Fora itself, a spit of land that extends a few hundred meters into the lake, is the site of our base camp. Each of the three principal areas of fossil-bearing sediments has its own natural boundaries; when these areas are seen from the air, they show up as pale patches among the darker volcanic terrain. For reference purposes they have been divided into smaller units that are identified by number on the project maps and that are readily distinguished in the field by their vegetation, dry rivers and the like.

In studies such as these it is of paramount importance to develop a chronological framework that allows the fossil finds to be placed in their correct relative positions. The construction of such a framework is the responsibility of the project geologists and geophysicists. The geology of East Turkana is straightforward in its broad outlines but extremely complex in many of its local details. Among the factors responsible for its complexity are abrupt lateral shifts in the composition of the exposed sedimentary strata, discontinuities, faulting that involves rather small displacements and above all the absence from many of the sediments of volcanic tuff: layers of ash that play a key role in correlating the strata.

The difficult work of geological mapping and stratigraphic correlation has been carried out by many of our colleagues but mainly by Bruce Bowen of Iowa State University, working in collaboration with Ian Findlater of the International Louis Leakey Memorial Institute for African Prehistory in Nairobi, Kay Behrensmeyer of Yale University and Carl F. Vondra of Iowa State.

FOSSIL-RICH AREAS in East Turkana appear as pale patches in a satellite image of the region. The deltas of the Omo River, which rises in Ethiopia, appear at the northern end of Lake Turkana (formerly Lake Rudolf). The short, narrow promontory that juts out from the eastern shore of the lake is Koobi Fora, where the base camp for project fieldworkers is situated. Imagery is from the satellite *Landsat 2*.

The complexities have nonetheless prevented the accurate placement of some of the most important early hominid fossils in the stratigraphic framework the geologists established. In such cases we have provisionally assigned the specimens to temporal positions on the basis of criteria other than stratigraphic ones.

The fossil-rich sediments are underlain by older rocks of volcanic origin. The sediments themselves are of various kinds, laid down in such different ancient environments as stream channels and their associated floodplains, lake bottoms, stream deltas and former lakeshores. For the most part the strata dip gently toward the present Lake Turkana. In the past, extensions of various deltas and coastal plains frequently built out westward into the former lake basin; these intrusions alternated with periods when the lake waters intruded eastward. The result is a complex interdigitation of lake sediments and stream sediments.

The major basis for correlating the various strata is the presence of distinctive strata consisting of tuffs; the volcanic material has periodically washed into the lakeshore basin from the terrain to the north and to the east. Some of the tuff beds are widespread and some are not. The uncertainties of correlation between tuff layers in different locations are greatest in the Koobi Fora and Allia Bay areas; these are the areas farthest from the erosional sources of the volcanic ash. At the same time volcanic rocks can be dated by means of isotope measurements, which makes the ash strata particularly important.

Jack Miller of the University of Cambridge and Frank J. Fitch of Birkbeck College in London have conducted most of the dating studies of the tuff (based on the decay of radioactive potassium into argon). Independent chronological data are also available from studies of the magnetic orientation of some particles in the volcanic ash and studies of fission tracks in bits of zircon in the ash. The various measurements are by no means unequivocal, but it can be stated as a generality that the location of a fossil find below one such layer of tuff and above another at least establishes a relative chronological position for the fossil even if its precise age remains in doubt.

There are five principal tuff marker layers. The earliest, which provides a boundary between the Kubi Algi sediments below it and the Koobi Fora sediments above it, is the Surgaei tuff. The next layer of tuff divides the lower member of the Koobi Fora sedimentary formation approximately in half; this is the Tulu Bor tuff, 3.2 million years old. The next layer, the KBS tuff, is named for the exposure where it was first recognized: the Kay Behrensmeyer site. It marks the boundary between the lower and upper members of the Koobi Fora Formation, and its exact age is debated.

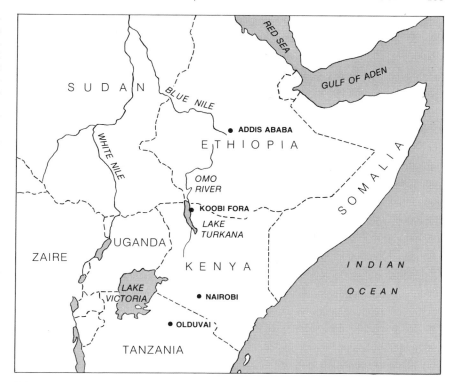

KOOBI FORA REGION of East Turkana lies near the border between Kenya and Ethiopia. The fossil material collected there is brought back to Nairobi for preparation and analysis.

In 1970 Miller and Fitch ran a potassium/argon analysis of the KBS tuff that showed it to be 2.61 (±.26) million years old. They have recently made new calculations indicating that the tuff is 2.42 (±.01) million years old. At the same time an age determination based on other KBS samples, done by Garniss H. Curtis of the University of California at Berkeley, yields two much younger readings: 1.82 (±.04) million and 1.60 (±.05) million years.

Above the anomalous KBS tuff the next marker layer is the Okote tuff, which divides the upper member of the Koobi Fora sediments approximately in half. The Okote tuff is between 1.6 and 1.5 million years old. The fifth and uppermost marker layer, roughly indicating the boundary between the Koobi Fora sediments and the overlying Guomde sediments, is the Chari-Karari tuff; it is between 1.3 and 1.2 million years old.

We present the stratigraphy in such detail because most of the early hominid fossils discovered thus far in East Turkana are sandwiched between the Tulu Bor tuff at the most ancient end of the geological column and the Chari-Karari tuff at the most recent end. Twenty-six specimens, including a remarkable skull unearthed in 1972, designated KNM-ER 1470, come from sediments that lie below the KBS tuff. (The designation is an abbreviation of the formal accession number: Kenya National Museum–East Rudolf No. 1470.) Another 34 specimens, including several skulls, come

from sediments that lie above the KBS tuff but below the Okote tuff. Unfortunately for those interested in measuring the rates of evolutionary change in hominid lineages, the difference between the oldest and the youngest proposed KBS-tuff dates is some 1.3 million years.

That span of time far exceeds the one allotted to the whole of human evolution not many years ago. Even by today's standards the KBS-tuff discrepancy is enough to allow uncomfortably different evolutionary rates for various hypothetical hominid lineages. So far evidence of other kinds has not resolved the issue. For example, our colleagues John Harris of the Louis Leakey Memorial Institute and Timothy White of the University of California at Berkeley, who have conducted a detailed study of the evolution of pigs throughout Africa, suggest that the more recent date for the KBS tuff would best suit their fossil data. At the same time the fission-track studies of zircons from the KBS tuff indicate that the older dates are correct. For the time being we must accept the fact that the KBS tuff is either about 2.5 million years old or somewhere between 1.8 and 1.6 million years old.

Relative and absolute chronology apart, other kinds of investigation are increasing our knowledge of the different environments that were inhabited by the East Turkana hominids. For example, most of the hominid specimens can in general be placed either in the genus *Australopithecus* or in the genus *Homo*. Behrensmeyer, Findlater and Bowen

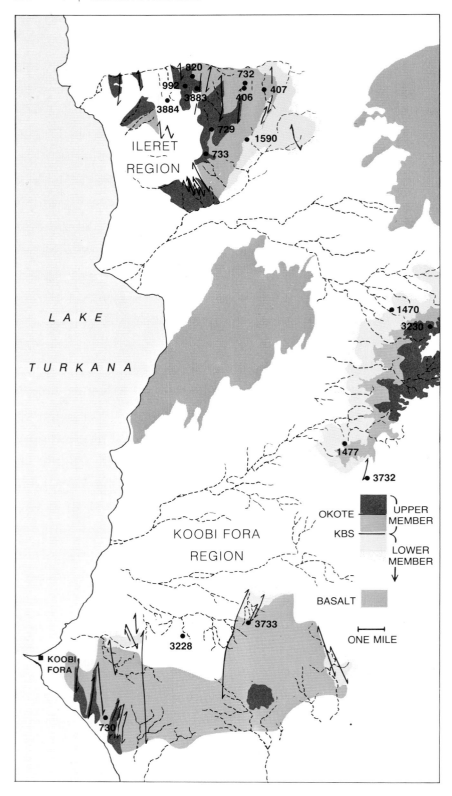

EAST TURKANA FOSSILS are found in a complex interdigitation of sedimentary rocks, some of lakeshore and delta origin and others of streambed origin, that were laid down during alternating periods of lake transgression and land buildup. The major geological marker layers are beds of volcanic tuffs that have been washed from the east and north across parts of the region. Two of the three main fossil-bearing areas are the Ileret region (*top*) and the Koobi Fora region (*bottom*). Allia Bay is not shown. The map is based on the work of Ian Findlater; numbers identify some of the hominid-fossil finds in both regions. Geological faults throughout the region are identified by conventional symbols; also identified (*color*) are sedimentary strata of the upper member and part of the lower member of the Koobi Fora Formation. These strata are separated by two volcanic-tuff marker beds: the Okote complex and the KBS complex.

are engaged in microstratigraphic studies that have enabled Behrensmeyer to associate many of the specimens with a specific environment of sedimentary burial. Preliminary analyses indicate that the specimens identified as *Homo* were fossilized more commonly in lake-margin sediments than in stream sediments whereas the specimens identified as *Australopithecus* are equally common in both sedimentary environments. Facts such as these promise to be of great help in reconstructing the lives of early hominids. In this instance the chance that an organism will be buried near where it spends most of its time is greater than the chance that it will be buried farther away. Thus Behrensmeyer has hypothesized that in this region of Africa early *Homo* exhibited a preference for living on the lakeshore.

Because of the unusual circumstances in this badlands region it will be useful to describe how the hominid fossils have been collected. The initial process is one of surface prospecting. The Kenyan prospecting team is led by Bwana Kimeu Kimeu; its job is to locate areas where natural erosion has left scatterings of mammalian bones and teeth exposed on the arid surface of the sedimentary beds. Kimeu is highly skilled at recognizing even fragmentary bits of hominid bone in the general bone litter present in such exposures.

Once the presence of a hominid fossil is established by the prospectors one of the project geologists determines its position with respect to the local stratigraphic section and records the location. Thereafter one of two procedures is generally followed. If the bone fragment has been washed completely free from the sedimentary matrix that held it, the practice is to scrape down and sieve the entire surrounding surface area in the hope of recovering additional fragments. As the scraping is done a watchful eye is kept for fragments that might still be in situ, that is, partly or entirely embedded in the rock.

If the initial discovery is a fossil fragment in situ, the procedure is different. Excavation is begun on a near-microscopic scale, the tools being dental picks and brushes. The Turkana hominid fossils are often so little mineralized that a preservative must be applied to the bone as excavation progresses in order to keep it from fragmenting further. Indeed, sometimes the preservative fluid must be applied with painstaking care because the impact of a falling drop can cause breakage.

After excavation each site is marked by a concrete post inscribed with an accession number provided by the Kenya National Museum. The next task, usually undertaken in the project laboratory in Nairobi, is piecing together the specimens. This is rather like doing a three-dimensional jigsaw puzzle with many of the pieces missing and no picture on the box. Any adhering matrix is now removed under the microscope, most often with an air-powered miniature jackhammer. (Cleaning with acid, a common laboratory method, is out of the question because the fossil bone is less resistant to the acid than the matrix.)

Finally, the hardened pieces of bone are reconstructed to the extent possible by gluing adjacent fragments together.

Can the East Turkana collection be considered representative of the hominid populations that occupied the area more than a million years ago? Taphonomy, the relatively new discipline that

attempts to define the processes whereby communities of plants and animals do or do not become preserved as fossils, is beginning to provide some helpful answers to the question. The biases that affect fossil samples are many. For example, circumstances may result in the preservation of only some parts of

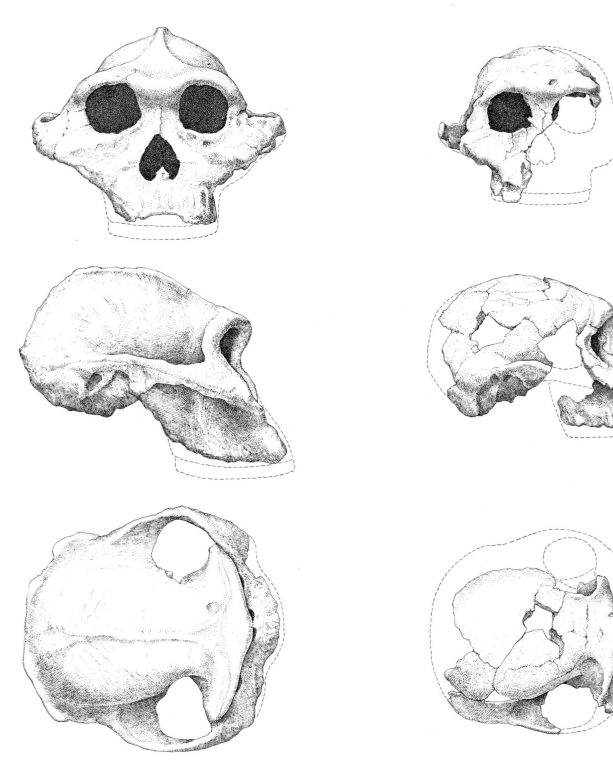

HOMINID FOSSILS found early in the process of collecting in East Turkana are illustrated. At the left is Kenya National Museum–East Rudolf (KNM-ER) accession No. 406, a robust cranium with well-preserved facial bones. At right is KNM-ER 732, a fragmented cranium that has little of the face preserved and is less robust than KNM-ER 406. Both specimens are placed in the genus *Australopithecus*.

certain individuals. Or a particular specimen may be severely deformed by pressure during its long burial. One bias that is easy to recognize in the East Turkana collection is a disproportionate number of lower jaws of the early hominid *Australopithecus robustus*. This hominid had powerful jaws and unusually large

teeth; its lower jawbone is particularly massive. The relative abundance of these jaws and teeth in the East Turkana sediments probably results more from their mechanical strength, and thus their enhanced ability to survive fossilization, than from any preponderance of *A. robustus* individuals in the population.

Another example of bias in the hominid-fossil collection is the disproportionate representation of different parts of the skeleton. Teeth are by far the hardest parts, and so it is not surprising to find that teeth account for the largest fraction of the East Turkana sample. In contrast, vertebrae and hand and foot

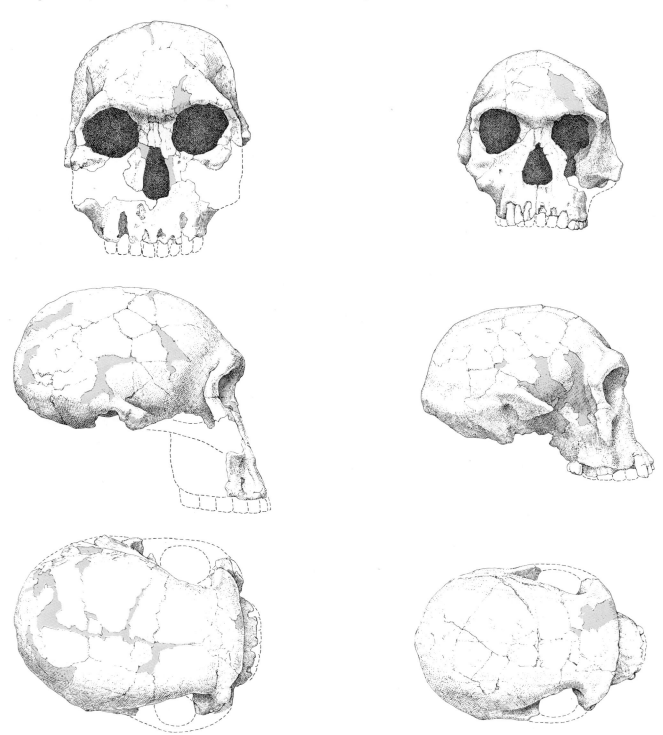

HOMINID FOSSILS OF DIFFERENT AGE are KNM-ER 1470, at left, and KNM-ER 1813, at right. The first cranium comes from the lower member of the Koobi Fora Formation; it cannot be less than 1.6 million years old and may be more than 2.5 million years old. It has a cranial capacity of about 775 cubic centimeters, compared **with the *Australopithecus* average of about 500 c.c. The second cranium is provisionally assigned to the upper member of the Koobi Fora Formation, suggesting that it is no less than 1.2 million and may be more than 1.6 million years old. Its cranial capacity is about 500 c.c. It resembles other African hominid fossils 1.5 to two million years old.**

bones are rarely found. Can this bias be attributed to the destructive processes associated with burial and exposure alone? It seems only logical to take into account a third process: carnivore and scavenger feeding on the hominid bodies before the sediments covered them.

What fraction of the hominid population in East Turkana more than a million years ago might our fossil collection represent? The answer, based on modern population studies of wild dogs and baboons, is that the fraction is extremely small. Assuming an appropriate interval between generations, if the hominid population density was low, as it is among living wild dogs, the collection represents two ten-thousandths, or .02 percent, of the original population. If the hominid population density was high, as it is among baboons, the fraction is very much smaller: two ten-millionths, or .00002 percent. If we ask further what fraction of the ancient popu-

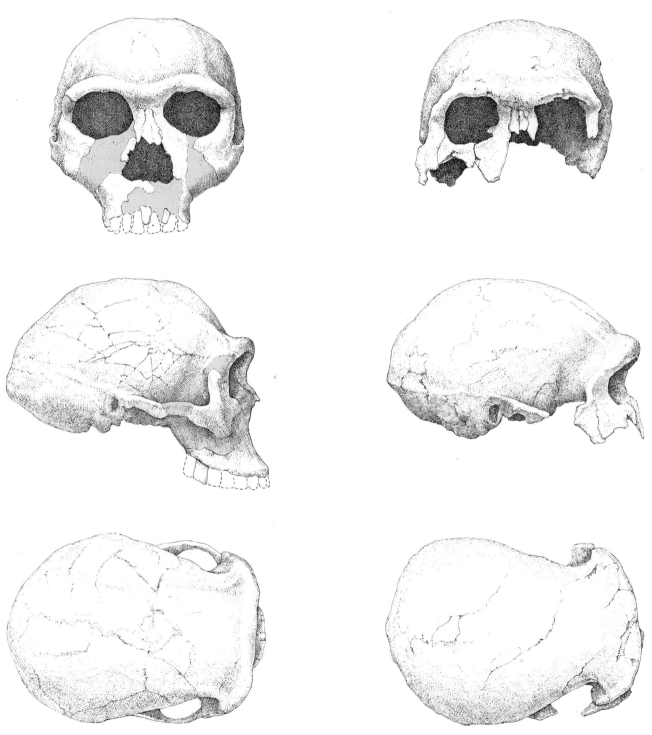

TWO SKULLS OF THE GENUS HOMO from the East Turkana fossil beds represent the early human species *Homo erectus*, a fossil hominid first discovered in Java and China. These are KNM-ER 3733 and KNM-ER 3883. Both are more than 1.5 million years old, which makes them a million years older than the specimens from Chi-na. Their great age strongly suggests that *H. erectus* first evolved in Africa. KNM-ER 3733 has a cranial capacity of about 850 c.c.; the cranial capacity of KNM-ER 3883 has not yet been measured but is probably about the same. Some specimens of *H. erectus* from Africa and Java have cranial capacities that are greater than 1,000 c.c.

lation is represented by the relatively complete skulls in the collection, it may be smaller still: it is between a hundred-thousandth and a hundred-millionth of the total. The second figure is the equivalent of someone's selecting two individuals at random to represent the entire population of the U.S. today. It is on this small sample that our hypotheses concerning hominid evolution must be based.

In developing such hypotheses we must keep in mind a number of fundamental questions. One question is: How many different species of early hominids were there in East Turkana? Another question is whether those species, whatever their number, existed over long periods of time or were replaced by other species. Again, do any or all of the species show signs of evolution during this interval of perhaps 1.5 million years or perhaps only 700,000 years? If there was any evolutionary change, what was its nature? How did the early hominids feed themselves? Were they relatively low-energy herbivores or relatively high-energy omnivores? If indeed several species were present at the same time, did each occupy a distinct ecological niche? What kept the niches separate?

The questions do not end here. Other questions, more specifically anatomical, also call for answers. Do the hominid fossils possess any morphological attributes that might be correlated with the archaeological record of tool use and scavenger-hunter behavior in the area? For example, can we detect any evidence of significant brain evolution during the period? Are there any morphological changes suggestive of altered patterns of locomotion or hand use that might shed light on the origins of certain unique human attributes? (For the purposes of this discussion we define these attributes as including not only walking upright and making use of tools but also an enlarged brain and the ability to communicate by speaking.)

These questions and many more can be answered only after the first question in the series is disposed of. Basically taxonomic in nature, it asks how many species are represented in the East Turkana fossil-hominid assemblage. We have already said that in general two genera were present: *Australopithecus* and *Homo*. How may the two be subdivided?

The answer to this basic and far from simple question is not an easy one. Conspiring against a clear-cut response are such factors as the smallness of the sam-

ple, the fragmentary condition of the individual specimens, the fact that even among individuals of the same species a large degree of morphological variability is far from uncommon and, under this same heading, the fact that a great deal of variation is often found between the two sexes of a single species. Also not to be neglected is the fallibility of the analyst, who is prone to human preconceptions. For example, the very order of discovery of the East Turkana hominids has affected our hypotheses, and we have had to chop and change in order to keep abreast of later discoveries.

It is illuminating in this connection to review the sequence of hominid discoveries in East Turkana. The first specimens to be identified were individuals of the species *Australopithecus robustus*. Fossil representatives of this species were first found at sites in South Africa decades ago. Characteristically they are large of face and massive of jaw; the molar and premolar teeth are very large, although the incisors and canines are small, about the same size as the front teeth of modern man. Although the facial skeleton is large, the brain case is relatively small: the average cranial capacity is about 500 cubic centimeters, compared with the modern human aver-

ILERET

AGE (MILLION YEARS)	TUFF COMPLEX	1/1A	3	5	6/6A	7A	8	10	11	12	15
GUOMDE FORMATION				•3884	× 999						
1.22 – 1.32	CHARI						– – – –		– – –	– – –	– – –
		• 725 × 739 • 728 × 741 • 805 × 993 • 3883	• 992 • 1467 × 740		• 731 • 1466	• 404			• 726 × 1465		
UPPER MEMBER, KOOBI FORA FORMATION 1.48 – 1.57	MIDDLE/ LOWER	× 1463		– – –	• 818 • 2593		• 729 • 807 • 733 • 808 ×803 • 806 • 809		– – • 1468	– – –	– – –
		• 819 • 820 • 1817 • 2595	• 1819		• 727 • 2592 • 801 • 3737 • 802 × 1464 • 1170 × 1823 • 1171 × 1824 • 1816 × 1825 • 1818			• 406 • 407 • 732 × 815		× 1591 × 1592	
2.42 (FITCH, MILLER) 1.6 – 1.8 (CURTIS)	(KBS EQUIVALENT)	– – – –								• 1593	• 2597 • 2599 • 2598 × 2596
LOWER MEMBER, KOOBI FORA FORMATION										• 1590	

INVENTORY OF FOSSIL HOMINIDS from East Turkana appears on this chart: specimens, identified by accession numbers, are listed according to the numbered areas where they were found. Their positions do not indicate any relative temporal position other than a location between the dated tuff marker layers (or equivalent markers). Such a position means that the specimen is older than the known

age of 1,360 c.c. Because the chewing muscles were evidently of a size commensurate with the large cheek teeth and massive jaws, many *A. robustus* individuals have not only extremely wide-flaring cheekbones but also a bony crest that runs fore and aft along the top of the brain case to provide a greater area for the attachment of chewing muscle.

Specimens of *A. robustus* have also been found in East Africa, most notably by Louis and Mary Leakey at Olduvai Gorge. The East African examples are on the whole even larger than those from South Africa, and their cheek teeth are more massive. A well-known example is "Zinjanthropus," an Olduvai cranium now accepted by most scholars as being closely related to the *A. robustus* specimens from South Africa. (Some scholars, it should be noted, still assign "Zinjanthropus" to a related species of *Australopithecus, A. boisei.*) Such taxonomic niceties aside, the fact is that the East Turkana deposits have been found to contain a good number of fossils that can be placed in this hominid species.

In the early investigations at East Turkana the skulls of certain smaller and less robust hominids were also discovered. Indeed, one such cranium, deformed by crushing, turned up near (although not in the same stratigraphic horizon as) a robust *Australopithecus* cranium: KNM-ER 406. When this crushed specimen was first discovered, it could not easily be given a taxonomic position. The finding of a second gracile (as opposed to robust) specimen, however, suggested to us that male-female dimorphism might account for both kinds of cranium. In the second specimen most of the right side of the brain case and facial skeleton and part of an upper-jaw premolar tooth and the roots of the molars were preserved. It is evident that although this individual is substantially less robust than KNM-ER 406, its premolar and molar teeth were only a little less massive than those of the robust one. If among the species *A. robustus* the morphological differences between males and females were as great as they are among gorillas, then the robust, crested specimens from East Turkana could be males and the more gracile specimens could be females.

The age of these *Australopithecus* specimens is substantially greater than that of any previously uncovered in East Africa (the age of the South African specimens remains in question), but their discovery presented no taxonomic problems. This happy state of simplicity came to an end in 1972 with the discovery of the cranium KNM-ER 1470. Bwana Bernard Ngeneo came across it on an exposure of older sediments belonging to the lower member of the Koobi Fora Formation. When he found the specimen, all that could be seen was a scattering of bone fragments on the rock surface. The fragments were relatively fragile, which led us to assume that they had been washed out of the matrix quite recently.

The specimen KNM-ER 1470 is a large, lightly built brain case with a considerable amount of the facial skeleton preserved. Our colleague Ralph L. Holloway, Jr., of Columbia University has determined that its cranial capacity is about 775 c.c. The facial skeleton is very large, and the proportions of the front and cheek teeth are indicated by the preserved tooth sockets and by both the sockets and the broken roots of the molars. The proportions are the reverse of those for *A. robustus;* the incisors and canines are very large and the premolars and molars are only moderately large. Even though the tooth size suggests a formidable chewing apparatus, the brain case shows no sign of a crest for the attachment of heavy chewing

KOOBI FORA

TUFF COMPLEX	118	129	130	131	105	117	116	103	104	119	127	121	123	124
KARARI		– – –												
(dark band)								×1807						
OKOTE			•3230		– – –	– –	– –	×737	– – –					
	•1648		•1805 •1806		•405 ×738 •1477 ×1476 •1478 ×3736 •1479 •1480 •2607			•403 •730 •734 •1515 •1820 ✖1808 ×736	•164 •1804 •810 •3733 •811 ×813 •812 ×997 •814 •816 •998	•1509	•1507 •1508 •1814	•1506 ×1809	•1501 ×1503 •1502 ×1504 •1811 ×1505 •1813 ×1810 •1821 ×1822 ✖1812	•817
KBS		•417	•1462 •1800 •2601 •2660 ×1500	•1469 ×1471 •1470 ×1472 •1474 ×1473 •1482 ×1475 •1801 ×1481 •1802 •1803 •1873	•3731 •3732 •3734	– – –	×3735	•2602 •2604						
TULU BOR (3.18 MILLION YEARS)	– –		– – –	– – –	– – –	•2603 •2605 •2606								

PLACEMENT PROVISIONAL (under columns 121–123)

• SKULL, SKULL FRAGMENTS, JAWS OR TEETH
× OTHER BONES
✖ BONES OF BOTH CLASSES

age of the tuff layer above it and younger than the known age of the tuff below. Numbers in color identify fossils illustrated on pages 195 through 197. Two geological columns define sedimentary strata; note ages of the tuff marker layers. Broken lines show correlation between Ileret and Koobi Fora tuffs. Placement of specimens listed at the far right is provisional because marker layers are absent there.

muscles. Similarly, the cheekbones, although incomplete, do not suggest the same great width of face that is characteristic of *A. robustus*.

Much has been written about the significance of KNM-ER 1470. We believe that certain hominid specimens found at Olduvai Gorge in broken and fragmentary condition are examples of the same kind of skull. If it is necessary to decide on a taxonomic term for these hominids, the species name may well turn out to be *habilis*. (*Homo habilis* is the name that was given to an early species of the genus *Homo* by Louis Leakey and his col-

leagues John Napier of the University of London and Phillip V. Tobias of the University of the Witwatersrand. The name was not accepted unanimously by other students of fossil man and has even caused heated argument.)

We ourselves cannot agree on a generic assignment for KNM-ER 1470. One of us (Leakey) prefers to place the species in the genus *Homo,* the other (Walker) in *Australopithecus.* The disagreement is merely one of nomenclature; we are in firm agreement on the evolutionary significance of what are now multiple finds. Since 1972 two additional

partial skulls of this large-brained, thin-vaulted kind have been found in association with strata assigned to the lower member of the Koobi Fora Formation.

It was at about the time of the discovery of KNM-ER 1470 that we and our colleagues began to disagree as to the taxonomic position of certain well-preserved lower jaws from the East Turkana region. The initial source of disagreement was a small cranium: KNM-ER 1813. It is the cranium of a small-brained hominid with the average *Australopithecus* cranial capacity: 500 c.c. It has a relatively large facial skeleton and

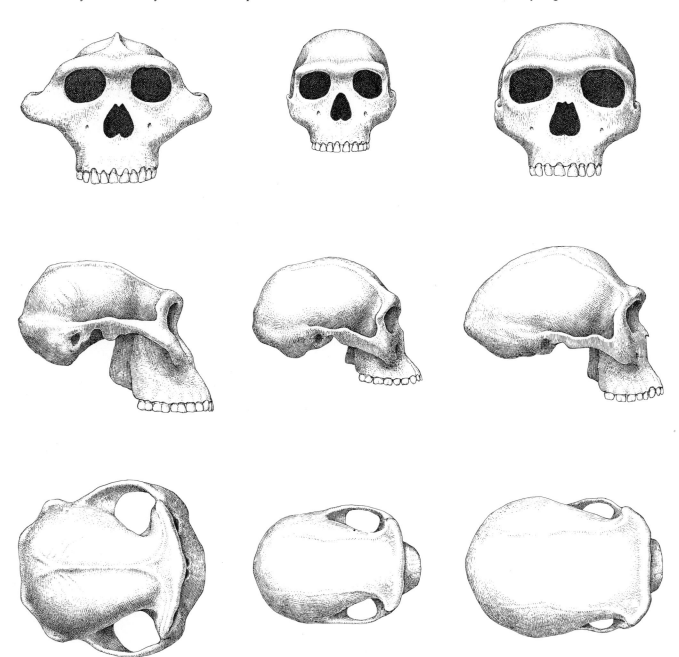

THREE FORMS OF HOMINID are represented among the fossils found in the upper member of the Koobi Fora Formation. The illustration shows them as they would appear if they were restored; lower jaws are omitted. The three may be assigned to particular species in five different ways; all three may belong to a single highly variable species, or they may belong to two species in three possible combinations, or each form may be a valid species on its own. The authors suggest that the three-species hypothesis is the most probable of the five.

SURFACE-SCRAPING PARTY of the Koobi Fora Research Project is seen at work in the Ileret area of East Turkana. The discovery of a hominid jawbone on the surface has led to the collection and screening of loose topsoil and rock in the hope of locating additional fossil fragments. Geologists in the party simultaneously record the stratigraphic position of the exposure and set up a site marker.

FOSSIL JAWBONE is photographed where it was found, exposed on the surface by erosion of the sedimentary rock that once surrounded it. This is the fragmented mandible of a robust member of the genus *Australopithecus;* it was preserved through chance burial in sediments almost two million years ago. Mandibles such as this one are so sturdy that a disproportionate number have survived as fossils.

palate. The upper teeth preserved with the palate are comparatively small, however, and bear a striking resemblance to the teeth of one of the *Homo habilis* specimens from Olduvai, OH-13. It happens that in the initial controversy over the original *H. habilis* specimens, OH-13 represented a species that even skeptics agreed was nearly, if not actually, identical with the species *Homo erectus,* a member in good standing of the genus *Homo* that was first recognized in Java and northern China.

The resemblances between KNM-ER 1813 and OH-13 go further than their teeth. In all the parts that can be compared—the palate as well as the teeth, much of the base of the skull and most of the back of the skull—the two specimens are virtually identical. This leads us to believe the usual reconstructions of OH-13, which have assumed that the specimen had a large cranial capacity and an *erectus*-like skull, are in error. That is not all. The mandible of OH-13 was preserved; its small size and the details of the teeth were major components of the evidence leading to the conclusion that *habilis* was near the *erectus* line. The comparable lower jaws we have found at East Turkana, we can now see, make it clear that the OH-13 mandible could just as well have been hinged to a small-brained, thin-vaulted skull like that of KNM-ER 1813.

To make a final point about these enigmatic East Turkana specimens, we believe there are strong resemblances between them and some of the smaller *Australopithecus* specimens from sites in South Africa, specimens that are usually placed in the gracile species *A. africanus*. Faced with so many possibilities, we argue for caution in the making of taxonomic judgments. Such caution should prevail not only when the evidence in hand is a few isolated teeth but also when the evidence is more generous: lower jaws and upper jaws with the teeth still in place.

In the 1975 season we discovered a remarkably complete cranium: KNM-ER 3733. The find showed unequivocally that a member of our own genus was present in East Turkana in the early strata of the Koobi Fora upper member were formed. The skull bears a striking resemblance to some of the *Homo erectus* skulls found in the 1930's near Peking and is certainly a member of that species. The brain case is large, low and thick-boned. Its principal part is formed by the projecting occipital bone, and its cranial capacity is about 850 c.c. The brow ridges jut out over the eye orbits, and a distinct groove is visible behind them where the frontal bone rises toward the top of the vault.

KNM-ER 3733 has a small face tucked in under the brow ridges. The sockets of some of the upper teeth are preserved. The missing front teeth were relatively large, but the back teeth (some of them still in place) are of only modest proportions. The third molars are among the missing teeth, but the evidence is that they were quite reduced in size. They had been erupted long enough for them to wear grooves in the second molars in front of them, yet the bone forming the sockets indicates that their roots were very small. The point is important because a diminution in the size of this tooth is a common phenomenon in modern human populations.

This fine example of *Homo erectus* from East Turkana predates the example of the same species found at Olduvai Gorge by half a million years and is about a million years older than the examples from northern China. And KNM-ER 3733 is not alone. Another *erectus* specimen was found shortly afterward in Area No. 3 of the Ileret fossil beds. This is KNM-ER 3883. The Ileret specimen is from approximately the same geological horizon as KNM-ER 3733. It has much the same cranial conformation, but its brow ridges, facial skeleton and mastoid processes are somewhat more massive. The cranial capacity of KNM-ER 3883 has not yet been determined, but there is no reason to expect that it will be much different from that of 3733. The similarity of the two East Turkana specimens to specimens from far away that are very much younger strongly suggests that *Homo erectus* was a morphologically stable species of man over a span of at least a million years.

Leaving aside the problems presented by various fragmentary specimens, how can the new hominid finds from East Turkana be assessed taxonomically? One might simply suggest a series of normal taxonomic assignments in the light of what we see at present, acknowledging that as in the past the assignments are likely to be changed. We shall not do so here because we now view the problem of taxonomic assignment in a slightly different way.

Considering only the fossils from the upper member of the Koobi Fora Formation, we think we recognize specimens that might be assigned to three different species. At the same time we may have seriously misunderstood the quantity and quality of variation in any one of the three species. In order to acknowledge this we should like to consider the following possibilities:

1. The three forms are only artifacts of our imagination. Only one hominid species was present, and what we take to be distinct types are merely morphological variants within that species.

This, of course, is the single-species hypothesis. Its strongest proponents have been Loring C. Brace and Milford H. Wolpoff of the University of Michigan. Simply put, the hypothesis states that ever since the human attributes of upright walking, a prolonged childhood, a large brain and small canine teeth became established there has been only one hominid species. What brought this about, those who favor the hypothesis aver, was culture. Once culture became the human domain—and evidence for this can be sought in the fossil bones by looking for the basic human anatomical attributes that developed along with culture—the human ecological niche became so large that the species with culture always had the edge over any other species in competition for the available resources.

2. Two of the three forms represent one species; the third represents a second species. In this hypothesis the separate species is *Homo erectus*. Both of the other forms, then, must be members of a single, highly variable and sexually dimorphic species. The very robust specimens are males and the very gracile ones are females.

To accept this version of the dimorphic hypothesis one must postulate a great deal of variability. The difference in cranial capacity between the robust and the gracile specimens is within the limits seen in living populations of sexually dimorphic apes. The difference in tooth size, however, is outside the observed limits among living apes.

3. Two of the three forms constitute one species, as above, except that the robust forms represent one of the two species and the *Homo erectus* specimens and the gracile forms together represent another highly variable species.

In this dimorphic hypothesis the principal postulated variation is in the size and shape of the brain case. The admissibility of the hypothesis hinges on accepting the fact that among early humans the cranial capacity of the females was roughly half that of the males.

4. Two of the three forms constitute one species, as above, except that the gracile forms represent one of the two species and the highly variable second species consists of the robust forms and the *Homo erectus* specimens.

In this dimorphic hypothesis the principal postulated variations involve the brain case, the jaws and the teeth. The admissibility of the hypothesis hinges on accepting the fact that small-brained but large-jawed forms and large-brained but small-jawed forms can be placed together in the same species.

5. The three forms represent three separate species.

Having listed the five possible hypotheses, we can now assess the probability of each being correct. We shall do so bearing in mind both the fossil evidence and what is known about the variability of living primate populations. We think the probability that the single-species hypothesis is correct is very low. First, the hypothesis involves accepting the fact that there is an enormous amount of intraspecies variability. Second, we think, along with others, that the adaptations apparent in the skulls of both extreme forms (*Homo erectus* and *Australopithecus robustus*) are different. In *H. erectus* the size of the brain case seems to overwhelm the chewing apparatus, as it does in living man. In *A. robustus* the opposite is true.

For the same reasons we think the

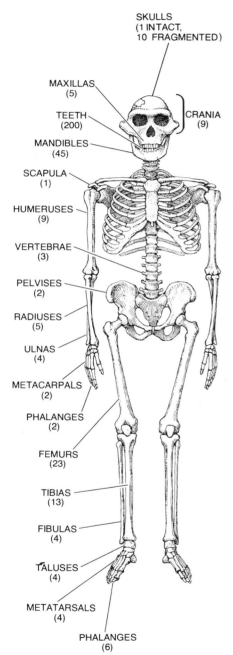

DISPROPORTION in preservation of the remains of fossil hominids unearthed thus far in East Turkana is assessed in this illustration. Teeth, which are the hardest of all body parts, are the most numerous remains. Mandibles, most of them representing the robust species of *Australopithecus*, come next. Rarest of all are pelvic bones, the bones of hands and feet, vertebrae and the bones of the lower arms.

probability that the fourth hypothesis is correct is very low. In addition difficulties other than anatomical ones stand in the way. One must also ask why no specimens of *A. robustus* are found in Java and China, where *H. erectus* specimens are comparatively abundant. If the answer is that such specimens were present but for some reason were not fossilized or have not yet been unearthed, then how is it that *A. robustus* is the commonest hominid in the East African fossil record?

The third hypothesis is an attractive one, but we think the probability of its being correct is low. It would be difficult enough to account for the enormous dimorphism in brain size without having to supply an answer to an additional question: Why are these gracile forms of *Australopithecus* not found as fossils in Java and China?

The second hypothesis, we think, is more likely to be correct than the third. Although the variability in the dimensions and proportions of the teeth cannot be matched in living dimorphic populations, there is some hint that dental dimorphism might have been greater in extinct hominoids: the superfamily that includes both the hominids and the apes. Louis de Bonis of the University of Paris has collected a series of Miocene hominoid jawbones in Macedonia. His sample is from one small area; if the mandibles all belong to the same species, they represent a degree of dental dimorphism greater than that found in living anthropoid apes.

The hypothesis with the best chance of being correct, we believe, is the last of the five: that three species are present. Sorted out in this way, none of the specimens within each group shows more variability in brain size and chewing apparatus than we see among living anthropoids. Although the fossil record from the lower member of the Koobi Fora Formation is far less rich than that from the upper member, a similar three-species hypothesis could also be advanced with respect to the specimens found in it. In this hypothesis the third species in addition to the robust and gracile ones would be represented by the *habilis* specimens.

Several consequences follow from our probability assessments. For example, in our view the demonstration at East Turkana that *Homo erectus* was contemporaneous with some of the largest representatives of *Australopithecus robustus* amounts to a disproof of the single-species hypothesis. We believe both *H. erectus* and *A. robustus* had essentially human characteristics. If this is the case, it follows that they occupied separate ecological niches. It would seem either that one of the species did not possess culture and yet still developed human characteristics or that the argument that the advantage of culture

would give the cultured hominid dominance within a very wide ecological niche is flawed.

We prefer the first alternative, and we would nominate *Australopithecus* for the role of the hominid without culture. Accepting this alternative requires that we keep searching for the natural-selection pressures that have been responsible for producing the basic human attributes.

Where did *Homo erectus* come from? Some have suggested that the species arose in Asia and migrated to Africa. This seems to us an unnecessarily complicated hypothesis. For one thing, it neglects *habilis*. Worse, it implies that a population of these large-brained hominids, who presumably made the stone tools found in the early East Turkana strata, evolved independently in Africa at the right time to fit into an ancestor-descendant relation with *Homo erectus* and then came to an abrupt halt, without playing any further part in human evolution.

It is our view that the *habilis* populations are directly antecedent to *Homo erectus*. If the earlier range of dates for the strata where the *habilis* specimen KNM-ER 1470 was found proves to be correct, then the transition from *habilis* to *erectus* could have been a gradual one, spanning a period of well over a million years. If the later dates are correct, then the transition must have been very rapid indeed.

We have concentrated here on giving an overview of the hominid record in East Turkana to the neglect of other work in progress that promises to answer some of the questions we have raised. The Koobi Fora Research Project is continuing its field activities, and a number of special studies are also under way outside Africa. Michael H. Day of St. Thomas's Hospital Medical School in London is examining the fossil limb bones and associated parts from East Turkana to see what can be learned about the various species' capacity for upright walking. Bernard A. Wood of the Middlesex Hospital Medical School in London is conducting a full analysis of the skulls and jaws in order to document the extremes and means of variability and dimorphism. Holloway is studying casts of the inside of brain cases in an effort to trace the evolution of the brain. One of us (Leakey) continues as codirector of the research project and the other (Walker) is studying the biomechanics of hominid mastication, examining the fossil teeth from East Turkana with the scanning electron microscope in an attempt to deduce dietary habits from the patterns of tooth wear. These studies and others aim at reconstructing as much as can be reconstructed about the biology of these very early hominids in the hope of determining just what it was that made us human.

16 The Neanderthals

by Erik Trinkaus and William W. Howells
December 1979

They flourished from western Europe to central Asia between 75,000 and 35,000 years ago. The differences between them and later peoples are not as great as was once thought but still call for an explanation

The Neanderthals were first recognized in 1856, when workmen uncovered fossil human bones in the Neander Valley near Düsseldorf in Germany. At the time and for some time thereafter the idea of early men, of men different from those now living, was so unfamiliar that the Neanderthals were regarded either as a freakish variant of modern men or as beings that were not quite human. They came to be classified not as members of our own species, *Homo sapiens,* but as a separate species, *Homo neanderthalensis.*

Today the Neanderthals cannot even be regarded as particularly early men. They arose long after other members of the genus *Homo* and longer still after the hominid genus *Australopithecus.* The Neanderthals belong to a rather late stage of the Pleistocene epoch. Indeed, their lateness is the main reason so much is known about them.

In recent years this knowledge has been simultaneously extended and refined. To look broadly at the new picture, the Neanderthals appear on the scene as competent hunters of large and small game, taking their prey in ways that might seem primitive to us but would nonetheless be familiar. They were able to deal with the rigors of a cold climate during the last phase of the Pleistocene. They flourished from western Europe to central Asia. They must have used animal skins for clothing and shelter, because there is clear evidence that earlier people had used them.

They took shelter in caves, where most of their bones have been found. The reason few Neanderthal bones are uncovered elsewhere, however, is that caves preserve bones as open-air sites rarely do. Neanderthals lived in the open as well, as is indicated by open-air sites with masses of the kind of stone tools that are associated elsewhere with Neanderthal bones. Moreover, hearths and rings of mammoth bones at certain sites point to their occupants' living in skin tents. Indeed, it is probable that Neanderthals lived more in the open than they did in caves.

Most of the Neanderthals' tools were flakes of flint, struck from a "core" and trimmed into the projectile points, knives and scrapers that make up the Mousterian (or Middle Paleolithic) tool complex. This complex is not a uniform assemblage of tools everywhere it is found but manifests itself in local variations on a theme of similar manufacture. François Bordes of the University of Bordeaux has distinguished five such "subcultures" in France alone, and other variant assemblages, all loosely classified as Mousterian, stretch off to the east through central Europe and into Asia. The Mousterian culture was a long-lived one appropriate to this late period in the cultural evolution of the Paleolithic.

It is important to note that whereas all Neanderthals made Mousterian tools, not all Mousterian toolmakers were Neanderthals. The Mousterian tool complex is a general level of achievement in the making of stone tools rather than an expression of a specifically Neanderthal intellect and skill.

From about 40,000 years ago in eastern Europe and about 35,000 in western Europe the Mousterian tool assemblages were succeeded by those of somewhat more varied cultures that belong to what is designated the Upper Paleolithic. When these later tools are found with human fossils, the bones are those not of Neanderthals but of anatomically modern human beings. The basic innovation in the tools is that the flakes struck from the core were long, narrow blades. This made it easy to vary the final form of the tools and thus to have a larger assortment of tools. The innovation was also more economical of flint, often a scarce raw material.

In some early manifestations of the Upper Paleolithic certain technical ideas first seen in the Mousterian persist. In others, such as the Aurignacian, the break from the Middle Paleolithic is more complete. The Upper Paleolithic also introduces art: cave paintings, engravings on bone, statuettes of bone and stone, and such personal decoration as strings of beads. The Middle Paleolithic is devoid of such expressions except possibly for a few rock carvings. The Neanderthals did, however, bury their dead and place grave offerings with them. Goat horns have been found in a boy's grave in central Asia and flowers (identified from their pollen) in a burial at Shanidar Cave in Iraq.

In spite of such distinctions it would be unwarranted to decide on the basis of the Neanderthals' tools that their way of life differed radically from that of hunting peoples living into our own times. If the Neanderthals' stone implements were limited to flakes technically inferior to those of the Upper Paleolithic, the same is true of tools made over a period of perhaps 30,000 years by members of one modern population: the Australian aborigines. Again, whereas the Eskimos have had tools of great refinement and variety, comparable in their development to those of the Upper Paleolithic, the first people to occupy the New World certainly did not. It therefore seems safest to speculate that the Neanderthals were formed into hunting bands similar to those of recent hunting peoples, probably linked loosely into tribal groupings, or at least groups with a common language. To judge by the wide distribution and homogeneity of Neanderthal remains, the Neanderthals formed a distinct and major human population that was not a particularly

FRONT AND SIDE VIEWS of the skull of an adult male Neanderthal appear in the photographs on the opposite page. The skull is Shanidar 1, which was discovered at the Iraq cave site of the same name in 1957 by Ralph S. Solecki of Columbia University and his colleagues. The left side of the individual's head had suffered an injury of the eye socket and the bone around it that had healed before his death. Specimen is in the Iraq Museum in Baghdad. Photographs were made through the courtesy of Muayed Sa'id al-Damirji, Director General of Antiquities.

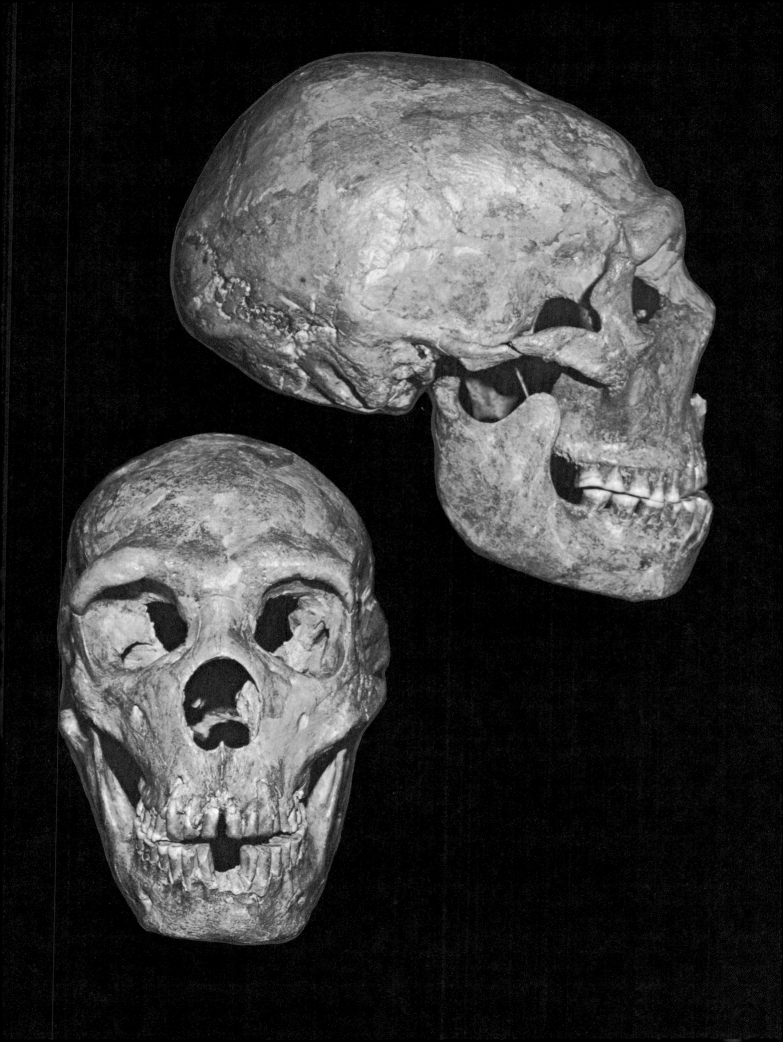

sparse one. Finally, whereas the organization of human populations in the Middle Paleolithic is necessarily a subject of speculation, the evidence of the fossil remains is concrete. Here the question becomes not what is meant by "Mousterian" but what is meant by "Neanderthal."

Although the Neanderthals were first recognized well over 100 years ago, the evolutionary significance of the original Neanderthal discovery and of other human remains uncovered at Paleolithic sites was not apparent until the turn of the 20th century. At that time a number of sites in Europe yielded new Neanderthal fossils, most of them partial skeletons. Among them were the remains of a man aged between 40 and 50 uncovered in a cave near the village of La Chapelle-aux-Saints in France. With this skeleton as a point of departure Marcellin Boule, the leading French anthropologist of his day, published in 1913 a monograph that reviewed all known Neanderthal remains. Boule's monograph included what soon became the standard description and interpretation of the Neanderthals.

At the time no older human fossils were known except those of "Java man," or "Pithecanthropus," which Boule did not regard as being human. He therefore inserted the Neanderthals taxonomically somewhere between chimpanzees and modern men. In this framework the differences between the Neanderthals and modern men tended to make the Neanderthals seem apelike. Boule was further misled, both by anatomical views then current and by his own evolutionary preconceptions, into seeing the Neanderthals as being somewhat stooped and having knees slightly bent and feet rolled in such a way that the outer edge of the foot, rather than the sole, formed the walking surface. Other experts either did not disagree or enthusiastically agreed, and Boule's view of the Neanderthals as an aberrant branch of humanity prevailed. He gave his stamp of approval to their classification as a species distinct from and not ancestral to *Homo sapiens.*

Some two decades later there was a reaction against this view as many other human fossils were discovered, notably in Java and China. Most of these fossil forms were earlier and more primitive than the Neanderthals. Some anthropologists now inserted the Neanderthals not between men and apes but between modern man and such probable ancestors as the Indonesian and Chinese fossils now classified as *Homo erectus.* The Neanderthals were thus viewed as a stage of human evolution, well up on the scale of time and development: a "Neanderthal phase." By extension some earlier fossils were considered to be representative of the same phase in other areas. For example, the Broken Hill skull from Zambia ("Rhodesian man") was classified as an African Neanderthal and the Solo skulls from Java were classified as Eastern Neanderthals. The classification was based on their common possession of large, bony brow ridges and a low-vaulted brain case. In addition two important human populations, relatively recent but still making Mousterian tools, were interpreted as being in transition between the Neanderthals and modern men. These populations were represented by skeletal samples from the caves at Mugharet es-Skhūl and Jebel Qafzeh in Israel.

In this picture the populations of the earliest anatomically modern men arose separately from such immediate predecessors in various parts of the Old World and in association with cultural advances such as those of the Upper Paleolithic of Europe. All these predecessors were taken to be "Neanderthals" of one kind or another, and in contrast to Boule's view most if not all of them were accepted as being the ancestors of various living peoples.

Today a more noncommittal attitude is conveyed by the practice of classifying the Neanderthals as a subspecies within our own species. Thus they are commonly referred to as *Homo sapiens neanderthalensis,* and all living human beings are referred to as *Homo sapiens sapiens.* Such taxonomic distinctions, however, are merely an aid to grouping related individuals. They are not particularly useful as a guide in exploring the actual relations between the Neanderthals and modern men. To arrive at an

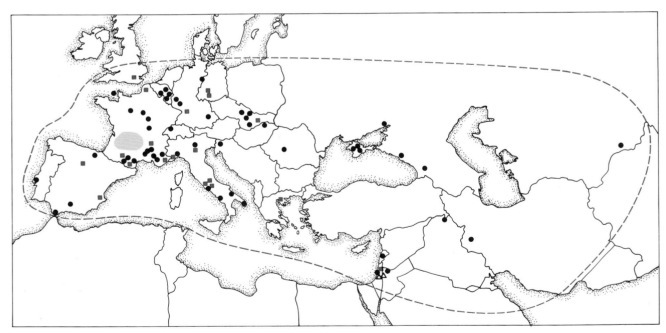

SPATIAL DISTRIBUTION of sites where Neanderthal fossils have been found is plotted on this map. The westernmost site is in Portugal and the easternmost is in Uzbekistan in Soviet Central Asia. The greatest concentration of Neanderthal remains is in the western Massif Central of France (*colored area*), where at least 10 early Neanderthal sites and 25 later Neanderthal ones are situated. Elsewhere on the map early sites appear as colored squares and later sites as black dots. The 19 early sites have yielded the partial remains of some 75 individuals and the 52 later sites the remains of at least 200 more, ranging from a few isolated teeth to complete skeletons. Two open triangles in the Levant locate Mugharet es-Skhūl and Jebel Qafzeh; some of the 30 fossils there were formerly classified as being Neanderthal.

understanding of what the term Neanderthal means one should ignore past controversies and take account of everything that is known today. For example, one can systematically examine the large corpus of Neanderthal fossils in the light of present knowledge of the anatomical functions of bone and muscle. Furthermore, the fossils can be examined against a fuller chronological background based on carbon-14 dating and on recent archaeological work. When this is done, the picture that emerges is quite clear. It reveals a human population complex with a special pattern of anatomical features that extends without interruption from Gibraltar across Europe into the Near East and central Asia. That population complex occupies a span of time from about 100,000 years ago (or at least before the beginning of the last Pleistocene glaciation) down to 40,000 or 35,000 years ago (depending on the locality). Within that space and time only remains recognizable as belonging to this population complex have been found.

The Neanderthal anatomical pattern, or combination of skeletal features, can now be distinguished from that of modern human populations and from the patterns of the European Upper Paleolithic and the Near Eastern late Mousterian. The Neanderthals can also be consistently differentiated from the human beings who lived at the same time in Africa and eastern Asia. Although some of the pattern's individual features grade into those of neighboring populations, its important aspects appear to be distinctively Neanderthal. Moreover, the Neanderthal population is at least as homogeneous as the human populations of today. The people of this anatomical pattern have often been called "classic Neanderthals." In our own view they are the only Neanderthals. To apply the term to specimens of any other time and place is only to invite confusion.

The anatomical pattern must be carefully defined. To begin with, the Neanderthal skull and skeleton exhibit a specific overall pattern. Compared with its modern counterpart the long Neanderthal skull is relatively low but not exceptionally so. The low cranium and the prominent brow ridges give an appearance resembling that of *Homo erectus,* and they are probably derived from such ancestry. Here, then, is the basis for the belief in a "Neanderthal stage" between *Homo erectus* and modern man. The brain encased in the Neanderthal skull, however, was on the average slightly larger than the brain of modern men. This anatomical feature is undoubtedly related to the fact that the musculature of the Neanderthals was more substantial than that of modern men; it does not suggest any difference in intellectual or behavioral capacities.

The Neanderthal face is unique. A

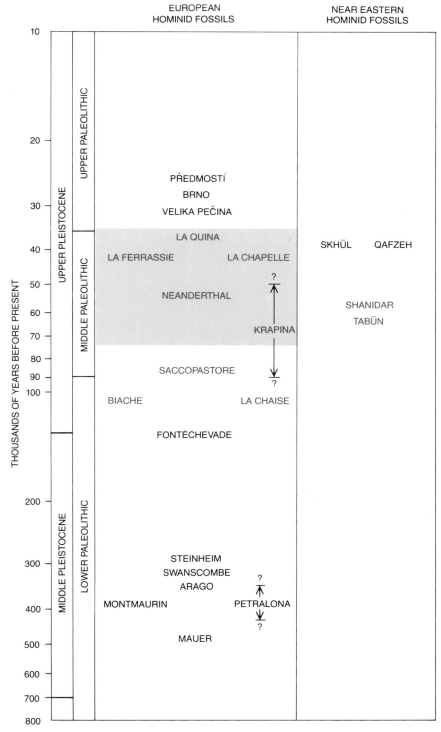

TEMPORAL DISTRIBUTION of the Neanderthals is shown on this chart, which extends from 10,000 years ago to 800,000. The time scale is logarithmic, which expands the space available for the Middle and Upper Paleolithic and the Upper Pleistocene. The last glacial phase of the Upper Pleistocene lasted from 80,000 years ago to 10,000 and was interrupted by a warm interval 35,000 years ago. Although many Neanderthal sites in Europe are not precisely dated, most are between 75,000 and 35,000 years old (*colored band*). The oldest of the fossils from Krapina are slightly older than other European Neanderthals, but most are contemporaneous. These Neanderthal site names appear in color, as do others more than 80,000 years old containing fossils that can be classified as early Neanderthals: Saccopastore, Biache and La Chaise. Still earlier European fossils, from Fontéchevade to Mauer, show varying degrees of affinity with both the Neanderthals and *Homo erectus*. The Upper Paleolithic sites of Velika Pečina, Brno, Předmostí, Skhūl and Qafzeh all contain human fossils of the modern type.

prominence down the midline brings the nose and the teeth farther forward (with respect to the vault of the skull) than they are in any other human fossil, either older or younger. The cheek arches slope backward instead of being angled, as they are in modern "high cheekbones." The forehead slopes instead of rising abruptly as it does over the tucked-in face of modern men. The Neanderthal midfacial prominence may be related to the teeth. The dentition is positioned so far forward with respect to the face that in a profile view there is a gap between the last molar (the wisdom tooth) and the edge of the ascending branch of the mandible, or lower jawbone. This is something seldom seen except in Neanderthals. A distinct bony chin, supposedly a hallmark of modern men, is variably developed among Neanderthals. Its prominence may have been largely obscured by the forward position of the lower teeth with respect to the mandible below them.

The spectacular forward position of the teeth in the Neanderthal skull remains unexplained. The cheek teeth were not significantly larger than those of modern men. The front teeth were somewhat bulkier than is common today, with the result that the arch at the front of the jaws is broader and opener. C. Loring Brace of the University of Michigan has suggested that the front teeth were regularly employed for something more than routine biting: for holding objects or perhaps for processing skins. Indeed, the crowns of the

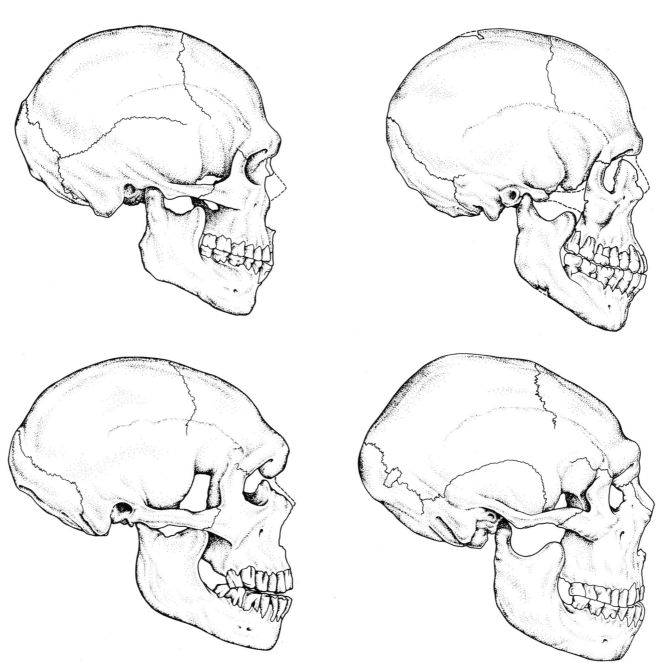

FOUR FOSSIL SKULLS are shown in profile, all slightly restored. The top two, anatomically modern, are Předmostí 3 from Czechoslovakia and Qafzeh 9 from Israel. The bottom two, both Neanderthals, are La Ferrassie 1 from France and Shanidar 1 from Iraq. (A profile photograph of the latter is on page 205.) Compared with the modern skulls the Neanderthal skulls are long, low and massive and their faces project, particularly around the nose and teeth. The anatomically modern skulls have a higher and rounder brain case, and their nose and teeth are more in line with their eye sockets. All should be compared with the Neanderthal precursor illustrated on opposite page.

front teeth of elderly Neanderthals are worn down in an unusual rounded way. Brace's hypothesis is that with the appearance of better tools in the Upper Paleolithic such uses of the front teeth became obsolete, and so the front teeth and jaws became smaller. It would seem, however, that the Mousterian tools were not so inferior as to account for the difference. Furthermore, some Mousterian toolmakers, the Skhūl people, already had front teeth that were like those of Upper Paleolithic populations in size and form.

The Neanderthal face as a whole is large, although it is not as large as the face in earlier members of the genus *Homo*. The front part of the upper jaw was generously proportioned; it accommodated the relatively long roots of the front teeth, particularly the canines. The nasal cavity and the rounded eye sockets are capacious. The sinus cavities are also large. For example, the frontal sinuses fill the brow ridges from above the nose out to the middle of the eye sockets with multichambered "cauliflower" cavities. They do not, however, reach up into the frontal bone above the brows, as is the case in earlier members of the genus *Homo*. In modern skulls the frontal sinuses are flattened, often extend above the brows and are quite irregular in size and shape.

In order to explain the Neanderthals' large, projecting face scholars have invoked a variety of causes, most often adaptation to cold. Carleton S. Coon and others have suggested that the Neanderthal midfacial prominence was such an adaptation. The nasal cavities were placed well away from the temperature-sensitive brain, and at the same time the enlarged size of the cavities may have provided additional space for warming inhaled air. The same Neanderthal facial shape, however, is found in Europe before the last onset of subarctic glacial cold and also in the Near East, where subarctic conditions never arrived. The unique Neanderthal facial configuration is more probably the result of a combination of factors: a highly complex interaction of forces from the chewing apparatus, a response to climatic conditions and a variety of other factors as yet undetermined. Sorting out these factors is one of the principal goals of current research on the Neanderthals. Meanwhile no coherent adaptive explanation for the total Neanderthal cranial pattern has been offered.

For the rest of the skeleton the situation is different. The postcranial skeleton, after all, is the structure that enables a large animal to maintain an erect posture, and *Homo* is a large animal. It is also the structure that allows the muscles to place enormous stresses on the bones while driving the body through the complex characteristic

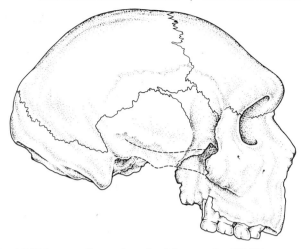

PETRALONA SKULL, from Greece, is undated but may be 400,000 years old. It shows a number of features reminiscent of *Homo erectus*: a low, wide brain case, a large, heavily built face and a large area at back of the skull for attachment of strong neck muscles. Although it has no specific Neanderthal character, it represents Neanderthal ancestral stock in Europe.

movements of the animal. A 150-pound man standing still needs to support only that much weight with his legs, but if he starts to run, the forces generated by muscle contraction and momentum are greatly increased and amount to several times the weight of his body. The bones of the skeleton must sustain these great stresses structurally, and the tendons must be attached to the bones strongly enough to produce the desired motion (or resistance to motion) effectively and efficiently. The tendon attachments leave characteristic marks on the bones that are clues to the power and action of the muscles. Equally significant, living bone under habitual stresses will reshape itself, within limits, to accommodate the stresses more efficiently.

There are many differences between the skeletons of Neanderthals and those of modern men. Some of the differences were formerly interpreted (as, for example, they were by Boule) as evidence of Neanderthal primitiveness. They were easily misinterpreted, and it was not possible to judge which might be more significant and which less so. Today, systematically examined in the light of functional anatomy, the skeletal differences present a coherent and satisfying picture.

To summarize, Boule and others were mistaken: Neanderthals were not less human than modern men, nor were their heads hung forward, their knees bent and their feet rolled over. A touch of arthritis in the neck bones of the La Chapelle-aux-Saints Neanderthal helped to lead Boule astray, but what misled him most was faulty anatomical interpretation. It is now clear that the Neanderthals had the same postural abilities, manual dexterity and range and character of movement that modern men do. They nonetheless differed from modern men in having massive limb bones, often somewhat bowed in the thigh and

forearm. The skeletal robustness evidently reflects the Neanderthals' great muscular power. Everything indicates that for their height both Neanderthal men and Neanderthal women were bulkily built and heavily muscled. Furthermore, signs of this massiveness appeared early in their childhood.

Many skeletal parts testify to this conclusion. For example, the talus, or anklebone, differs slightly in shape from the modern human talus. This difference was once taken to be a sign of primitiveness. It consists, however, only in just such an expansion of the joint surfaces at the ankle as would resist greater stress under load. The bones of the foot arches and the toes show stronger tendon attachments for the muscles that support the arches and propel the body in walking and running. The finger bones show similar attachments for the tendons of the powerful muscles that flexed the fingers. They also show an enlargement of the tuberosities that supported the pads at the fingertips. Both features indicate a much stronger grip than that of modern men, but there was nothing gorillalike in it; the control of movement was evidently the same as ours.

This kind of refined control, coupled with great power, also appears in a curious feature of the scapula, or shoulder blade. The feature has long been recognized but not explained. In modern individuals the outer edge of the scapula usually has a shallow groove on the front, or rib, surface. In Neanderthals a deeper groove characteristically appears on the back surface. This feature seems to reflect the strong development in Neanderthals of the teres minor muscle that runs from the scapula to the upper end of the humerus, or upper-arm bone. Part of the action of the muscle is to roll the arm (with the hand) outward. This action would have balanced and counteracted the major muscles that pull the arm down. In throwing or pounding motions, for example, all these muscles roll the arm inward. In balancing this tendency the teres minor muscle would have made possible a finer control of the arm and the hand in throwing a spear or retouching a flint tool, without any loss in the great muscular strength of the limb.

Other Neanderthal body parts repeat the theme. For example, quantitative analysis of the shape and cross-sectional area of the upper and lower bones of the leg shows that the difference between Neanderthal and modern human bones can also be explained in terms of resistance to the higher stresses of weight and activity in Neanderthals.

One difference still calls for explanation. In Neanderthals the pubic bone, at the front of the pelvis, has a curiously extended and lightened upper branch that forms a part of the rim of the pelvis. This is true of every Neanderthal specimen, male or female, from Europe and the Near East, in which the fragile bone is preserved. Possibly the feature is an adaptation for increasing the size of the birth canal in females. That would have allowed easier passage of an infant's head (which was presumably large) at birth. The presence of the same feature in males as well as females might be explained in terms of close genetic bonding between the two sexes. In any event it is not a trait that lends itself to explanation in terms of patterns of muscle action and movement. The peculiarity seems to be a significant Neanderthal anatomical marker. The pubic bones from Skhūl and Qafzeh are modern in form, as are those of the earliest Upper Paleolithic fossils from Europe. How the feature originated remains an unanswered question, because the pubic bones of earlier members of the genus *Homo* have not been preserved.

It now seems likely that the Neanderthals' antecedents can be traced in at least one section of the Neanderthal range, namely western Europe. Early human fossils from Europe are still few and fragmentary, but their number has increased greatly in recent years. Moreover, certain important specimens were formerly held to be more "progressive," or more modern in appearance, than Neanderthal specimens. These specimens now appear, on reexaminations that include multivariate statistical comparisons, to ally themselves more closely with the Neanderthals than with any other known human type. The specimens include the well-known Swanscombe skull from England, the Fontéchevade skull from France and the Steinheim skull from Germany.

In summarizing the evidence for the origins of the Neanderthals one can begin with the Petralona skull from Greece. It is of uncertain age but is probably as much as 400,000 years old. It shows no specifically Neanderthal character, looking more like an advanced *Homo erectus*. Next in line are the early jaws from Montmaurin in France and from Mauer (near Heidelberg) in Germany. Neither jaw shows any sign of a forward extension of the face such as the Neanderthal postmolar gap. A facial skeleton and two mandibles from Arago, a site in the French Pyrenees, are some 300,000 years old. Like the Swanscombe and Steinheim skulls, the Arago fossils show some suggestions of Neanderthal form, but the projection of the face and the tooth row is not as strongly developed.

From the last interglacial period, which began about 130,000 years ago, come several fossils that show clear signs of the Neanderthal pattern. They include the rear half of a skull, recently found at Biache in northeastern France, that has the lowness of the vault and the protruding rear characteristic of the Neanderthals. A mandible from another French site, the cave of Bourgeois-Delaunay near La Chaise, shows the typical Neanderthal position of the teeth. Two other skulls, from Saccopastore in Italy, clearly approach the typical Neanderthal pattern. Toward the end of the last interglacial the Neanderthal physique is seen in its complete development among the Krapina people. Found in a rock-shelter at Krapina in Yugoslavia, these fossil remains are believed to have accumulated over a considerable span of time both before and after the onset of the last glaciation. All the typical Neanderthal traits are visible: the shape of the skull, the projection of the face, the form of the limbs and the peculiarities of the shoulder blade and the pubic bone.

Why this physical pattern evolved can only be guessed at. The activity of the Neanderthals' massive muscles would have supplied their chunky body with more heat in a chill climate, but the pattern existed before the cold of the last glacial period began in Europe, and it was also present in the more temperate Near East. The robust physique was undoubtedly inherited from populations of *Homo erectus*. Those early men had a massive skull, and the few *H. erectus* limb bones that have been discovered are also massive. Such a heritage does not, however, explain the details of the Neanderthal pattern, particularly the skull pattern.

Whatever the origins of the Neanderthal physique, the fact that it was successful as an evolutionary adaptation is evident from its long stability. From the time of its full establishment perhaps 100,000 years ago down to 40,000 or 35,000 years ago this physical pattern continued without any evidence of evolutionary change. One possible exception is tooth dimensions. The teeth of the Krapina people are larger than those of more recent Neanderthals, which suggests that over a period of time there was a reduction in Neanderthal tooth size. In their details, however, the Krapina teeth are typically Neanderthal, and

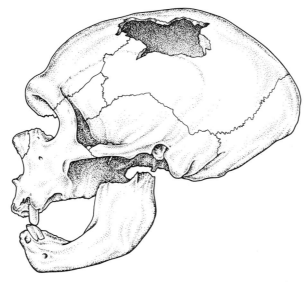

"BRUTISH" NEANDERTHAL, the "Old Man" from La Chapelle-aux-Saints in France, was discovered in 1908. The base of the skull was altered by arthritis and damaged after burial; this led to incorrect conclusions about the head posture of Neanderthals in general. Most of the teeth had been lost before death, so that the lower jaw acquired an abnormal rounded contour. This specimen is one of the longest Neanderthal skulls known and has one of the most projecting faces. Because the "Old Man" was for a long time the most complete and best-described Neanderthal specimen it became the stereotype of subspecies *Homo sapiens neanderthalensis*.

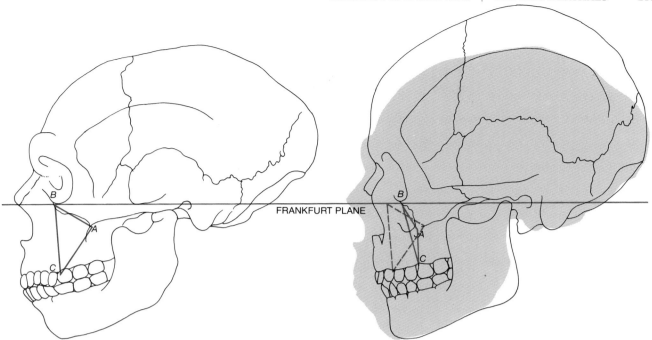

PROJECTING FACE of a Neanderthal (*left*) is annotated by the triangle connecting *A*, *B* and *C*. The forward edge of the first molar tooth (*C*) is well ahead of the lower edge of the cheekbone (*A*) and almost directly below the upper end of the cheekbone (*B*). The horizontal line passing above the triangle defines the Frankfurt plane, a standard orientation. The face of an anatomically modern skull (*right*), with point *B* and the Frankfurt plane superposed on a silhouette of the Neanderthal skull, has a tucked-in appearance; point *A* lies above the first molar tooth, and all three points in the triangle are nearly in the same vertical plane. The Neanderthal specimen used for this comparison is an idealized restoration of the "Old Man" of La Chapelle-aux-Saints. This specimen is shown unrestored on the opposite page.

the evidence of a trend toward a reduction of tooth size elsewhere among the Neanderthals is equivocal. It is possible that the Krapina people merely represent an extreme of tooth size, as the Australian aborigines do among modern populations.

After a stability lasting for perhaps 60,000 years the Neanderthal physical pattern was rapidly replaced by one similar to that of modern men. The first anatomically modern groups showed little difference from the Neanderthals in size. For example, the change in the teeth came not in average size but in details of form. The modern reduction in tooth size began later and has continued down to the present day. In general the anatomically modern people of the Near Eastern late Middle Paleolithic (Skhūl and Qafzeh) and the European early Upper Paleolithic had large bones and robust skulls. Fugitive signs of Neanderthal features appear in some of the Skhūl craniums, but they are rarely found in the Qafzeh group or in Upper Paleolithic specimens. The Neanderthal complex of traits is simply not there; these were ordinary robust representatives of modern humanity, like the Polynesians and northern Europeans of today. Indeed, the Upper Paleolithic skulls are specifically like those of later Europeans, or Caucasoids.

These are not subjective judgments: recent studies based on refined measurements, in particular those of Christopher Stringer of the British Museum (Natural History), make the separation between the cranial pattern of Neanderthals and that of early anatomically modern human beings quite clear. Moreover, both populations have skulls that are distinct from those of other fossil hominids. Other skeletal details, including the features of the scapula and the pubic bone mentioned above, fortify the distinction between the Neanderthals and their successors.

A transition from Neanderthals to their immediate successors undoubtedly took place, but there is little evidence for the actual course of events. The problem is dating. The period of time falls near the limit of accuracy of carbon-14 dating, and in any case samples suitable for carbon-14 analysis have been meager. The problem is particularly difficult in the Near East, and it is only beginning to be solved by the work of Arthur J. Jelinek of the University of Arizona and others. From what is known the most recent Neanderthals at Tabūn Cave in Israel and Shanidar in Iraq are at least 45,000 years old, but they may be considerably older. On the basis of archaeological comparisons the undated Skhūl skeletons are later, being probably no more than 40,000 years old and possibly younger. The Qafzeh remains are undated at the moment, but it seems reasonable to suppose they are about the same age as the Skhūl ones.

We shall therefore assume for the moment, but with no great confidence, that in the Near East anatomically modern men replaced the Neanderthals between 45,000 and 40,000 years ago.

In Europe the dating is a little clearer. Carbon-14 dates for sites that hold late Mousterian artifacts (taken as evidence of Neanderthal occupation) come as close to the present as about 38,000 years ago. A date of 35,250 years ago has been determined for the final Mousterian layer in the rock-shelter of La Quina in France. A frontal bone of modern form, found at Velika Pečina in Yugoslavia, has been dated to about 34,000 years ago. Meanwhile a series of carbon-14 dates obtained at sites across Europe place the beginnings of the culture level known as the Aurignacian (inferentially associated with people of modern physique) at about 33,000 years ago or slightly earlier. Hence in Europe the interval between Neanderthals and anatomically modern populations appears to have been extremely short.

What is important here is to contrast the departure of the Neanderthals with their arrival. The pace is totally different. From what is known about the arrival it can be seen as a gradual evolution. The departure can only be called abrupt; it probably took a tenth as much time as the arrival. Can the two transitions be assessed in the same terms? To answer the question one must accept some guidelines from current knowledge of evolutionary processes. These processes have been notably neglected in the two prevailing explanations of the Neanderthals' disappearance.

One of the two explanations, favored today by anthropologists in the U.S.S.R. and elsewhere in eastern Europe and by many in the U.S., is a restatement of the old "Neanderthal phase" hypothesis. In this view the Neanderthals evolved directly and on the spot into the anatomically modern people of the Upper Paleolithic. Its adherents see Neanderthal or "transitional" anatomical traits in the Skhūl population and in certain Upper Paleolithic specimens such as those from Brno and Předmostí in Czechoslovakia. One of those specimens has not only heavy brow ridges but also a lower jaw with the characteristic gap between the wisdom tooth and the jaw's ascend-

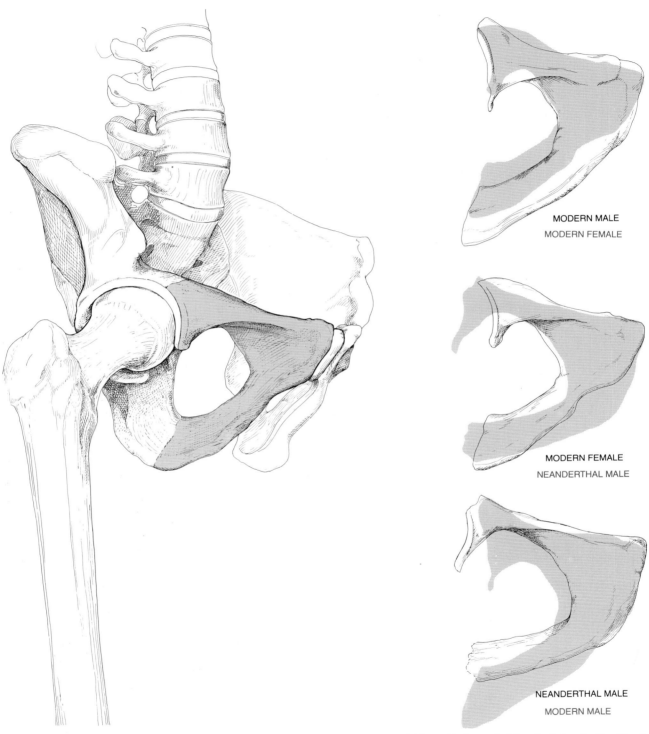

MODERN MALE
MODERN FEMALE

MODERN FEMALE
NEANDERTHAL MALE

NEANDERTHAL MALE
MODERN MALE

PUBIC BONES of a Neanderthal male and an anatomically modern male and female are compared. The pubic bone is the portion of the pelvis (left) that extends from the hip socket to the front midline; here the right pubic bone is seen from the front in all instances. First (top right) the pubic bone of a modern female is silhouetted in color. It is wider and less massive than the pubic bone of an anatomically mod-ern male (black). The pubic bone of a Neanderthal male, Shanidar 1, is silhouetted in color (center right). It is even wider and slenderer than that of a modern female (black). Finally (bottom right) the slenderness of the Neanderthal pubic bone is even more evident when it is compared with that of a modern male (silhouette in color). All Neanderthal pubic bones, male and female, show this characteristic.

ing branch. These features do not, how-ever, make up the full Neanderthal pattern. Both are found in some prehistoric skulls from Australia. Moreover, the brow ridges are not Neanderthal in form; they probably represent an extreme among the generally rugged skulls of Upper Paleolithic people in Europe. Finally, when the measurements of the overall skull shape and of several limb-bone features are subjected to statistical analysis, they indicate that the specimens lack any particularly close Neanderthal affinity. In general, then, fossil specimens from the early Upper Paleolithic, although they are robust and rugged, show no convincing sign of a total morphology that is transitional between Neanderthals and modern men. Nor do late Neanderthal fossils show signs of having begun an evolutionary trend in a modern direction.

The second interpretation ascribes the disappearance of the Neanderthals not to local evolution but to invasion by new peoples of modern form. If there were a cave containing the remains of killed Neanderthals in association with Upper Paleolithic tools, one might entertain the hypothesis of replacement. The hypothesis might also be supported if there were evidence of a homeland for the alleged invaders or for their migration route. Current archaeological and paleontological information, however, is far too fragmentary to support the hypothesis. One can only point out that human populations of modern, although not European, anatomical form certainly occupied the distant continent of Australia 32,000 years ago and probably 8,000 years earlier than that. Signs of even older modern men are found in sub-Saharan Africa. Hence if anatomically modern men sprang from a single original main population, a point that many dispute, then it was not a population of Neanderthals, since anatomically modern men were in existence elsewhere when the Neanderthals still inhabited Europe.

Any attempt to choose between these two interpretations is sterile unless evolutionary principles are taken into account. In evolutionary terms a significant change in a physical pattern, such as the one separating the Neanderthals from their Upper Paleolithic successors, normally comes in two steps. First the change arises as a consequence of new selective forces acting on the individuals of a particular population. Then the change somehow becomes established as the norm in all populations of the species. Since the skeletal differences among all living human populations are less than the differences between the living populations and the Neanderthals, there is little doubt that the new pattern has become established throughout *Homo sapiens* today.

More specifically, a widely distrib-

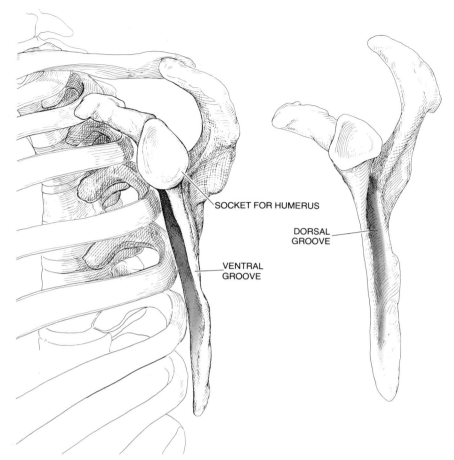

SHOULDER BLADES of a Neanderthal and an anatomically modern man are compared. These are left scapulas seen from the side. The modern scapula, at the left, shows a single groove on the ventral, or rib, side of the outer edge (*color*). This ventral-groove pattern is present in four out of five modern men; it is related to the development of a shoulder muscle, the teres minor, which in anatomically modern men connects the upper arm to the scapula by attaching to a small portion of the dorsal scapular surface (*see illustration below*). The Neanderthal scapula, at the right, has a single large groove on the dorsal, or back, side of the outer border (*color*). This dorsal-groove pattern appears in more than 60 percent of Neanderthal scapulas; the scapula illustrated is that of Shanidar 1. All of outer edge of the bone and part of the dorsal surface provided attachment for the teres minor muscle, indicating that it was well developed.

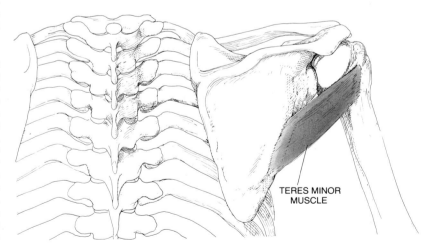

LINE OF ACTION of the teres minor muscle (*color*) is indicated in this dorsal view of the right scapula and part of the right upper arm bone, or humerus. When the muscle contracts, it pulls the humerus in toward the scapula, thereby strengthening the shoulder joint; at the same time it turns the upper arm, forearm and hand outward. All the major muscles of the shoulder that pull the arm downward, as in throwing or striking a blow, tend to turn the arm and hand inward. By countering this rotation the teres minor muscle gave Neanderthals more precise arm control.

uted species that interbreeds freely is likely to be relatively static in the evolutionary sense. It changes quite slowly because the lack of barriers to gene exchange between populations encourages a species-wide genetic homogeneity. Moreover, if the species has a common adaptation, and human culture can be so viewed, then selection will be similar for most of its features, which further promotes its general uniformity.

If, on the other hand, the species is more fragmented, perhaps by various degrees of geographical isolation, an increase in diversity is more likely. In small, isolated populations the substitution of genes under the pressure of natural selection will be faster. Only by some advantage in adaptation, for example a better adaptation to a new ecological situation or a better exploitation of the old one, can a new pattern in one population of a species become the dominant pattern of the species as a whole. The adaptive pattern may be propagated by the flow of genes to other populations or by the simple replacement of part or all of the old populations, with the new population expanding and successfully competing for existing resources. How is Neanderthal history to be seen in the light of such processes?

We have outlined two eventualities: either evolution throughout an entire species (or a large population) because of a common selective pressure or a faster evolution in one element of the population toward a specific adaptation, after which that element of the population replaces other elements because of its adaptive advantage. The two eventualities correspond generally to the two customary interpretations of why after a certain time men of the modern type came to the fore and the Neanderthals disappeared. The two interpretations also imply different degrees of complexity in the genetic basis of the observed anatomical differences. If, on the one hand, the Neanderthals evolved locally into early Upper Paleolithic people, one would expect that relatively few simply coded genetic traits were responsible for the anatomical differences in order for them to appear and spread across the Near East and Europe in a few thousand years. The rate of such evolution within local populations might have been accelerated by the influence of behavioral adaptations on certain aspects of growth, such as the robustness of limb bones; this characteristic is known to be sensitive, within limits, to patterns of individual activity. If, on the other hand, there was significant migration of non-Neanderthal peoples, together with interbreeding and replacement, then a far more complex set of circumstances of genetic substitution and change might be involved.

The fossils themselves furnish some hint of the complexity of the genetic basis for the anatomical differences. First, the Neanderthal pattern seems to have coalesced slowly during the late Middle Pleistocene and the early Upper Pleistocene. This suggests that the kind of complex genetic basis that could build up over many millenniums may well have been responsible for the Neanderthal pattern. Second, the fossil remains of Neanderthal children show that the characteristic Neanderthal morphology had developed by the age of five, and perhaps earlier. Since it is difficult to see how activity could seriously affect the developmental pattern of an infant, it appears likely that there was a complex genetic determination of the development of many details of the Neanderthal morphological pattern.

Broadly, then, the Neanderthal physical pattern evolved in 50 millenniums or more; thereafter it remained relatively constant for about another 50 millenniums. Then came the observed transition, within Neanderthal territory, to an essentially modern human anatomy in about 5,000 years. The various evolutionary and anatomical considerations seem to fit best a model that presents the evolution of populations of anatomically modern men (the early Upper Paleolithic people of Europe and the final Mousterians of the Near East) in partial isolation from the majority of the Neanderthals. These populations may have arisen from a strictly Neanderthal population or a non-Neanderthal one. At any rate they undoubtedly spread, absorbing and replacing various local Neanderthal populations across the Near East and Europe. The time and place of the establishment of these earliest modern people within the Neanderthal area are not yet known.

The main selective force that favored

MULTIVARIATE ANALYSIS of fossil and modern skulls compares 18 measurements in terms of size (*ordinate*) and shape (*abscissa*). An average of European Upper Paleolithic skulls (*open square*) is used as the point of departure. Farthest removed from the starting point in terms of shape and well removed in terms of size is the Middle Pleistocene skull from Petralona (*colored triangle at far right*). More removed in terms of size but nearer in terms of shape are the skulls of Rhodesian man (*colored triangle to right of center*) and a Near Eastern Neanderthal, from Amud (*colored triangle at top center*). The colored square near the center represents the average of European Neanderthal skulls; the Middle Pleistocene European fossil from Steinheim (*open colored circle*), although it is smaller than the Neanderthal average, is quite close to the Neanderthal average in shape. The Saccopastore skulls (*colored dot*), most recent of the early Neanderthals, are surprisingly close to the averages of two modern skull samples (*black dots*): Norwegians and Zulus. The same is true of two skulls from Qafzeh (*black triangle*) and another specimen from the Levant, Skhūl 5 (*black square*). The modern and the Levant specimens diverge only trivially from the European Upper Paleolithic average. This Penrose size and shape analysis was done by Christopher Stringer of the British Museum (Natural History).

the modern physique over the Neanderthal one also remains to be discovered. Was it perhaps climatic change? Actually the last glacial period reached its coldest point more than 10,000 years after the transition from the Neanderthal physique to the modern one, and there is no consistent correlation between the transition and any major climatic change. Was it ecology? Both kinds of men hunted the same game and presumably collected the same plant foods.

Was it cultural advance? Here we have the best evidence in the form of the stone tools. It is hard to see that the specific Upper Paleolithic tools have much of an advantage over the Mousterian ones, for hunting, for gathering or for any other subsistence activity. It is more likely that, as the Mousterian tools were beginning to suggest the Upper Paleolithic forms, there arrived a threshold in human subsistence patterns, and that the only indication of the threshold appears in the tools themselves. One might hypothesize that crossing this threshold made the bulky Neanderthal physique both unnecessary and too costly in its food requirements, thereby initiating a rapid reduction in body size and conceivably even a change in all the special Neanderthal traits. Alternatively the improvement in stoneworking techniques and the associated behavioral changes may have given a significant adaptive advantage to the less heavy-bodied Upper Paleolithic people.

It is interesting that between 40,000 and 35,000 years ago there was a marked increase in the complexity of the sociocultural system of these hominids. Soon thereafter various forms of art are a regular feature at archaeological sites, implying the existence of well-established rituals for various kinds of social behavior. Although ritual existed considerably earlier among the Neanderthals, as is indicated by their burial practices, a rapid increase in its complexity would suggest that some threshold had been reached in the evolution of the sociocultural system. The crossing of such a threshold may well have had significant influence on the biological evolution of these prehistoric human populations.

The problems remain: on the theoretical side the nature of the advantage giving rise to the transition and on the factual side the lack of datable fossils that would make the real story clear. Yet the importance of the Neanderthals is that so much is now known about them, incomparably more than is known about other human populations that lived at the same time. Reconciling this wealth of information with what is known about evolution in general presents by far the best opportunity for the scientific study of human development in the late Pleistocene.

BIBLIOGRAPHIES

I THE PHENOMENON OF EVOLUTION

1. Evolution

ON THE ORIGIN OF SPECIES. Charles Darwin. Facsimile Edition. Harvard University Press. 1964.

THE NATURE OF THE DARWINIAN REVOLUTION. Ernst Mayr in *Science*, Vol. 176, No. 4038, pages 981–989; June 2, 1972.

DARWIN AND NATURAL SELECTION. Ernst Mayr in *American Scientist*, Vol. 65, No. 3, pages 321–327; May–June, 1977.

2. Adaptation

ADAPTATION AND NATURAL SELECTION: A CRITIQUE OF SOME CURRENT EVOLUTIONARY THOUGHT. George C. Williams. Princeton University Press, 1966.

EVOLUTION IN CHANGING ENVIRONMENTS: SOME THEORETICAL EXPLORATIONS. Richard Levins. Princeton University Press, 1968.

ADAPTATION AND DIVERSITY: NATURAL HISTORY AND THE

MATHEMATICS OF EVOLUTION. E. G. Leigh, Jr. Freeman, Cooper and Company, 1971.

3. The Evolution of Multicellular Plants and Animals

EVOLUTIONARY PALEOECOLOGY OF THE MARINE BIOSPHERE. James W. Valentine, Prentice-Hall, 1973.

PLATE TECTONICS AND THE HISTORY OF LIFE IN THE OCEANS. James W. Valentine and Eldridge M. Moores in *Scientific American*, Vol. 230, No. 4, pages 80–89; April 1974.

THE EVOLVING CONTINENTS. Brian F. Windley. John Wiley & Sons, 1977.

PRINCIPLES OF PALEONTOLOGY, Second Edition. David M. Raup and Steven M. Stanley. W. H. Freeman and Company, 1978.

II EARLIEST TRACES OF LIFE

4. The Evolution of the Earliest Cells

ORIGIN OF EUKARYOTIC CELLS. Lynn Margulis. Yale University Press, 1970.

THE BIOLOGY OF BLUE-GREEN ALGAE. Edited by N. G. Carr and B. A. Whitton. University of California Press, 1973.

PRECAMBRIAN PALEOBIOLOGY: PROBLEMS AND PERSPECTIVES. J. William Schopf in *Annual Review of Earth and Planetary Sciences: Vol. 3*, edited by Fred A. Donath, Francis G. Stehli and George W. Wetherill. Annual Reviews, 1975.

BIOSTRATIGRAPHIC USEFULNESS OF STROMATOLITIC PRECAMBRIAN MICROBIOTAS: A PRELIMINARY ANALYSIS. J. William Schopf in *Precambrian Research*, Vol. 5, No. 2, pages 143–173; August, 1977.

5. Pre-Cambrian Animals

THE GEOLOGY AND LATE PRECAMBRIAN FAUNA OF THE EDIACARA FOSSIL RESERVE. M. F. Glaessner and B. Daily in *Records of the South Australian Museum*, Vol. XIII, No. 3, pages 369–401; July 2, 1959.

THE OLDEST FOSSIL FAUNAS OF SOUTH AUSTRALIA. M. F. Glaessner in *Sonderdruck aus der Geologischen Rundschau*, Vol. 47, No. 2, pages 522–531; 1958.

SEARCH FOR THE PAST. J. R. Beerbower. Prentice-Hall, 1960.

TIME, LIFE AND MAN: THE FOSSIL RECORD. R. A Stirton. John Wiley & Sons, 1959.

6. The Animals of the Burgess Shale

THE ENIGMATIC ANIMAL *OPABINIA REGALIS*, MIDDLE CAMBRIAN, BURGESS SHALE, BRITISH COLUMBIA. H. B. Whittington in *Philosophical Transactions of the Royal Society of London*, Series B, Vol. 271, pages 1–43; June 26, 1975.

FOSSIL PRIAPULID WORMS. Simon Conway Morris in *Special Papers in Paleontology, No. 20*. The Paleontological Association of London; December, 1977.

THE BURGESS SHALE (MIDDLE CAMBRIAN) FAUNA. Simon Conway Morris in *Annual Review of Ecology and Systematics: Vol. 10*, edited by Richard F. Johnston et al. Annual Reviews, 1979.

III   INTERPRETING FOSSILS

7. The Evolution of Reefs

REVOLUTIONS IN THE HISTORY OF LIFE. Norman D. Newell in *Uniformity and Simplicity: A Symposium on the Principle of the Uniformity of Nature*. The Geological Society of America, Special Paper 89, edited by Claude C. Albritton, Jr., 1967.

AN OUTLINE HISTORY OF TROPICAL ORGANIC REEFS. Norman D. Newell in *American Museum Novitates*, No. 2465; September 21, 1971.

REEF ORGANISMS THROUGH TIME. Proceedings of the North American Paleontological Convention, edited by Ellis Yochelson. Allen Press, 1971.

8. Fossil Behavior

BIOGENIC SEDIMENTARY STRUCTURES. Adolf Seilacher in *Approaches to Paleoecology*, edited by John Imbrie and Norman Newell. John Wiley & Sons, 1964.

PALEONTOLOGICAL STUDIES ON TURBIDITE SEDIMENTATION AND EROSION. Adolf Seilacher in *The Journal of Geology*, Vol. 70, No. 2, pages 227–234; March, 1962.

TRACE FOSSILS AND PROBLEMATICA. Walter Häntzschel in *Treatise on Invertebrate Paleontology: Part W*, edited by Raymond C. Moore. Geological Society of America and University of Kansas Press, 1962.

VORZEITLICHE LEBENSSPUREN. Othenio Abel. Verlag von Gustav Fischer, 1935.

9. Micropaleontology

ATLANTIC DEEP-SEA SEDIMENT CORES. David B. Ericson, Maurice Ewing, Goesta Wollin and Bruce C. Heezen in *Bulletin of the Geological Society of America*, Vol. 72, No. 2 pages 193–286; February, 1961.

CATALOGUE OF FORAMINIFERA. Edited by Brooks Fleming Ellis and Angelina R. Messina. American Museum of Natural History, 1940—.

ECOLOGY AND DISTRIBUTION OF RECENT FORAMINIFERA. Fred B. Phleger. Johns Hopkins Press, 1960.

INTRODUCTION TO MICROFOSSILS. Daniel J. Jones. Harper & Row, 1956.

PRINCIPLES OF MICROPALAEONTOLOGY. Martin F. Glaessner. John Wiley & Sons, 1947.

10. Dinosaur Renaissance

ECOLOGY OF THE BRONTOSAURS. Robert T. Bakker in *Nature*, Vol. 229, No. 5281, pages 172–174; January 15, 1971.

DINOSAUR MONOPHYLY AND A NEW CLASS OF VERTEBRATES, Robert T. Bakker and Peter M. Galton in *Nature*, Vol. 248, No. 5444, pages 168–172; March 8, 1974.

EXPERIMENTAL AND FOSSIL EVIDENCE OF THE EVOLUTION OF TETRAPOD BIOENERGETICS. Robert T. Bakker in *Perspectives in Biophysical Ecology*, edited by David Gates and Rudolf Schmerl. Springer-Verlag, 1975.

IV   SOME MAJOR PATTERNS IN THE HISTORY OF LIFE

11. Crises in the History of Life

BIOTIC ASSOCIATIONS AND EXTINCTION. David Nicol in *Systematic Zoology*, Vol. 10, No. 1, pages 35–41; March, 1961.

EVOLUTION OF LATE PALEOZOIC INVERTEBRATES IN RESPONSE TO MAJOR OSCILLATIONS OF SHALLOW SEAS. Raymond C. Moore in *Bulletin of the Museum of Comparative Zoology at Harvard College*, Vol. 112, No. 3, pages 259–286; October, 1954.

PALEONTOLOGICAL GAPS AND GEOCHRONOLOGY. Norman D. Newell in *Jounal of Paleontology*, Vol. 36, No. 3, pages 592–610, May, 1962.

TETRAPOD EXTINCTIONS AT THE END OF THE TRIASSIC PERIOD. Edwin H. Colbert in *Proceedings of the National Academy of Sciences of the U.S.A.*, Vol. 44, No. 9, pages 973–977; September, 1958.

12. Plate Tectonics and the History of Life in the Oceans

DYNAMICS IN METAZOAN EVOLUTION. R. B. Clark. Oxford University Press, 1964.

GLOBAL TECTONICS AND THE FOSSIL RECORD. James W. Valentine and Eldridge M. Moores in *The Journal of Geology*, Vol. 80, No. 2, pages 167–184; March, 1972.

A REVOLUTION IN THE EARTH SCIENCES: FROM CONTINENTAL DRIFT TO PLATE TECTONICS. A. A. Hallam. Oxford University Press, 1973.

V   HUMAN EVOLUTION

14. The Evolution of Man

PHYLOGENY OF THE PRIMATES: A MULTIDISCIPLINARY APPROACH. Edited by W. Patrick Luckett and Frederick S. Szalay. Plenum Press, 1975.

EVOLUTION AT TWO LEVELS IN HUMANS AND CHIMPANZEES. Mary-Claire King and A. C. Wilson in *Science*, Vol. 188, No. 4184, pages 107–116; April 11, 1975.

HUMAN ORIGINS: LOUIS LEAKEY AND THE EAST AFRICAN EVIDENCE. Edited by Glynn Ll. Isaac and Elizabeth R. McCown. W. A. Benjamin, 1976.

MOLECULAR ANTHROPOLOGY: EVOLVING INFORMATION MOLECULES IN THE ASCENT OF THE PRIMATES. Edited by Morris Goodman and Richard E. Tashian. Plenum Press. 1976.

ORIGINS. Richard Leakey and Roger Lewin. E. P. Dutton, 1977.

HUMAN EVOLUTION: BIOSOCIAL PERSPECTIVES. Edited by Sherwood L. Washburn and Elizabeth R. McCown. Benjamin/Cummings, 1978.

15. The Hominids of East Turkana

EARLIEST MAN AND ENVIRONMENTS IN THE LAKE RUDOLF BASIN: STRATIGRAPHY, PALEOECOLOGY AND EVOLUTION. Edited by Yves Coppens. F. C. Howell, Glynn Ll. Isaac and Richard E. F. Leakey. University of Chicago Press, 1976.

HUMAN ORIGINS: LOUIS LEAKEY AND THE EAST AFRICAN EVIDENCE. Edited Glynn Ll. Isaac and Elizabeth R. McCown. W. A. Benjamin, 1976.

KOOBI FORA RESEARCH PROJECT, Vol. 1: THE FOSSIL HOMINIDS AND AN INTRODUCTION TO THEIR CONTEXT 1968–1974. Edited by Mary Leakey and Richard E. F. Leakey. Oxford University Press, 1978.

13. Continental Drift and Evolution

VERTEBRATE PALEONTOLOGY. Alfred Sherwood Romer. University of Chicago Press, 1966.

THE AGE OF THE DINOSAURS. Björn Kurten. World University Library, 1968.

16. The Neanderthals

EVOLUTION OF THE GENUS *HOMO*. William W. Howells. Addison-Wesley, 1973.

NEANDERTHAL MAN: FACTS AND FIGURES. William W. Howells in *Paleoanthropology: Morphology and Paleoecology*, edited by Russell H. Tuttle. Mouton, 1975.

HUMAN EVOLUTION. C. L. Brace and M. F. Ashley Montagu. Macmillan, 1977.

LES ORIGINES HUMAINES ET LES ÉPOQUES DE L'INTELLIGENCE. Edited by J. Piveteau. Masson et Cie., 1978.

INDEX